Simulation and Control of Chaotic Nonequilibrium Systems

ADVANCED SERIES IN NONLINEAR DYNAMICS*

Editor-in-Chief: R. S. MacKay *(Univ. Warwick)*

Published

Vol. 12 Localization and Solitary Waves in Solid Mechanics
A. R. Champneys, G. W. Hunt & J. M. T. Thompson

Vol. 13 Time Reversibility, Computer Simulation, Algorithms, Chaos (2nd Edition)
W. G. Hoover & C. G. Hoover

Vol. 14 Topics in Nonlinear Time Series Analysis – With Implications for EEG Analysis
A. Galka

Vol. 15 Methods in Equivariant Bifurcations and Dynamical Systems
P. Chossat & R. Lauterbach

Vol. 16 Positive Transfer Operators and Decay of Correlations
V. Baladi

Vol. 17 Smooth Dynamical Systems
M. C. Irwin

Vol. 18 Symplectic Twist Maps
C. Gole

Vol. 19 Integrability and Nonintegrability of Dynamical Systems
A. Goriely

Vol. 20 The Mathematical Theory of Permanent Progressive Water-Waves
H. Okamoto & M. Shoji

Vol. 21 Spatio-Temporal Chaos & Vacuum Fluctuations of Quantized Fields
C. Beck

Vol. 22 Energy Localisation and Transfer
eds. T. Dauxois, A. Litvak-Hinenzon, R. MacKay & A. Spanoudaki

Vol. 23 Geometrical Theory of Dynamical Systems and Fluid Flows (Revised Edition)
T. Kambe

Vol. 24 Microscopic Chaos, Fractals and Transport in Nonequilibrium Statistical Mechanics
R. Klages

Vol. 25 Smooth Particle Applied Mechanics – The State of the Art
W. G. Hoover

Vol. 26 Geometry of Nonholonomically Constrained Systems
by R. H. Cushman, J. Śniatycki & H. Duistermaat

Vol. 27 Simulation and Control of Chaotic Nonequilibrium Systems
by W. G. Hoover & C. G. Hoover

*For the complete list of titles in this series, please visit
http://www.worldscientific.com/series/asnd

Simulation and Control of Chaotic Nonequilibrium Systems

With a Foreword by Julien Clinton Sprott

William Graham Hoover
Carol Griswold Hoover
University of California, Davis, USA

NEW JERSEY • LONDON • SINGAPORE • BEIJING • SHANGHAI • HONG KONG • TAIPEI • CHENNAI

Published by

World Scientific Publishing Co. Pte. Ltd.
5 Toh Tuck Link, Singapore 596224
USA office: 27 Warren Street, Suite 401-402, Hackensack, NJ 07601
UK office: 57 Shelton Street, Covent Garden, London WC2H 9HE

British Library Cataloguing-in-Publication Data
A catalogue record for this book is available from the British Library.

Advanced Series in Nonlinear Dynamics — Vol. 27
SIMULATION AND CONTROL OF CHAOTIC NONEQUILIBRIUM SYSTEMS

ISBN 978-981-4656-82-5

Dedicated to
Our Friends and Neighbors,
the Good People of Ruby Valley,
in Nevada, United States of America

Nonequilibrium phase-plane section
of the Hoover-Holian oscillator
showing local Lyapunov instability.

[Courtesy of Julien Clinton Sprott]

Foreword

> The distinction between past, present, and future is only an illusion, even if a stubborn one. —Albert Einstein

When two atoms in a gas collide, they are flung off in different directions but in perfect agreement with the laws of motion set down by Sir Isaac Newton (1642–1726) two hundred years before it was widely accepted that matter is made of atoms. Newton's Laws conserve mass, momentum, and energy. An additional fundamental property of Newton's Laws, stressed in this book, is their *time-reversibility.* A movie showing the collision of those two atoms will, when run backwards, look perfectly reasonable. The reversed collision accurately obeys the same Laws of motion.

Now imagine the sudden expansion of an isolated N-body gas, still governed by Newton's equations but also well-described by continuum mechanics. Macroscopic thermodynamics predicts the entropy increase in this process. And in accord with the Second Law of Thermodynamics the entropy can only increase. A reversed movie of an irreversible process like this has a fictional look. Thermodynamics says only that the reversed movie *is* fictional without saying "why". Unlike Newton's reversible mechanics irreversible thermodynamics contains Arthur Eddington's "Arrow of Time".

The renowned mathematical physicist, Jules Henri Poincaré (1854–1912), was also interested in reversibility and the general properties of dynamical systems described by ordinary differential equations like Newton's. In attempting to understand "why" the N-body gravitational problem is so much more complicated than the two-body problem and to win a prize for demonstrating the stability of the Solar System, Poincaré described what is now known as "chaos". Chaos amplifies small changes in the initial conditions so as to effect large changes in the subsequent trajectories, explaining how purely deterministic systems can exhibit apparently random unpredictable behavior. However, Poincaré's discoveries lay dormant for more than half a century awaiting the widespread availability of computers. Electronic computation enabled the prompt and accurate solution of such problems to be displayed with graphical elegance.

No one has done more to apply these recent developments to an understanding of the reversibility paradox than Bill Hoover. His books, *Molecular Dynamics* (1986), *Computational Statistical Mechanics* (1991), *Time Reversibility, Computer Simulation, and Chaos* (1999), along with his many published papers and lectures continually return to the topic of reversibility with ever increasing sophistication and insight. Now in this new book, in collaboration with his wife Carol, he sheds additional light on reversibility and a wealth of related issues in chaos, computation, and fractal geometry.

Colliding atoms provide a good metaphor for the way science is done. Some scientists work mostly alone. Others work in long-term, close collaborations, behaving a bit like molecules. All of us experience occasional redirections, gentle or profound, arising from our discussions with other scientists and our interpretations of their publications. Unlike atomic collisions, individual scientific interactions *are* irreversible. What one learns from a colleague cannot be unlearned. The sum total of many such interactions helps systematize and order our understanding of nature in a time-reversed analog of the Second Law of Thermodynamics.

My own interactions with Bill Hoover date back to 1995 when I received an email from him commenting on a paper I had written. As a converted plasma physicist, I was relatively new to the field of chaos and was proud to have discovered nineteen new chaotic systems, all of them simpler than the standard ones proposed by Ed Lorenz and Otto Rössler and widely adopted as *the* prototypical examples of chaos. I was slightly disappointed and embarrassed when Bill pointed out that one of my "new" systems – the only time-reversible one – was a special case of the Nosé-Hoover oscillator equations that he and Shuichi Nosé had previously studied. At the same time I was delighted that someone of his caliber had taken note of my work. It suggested that some of the other eighteen examples might also have some significance for physics. That innocuous two-body interaction inspired me to continue looking for new examples of chaotic systems with special properties, and that quest still continues twenty years later.

Years went by between our collisions, but we would occasionally find ourselves in an intense discussion, furiously trading emails and running computer programs late into the night to answer some question that we both found interesting. A recent such interaction occurred in 2013 when we discovered a system that was time-reversible with invariant tori for some initial conditions and irreversible with a strange attractor/repellor pair for other initial conditions. That work led to our first coauthored publication with the provocative title "Heat Conduction, and the Lack Thereof, in Time-Reversible Dynamical Systems: Generalized Nosé-Hoover Oscillators with a Temperature Gradient." To my delight, some of the ideas and figures from that paper appear in this book and on its cover.

Bill and Carol form a stable diatomic molecule, joyful that their collisions with their many colleagues and collaborators have changed the course of their own research. The coincidences and chaotic interactions throughout this shrinking world are mirrored in the varied topics of the Table of Contents. The academic life, with the stimulation of new students and the assimilation of others' points of view is indeed a Good Life, as is abundantly illustrated in this book.

The concept of reversibility is subtle, even difficult, having perplexed some of the greatest scientists of all time. So do not expect this book to throw back the veil and make it all "perfectly clear". Students, professors, researchers, and even armchair philosophers will find here much to ponder. Perhaps your own collision with this work of Bill and Carol Hoover will set you off on a delightful new trajectory and will inspire you to make your own contributions to this long-standing scientific mystery.

Julien Clinton Sprott
University of Wisconsin – Madison
Author of *Chaos and Time-Series Analysis*

Preface

The Aim, The Scope, and The Key Features of This Book

To provide a lively working knowledge of the thermodynamic control of microscopic simulations. To summarize the historical development of the subject, along with some personal reminiscences. To describe many computational examples, well-suited to learning by doing. To summarize our current understanding of the reversibility paradox.

We two have written this book as a form of "service", summarizing the fruits of our studies for others. Our intent is to simplify their own search for truth and to amplify their understanding of a paradoxical subject — the compatibility of *time-reversible* micromechanics with *irreversible* thermodynamics.

Together, we enjoyed the research support of the University of California for about 70 years. In retirement our support has continued in the form of generous pensions. This welcome largess, absent bureaucratic reviews, promotes more concentration on research rather than less. Our gratitude for this generosity and for our freedom to choose our own goals creates an obligation for us to document what we have learned.

Success in research lies in identifying and pursuing timely questions so as to develop, refine, and understand appropriate models of nature. We are fortunate to have the peace of mind and physical resources necessary to the academic study of such models. Such study and understanding is the antithesis to blind faith, fiction, and ignorance. Snow's "Two Cultures" spells out an important and particularly timely motivation for scientific work. Beyond the satisfaction and excitement of learning by doing and discovering, a second motivation for our current work is the pleasure of surveying and reviewing the literature for new insights. To our minds, the popular writings of Doyne Farmer, Richard Feynman, James Gleick, and David Ruelle have come closest to capturing the motivation and rewards for scientific work of the kind described here. There is the thrill of discovering, understanding, and refining fundamental truths. There is the service of passing them on.

In recent years new ideas have emerged for modelling atomistic systems. We have included a selection of these ideas for display and analysis in this book. With Clint Sprott we have investigated the phase-space-filling ergodicity of three- and four-dimensional thermostated oscillators with fourth-order and fifth-order, as well as adaptive hybrid Runge-Kutta integrators. These particular problems illustrate valuable lessons in numerical analysis. When time-reversible constraints or controls are imposed on numerical simulations of nonequilibrium steady states, strange attractors are the typical result. These attractors occupy a vanishing nonintegral-dimensional fraction of the equilibrium phase space. The farther from equilibrium the greater the dimensionality loss. We discuss the almost unlimited dimensionality loss which can occur in the phase-space attractors generated by atomistic ϕ^4 models of heat conduction.

We have reviewed the literature for new thermostat control methods and devote a chapter to comparing the old with the new. We continue to add to our understanding by quantifying instabilities through the Lyapunov exponents, along with their associated vectors, and report a selection of those results. Faster computers have made it possible to run the longer simulations necessary for an in-depth understanding of the phase-space structure of simultaneously time-reversible and sometimes-dissipative four-dimensional problems.

Nonequilibrium statistical mechanics is an intensely visual subject at both the atomistic and the continuum levels. Snapshots and movies of flows simplify the search for errors and help to solidify our understanding. Geometric descriptions of realistic models follow the rules of ordinary calculus as applied to the ordinary differential equations of atomistic mechanics and the partial differential equations of continuum mechanics. The fractal structures that can result from simulations are surprising and beautiful. We remember well the days of hand-drawn illustrations when Tolman's 704-page tome *The Principles of Statistical Mechanics*, with exactly *two* figures, was a popular academic representation of the state of the art in the Depression year of 1930 .

Progress has some difficult aspects. Typing on a MacBook Pro laptop is made clumsier both by the smaller keyboard and by advancing age. But the possibility of *enlarging* the copy, with an automatic spell-checker accenting unfamiliar words in red (words like "Gibbs" and "Boltzmann") more than compensates for the clumsiness by providing an overall efficiency gain. In working on our 2012 World Scientific book, *Time Reversibility, Computer Simulation, Algorithms, and Chaos*, both of our Dell desktop computers simultaneously gave up the ghost, enabling our transition to the use of laptops.

Computer simulation can be traced back to von Neumann's continuum studies and Fermi's atomistic work at Los Alamos. Fermi's one-dimensional studies of anharmonic chains revealed the inapplicability of Gibbs' statistical mechanics to these systems. Despite their anharmonicity the chains failed, at low energies, to explore a representative sample of their phase-space states. An understanding of chaos, exponentially sensitive dependence on initial conditions, has been helpful in mapping out the complexity necessary to an understanding of space-filling ergodicity. Fermi's chain evolutions retained memories of their initial conditions. Fermi found a very complex dynamical evolution region separating nearly-harmonic motion from ergodic-chaotic dynamics with all energy-surface states equally likely. Such complexity is typical of low-dimensional Hamiltonian chaos.

The early years of simulation were devoted to exploring simple models in order to evaluate the usefulness of statistical mechanics and to understand the mechanisms for strongly nonequilibrium processes. Beginning with Vineyard's studies of "radiation damage" – the formation of voids, interstitials, and cracks resulting from high-energy dissipative collision events in solids – simulation has broadened to include the description of new materials with complex structures and the biomolecules which are a part of our lives. Dissipation – the conversion of work to heat, is fundamental to an understanding of the Second Law of Thermodynamics. Computational controls of dissipative processes lead to contraction in phase space and to the formation of strange attractors.

Organization

This book is divided into self-contained Chapters, beginning with a description of atomistic mechanics and the problems to which it may be usefully applied. Macroscopic thermodynamics and continuum mechanics are introduced next, to contrast the less-detailed macroscopic description that applies to systems with many degrees of freedom. The details of simulation follow next, methods for solving the differential equations, and a catalog of different approaches to the description of temperature. Then models incorporating these details are used to illustrate nonequilibrium implementations. This is followed by a discussion of our understanding of the Second Law of Thermodynamics. We resolve the apparent conflict between the Second Law's irreversibility and the time-reversible nature of *both* forms of atomistic mechanics, conservative and dissipative. We consider the state of the art and forecast future developments in a final chapter, passing the baton to the researchers of tomorrow.

Atomistic Studies

40 years have passed since Les Woodcock's isokinetic studies, and 30 years since Shuichi Nosé's surprise introduction of Gibbs' canonical distribution into dynamical simulations. The field has broadened to include the hydrodynamic applications stressed here as well as the many applications we don't consider in detail, astrophysics, biology, chemistry, nuclear physics, and so on. So many types of boundary conditions and hydrodynamic constraints have evolved from the simple beginnings of the 1950s and 1960s that it would be both complex and unrewarding to attempt a comprehensive review.

Both stochastic and deterministic thermostats have been used. There are moral and pragmatic arguments that favor the deterministic approach. The numbers of "Old Guard" adherents of stochasticity are dwindling. The inherent portability, reproducibility, and the serendipitous link to Second-Law thermodynamics through the phase-space continuity equation all recommend determinism. Shuichi Nosé, through an apparent feat of magicianship, discovered a Hamiltonian approach to thermostating. A more nearly complete and useful understanding and generalization of his work came from the combined efforts of Bauer, Bhattacharya, Braga, Brańka, Bulgac, Dettmann, Evans, Kusnezov, Morriss, Patra, Travis, and Wojciechowski. By now much of their innovative work has been made available in widespread packaged software.

So far, Moore's Law (which declares that the speedup of computation with time is exponential) has applied equally well to simulation, enabling more-detailed mesoscopic and multiscale simulations as machines have progressed from kiloflops to megaflops to gigaflops to teraflops and petaflops. Exaflops are on the drawing boards.

Now it is ours to review and correlate some of this work. We were fortunate to meet most of the pioneers of simulations during our years at the Livermore Laboratory. That Laboratory, and its Los Alamos twin, were uniquely valuable in that esoteric theory and applied weapons engineering were in close proximity with symbiotic benefits to both. Bill's work situation, with his salary *set* by the University but *paid* by the Livermore Laboratory was ideally suited to research.

Large-scale laboratories are no longer necessary for research. Today it is possible to interact with one's colleagues without mailing a letter or scheduling air travel. Progress is faster too. Keeping track of it all while noting and avoiding what George Stell termed "Setbacks in Physics" (publishing erroneous work on previously-solved problems) is more than a full time job. It is also a source of great joy and continual learning.

Macroscopic Studies

Thermodynamics, while strictly phenomenological and empirical, has also a logical basis in Gibbs' statistical mechanics. The Second Law suggests maximizing Gibbs' entropy to find the equilibrium mechanical and thermal conditions, expressing pressure and temperature in terms of entropy.

The ideal-gas thermometer is uniquely well-suited to linking mechanics and thermodynamics. That thermometer exhibits a clear classical-mechanical relation linking ideal-gas pressure to the microscopic *and* macroscopic equilibrium *temperatures.* And temperature is the new variable that distinguishes thermodynamics from ordinary energy-based mechanics.

Temperature is itself a fascinating concept. *At equilibrium* it has infinitely-many possible definitions, all of them equivalent. *Away from equilibrium* there is not yet a consensus. In strongly nonequilibrium situations (fracture, plastic flow, shockwaves, ...) it is natural to define a *tensor* temperature, with different local values of transverse and longitudinal temperature. In weapons simulations at the Livermore Laboratory the physics requires simultaneous tracking of electronic, nuclear, and radiation temperatures. In three-dimensional rheological simulations (far-from-equilibrium shear, for instance) separate local values of $T_{xx} \neq T_{yy} \neq T_{zz}$ can be observed. From a formal standpoint, maximizing Gibbs' entropy S with respect to the distribution of energy E suggests the definition of temperature $T \equiv (\partial E/\partial S)_V$. More generally, but even *more* formally, Gibbs' canonical distribution of energy shows that *any* Cartesian degree of freedom, with its individual coordinate q and momentum p , when varied, gives the identity

$$kT\langle\ \nabla^2\mathcal{H}\ \rangle = \langle\ (\nabla\mathcal{H})^2\ \rangle\ .$$

The catch lies in the implementation of the change ∇ . Two differentiations of kinetic energy with respect to momentum give a constant. Two differentiations of potential energy with respect to a coordinate can give a rapidly-varying function whose many sign changes give divergent and negative "temperatures". Intuition, coupled with sound judgment is required to avoid such cul-de-sacs.

Without an explicit mechanical description detailing the "contact" between the system whose temperature is being measured and the ideal-gas thermometer doing the measuring it isn't possible to verify that Gibbs' distribution applies. The formal linkage between the first and second derivatives of the energy can be made the basis for computational thermostats based on Hamiltonian mechanics. Ergodicity – reaching all the energy states of the system-plus-thermometer model – can only be checked numerically. This limitation is responsible for a relatively-large theoretical literature with relatively-few useful conclusions. This book is an effort to systematize *useful* approaches to thermostating, demonstrating their utility through specific examples.

Computational thermostats have been grounded in Hamiltonian mechanics. This is quite natural, given that nonequilibrium state variables, density, velocity, energy, pressure, temperature, heat flux, and even more, are readily defined and measured in mechanical simulations. The many different approaches to boundary conditions and constraints are best evaluated from an æsthetic standpoint, weighing the benefits of the simulation relative to its cost, simplicity, and elegance. Deterministic time-reversible approaches are both elegant and useful. Their results are readily portable and reproducible once sufficient detail is provided.

Stochastic approaches mostly discard both these advantages. Stochastic simulations rely on pseudorandom number generators which are almost never documented. We believe stochastic approaches are inertial relics from the early Einstein-Langevin-Smoluchowski days of making simple models for Brownian motion. For joint fans of stochasticity *and* time reversibility we offered up on the Los Alamos arχiv a 2013 Ian Snook Memorial Challenge Prize. The goal was to formulate a time-reversed version of a prototypical two-seed random number generator. Within 24 hours Federico Ricci-Tersenghi (Rome) solved the problem and won the Prize.

Like stochastic thermostats, series expansions along the lines of the Mayers' virial expansion (pressure in powers of the density) or the Chapman-Cowling expansion of stress and heat flux as series in the nonequilibrium gradients, have mostly been abandoned in favor of direct simulation. New ideas and faster computers made possible Richard Wheatley's 2013 calculation of the eleventh and twelfth hard-sphere virial coefficients as well as the ninth and tenth for soft spheres.

Jaynes' appealing idea that nature chooses the distribution maximizing a *nonequilibrium* information entropy seems ill-conceived and likewise well worth abandoning. Even in the simplest Rayleigh-Bénard convection cells quite different solutions (one roll or two ; regular or chaotic) turn out to be equally "stable" despite their quite different local-equilibrium entropies and entropy production rates.

Ever since Lorenz' Butterfly-attractor work it has been well accepted that irreversible differential equations could generate multifractal strange attractors, with dimensionalities that varied from place to place. In 1987 it was a surprise to us to find that *time-reversible* nonequilibrium systems can likewise produce multifractal distributions in phase space. Rather than a smooth Gibbsian distribution, like $e^{-E/kT}$, typical *nonequilibrium* distribution functions from constrained or thermostated dynamics are singular everywhere. This means that Gibbs' relation for the entropy, $S \propto \langle \ln f(q,p) \rangle$, *diverges* for these nonequilibrium systems. As a corollary, the definition of temperature in terms of entropy, $T \equiv (\partial E/\partial S)_V$, is simply inapplicable away from equilibrium. It has gradually become apparent that the divergence of Gibbs' phase-space entropy and the consequent lack of a thermodynamic entropy-based definition of nonequilibrium temperature are both characteristic of deterministic nonequilibrium states.

Since publication of our *Time Reversibility, Chaos, Algorithms, Computer Simulation* book additional computational study of Aoki and Kusnekov's ϕ^4 model has revealed its utility in clarifying the puzzling relationship between time-reversible mechanics and time-irreversible thermodynamics. This simplest of models for heat conduction, an anharmonic chain very like Fermi's, but with attractive lattice sites for the particles, clearly illustrates the phase-space dimensionality loss associated with irreversible flows. With the simplest possible [Nosé-Hoover] thermostats, dimensionality losses exceeding those describing all the thermostat degrees of freedom have been observed. The ϕ^4 model holds out the promise of distinguishing far-from-equilibrium approaches to temperature where not all thermometers produce the same T . We emphasize the ϕ^4 model here as a useful bridge between thermodynamics and particle mechanics. This model is also a useful testing device for new thermostats.

As is our habit, we make the notation as simple as possible. We use no special fonts or notation to distinguish scalars, vectors, or tensors, relying on context to distinguish them. Likewise, we avoid ambiguous acronyms such as NEMD [which might mean Northeast Metal Detecting *or* New England Medical Design] and PROLA [Professional Loss Adjustment] . Rather than refer to numbered equations we prefer to repeat them where needed, to save the reader the unnecessary distraction of page-turning searches. Because the mass-dependence is simple and straightforward for all the classical models detailed here we typically choose the particle mass equal to unity. Likewise Boltzmann's constant is typically set equal to one.

We have added a few suggestive problems at the ends of Chapters. All of the computational figures in the book should be thought of as problems for students, some of them easy and some not. We would encourage students to consider the details required to replicate our work. A characteristic of useful knowledge is its reproducibility.

Future Research

Education and scientific research need an unfettered environment in order that progress can continue at the pace we have enjoyed since the early 1960s. In recent years there has been too much government control over funding and research directions at American universities. This has resulted in excessive lost time for professors and their students. Rather than matching their talents to selecting new research directions, they write grant proposals matched to research directions supported by government. Political interference with education curricula has resulted in the downgrading of standards. The resulting dearth of qualified students among our own citizens jeopardizes our global competitiveness. A further hazard to students is the tuition inflation also caused by the excessive overhead promoted by government grants and loans. Technological advances based on fundamental research can only continue if the burdens of our government's paternalistic control intervention are reduced.

Thanks to our Colleagues

The growing list of colleagues that have stimulated and helped us over the years is too long for completeness. Ken Aoki, Arek Brańka, Baidurya Bhattacharya, Estela Blaisten-Barojas, Aurel Bulgac, Carl Dettmann, Denis Evans, Daan Frenkel, Brad Holian, Masaharu Isobe, Julius Jellinek, Dimitri Kusnezov, Ben Leimkuhler, Marc Meléndez, Tetsuya Morishita, Puneet Patra, Mauricio Romero-Bastida, Janka Petravic, Harald Posch, Tamás Tél, Billy Todd, Karl Travis, Paco Uribe, Franz Waldner, Kris Wojciechowski, and Les Woodcock, were all generous with their mature and stimulating reflections on the still-gestating field of computational thermostats. Kris and Stefano Ruffo have both been very helpful to us in publishing papers that had somehow ruffled referees' idiosyncracies. We are particularly grateful to Clint Sprott, whose work we have long admired, and more recently shared, for agreeing to write his Foreword to the book.

Lakshmi Narayanan, our Editor, has been a constant source of inspiration and support. We are very grateful to her. We also mourn those on our lengthening list of colleagues who are deceased : John Barker, Hugh De Witt, Leo García-Colín, Edwin Jaynes, Joel Keizer, Don McQuarrie, Shuchi Nosé, George Rushbrooke, Harry Sahlin, Duward Shriver, Ian Snook, Tom Wainwright, and Fred Wooten. This work is in part a testimony to their continuing influence as well as to that of those who are still with us.

It is, or was, a pleasure for us to interact with all these colleagues and with many other like-minded souls. Though the mysteries of existence and consciousness are all around us, the satisfaction of conscious modelling of interesting aspects of that existence enriches it, and helps us to enjoy it. The free expression possible in books, as opposed to refereed journal articles, is also a welcome fringe benefit of "retirement". We are specially grateful to the University of California's retirement system for making it possible for us to continue our studies even when officially "unemployed".

William Graham Hoover and *Carol Griswold Hoover*
Ruby Valley, Elko County, Nevada USA, 31 December 2014

Contents

Chapter 1

An Overview of Atomistic Mechanics

Topics

Many-Body Mechanics / Controlling Mechanical Boundaries / Controlling Thermal Boundaries / Gibbs' Statistical Mechanics / Nosé-Hoover Temperature Control / Nonequilibrium Multifractal Distributions / Nonlinear Transport / Time-Reversible Thermostats and Thermometers / Background for Our Numerical Examples /

1.1 Newton's, Lagrange's, and Hamilton's Mechanics

Most of classical mechanics is devoted to the evolution of isolated systems with conserved energies. In this book we develop generalized versions of mechanics describing "open" systems, systems where work is done by external forces and heat is exchanged with external reservoirs. Classical mechanics, Newtonian, Lagrangian, and Hamiltonian, is the natural place to start. To begin we review the structure of Newton's 17th century approach to the subject. Newton's mechanics describes the time evolution of the coordinates $\{ x(t),\ y(t),\ z(t) \}$ defining the system of interest. These coordinates may change with the time t . The natural method for dealing with such changes is the calculus of differential equations. Newton invented (or discovered) calculus in order to treat the rates of change of coordinates in a quantitative way.

The *first* time derivatives of the coordinates define the "velocity" $v = (\dot{x}, \dot{y}, \dot{z})$, a vector with as many components as there are coordinates : $\{ v_x,\ v_y,\ v_z \}$.

$$\dot{x} = (d/dt)x = v_x \ .$$

We will often use a superior dot shorthand " $\cdot$ " to indicate a "comoving" time derivative, a time derivative following the motion. The *second* time derivative of each coordinate defines the corresponding acceleration a :

$$\ddot{x} = (d/dt)\dot{x} = (d/dt)^2 x = \dot{v}_x = a_x \; ; \; \ddot{y} = (d/dt)\dot{y} = (d/dt)^2 y = \dot{v}_y = a_y \; .$$

Newton's Second Law relates particles' accelerations to their masses $\{ m \}$ and to the forces imposed upon those masses :

$$\{ F = ma = m\dot{v} = m\ddot{r} \} \; .$$

Newton's First Law describes the special case $F \equiv 0$ and his Third "action-reaction" Law we will often set out to violate. The Second Law is useful.

Given initial values of all the coordinates and velocities and a recipe for the forces $\{ F \}$ giving the accelerations we can integrate the motion equations ,

$$\{ \dot{x} = v_x \; ; \; \dot{v}_x = a_x = (F_x/m) \} \; ,$$

into the future (or into the past) to find the particle trajectories $\{ x(t), \; y(t), \; z(t) \}$. Usually the forces in classical mechanics depend only on coordinates. In our generalizations we will often use forces which depend on velocities as well as coordinates.

Gravitational forces are proportional to particle mass and provide accelerations inversely proportional to the square of the separation :

$$F_r = ma_r = m(d/dt)v_r \propto -m/r^2 \; .$$

Likewise, electrical forces are proportional to particle charge, providing a second source for inverse-square forces. Both these results are empirical. Newton reasoned that the accelerations—the *second* time derivatives of the coordinates—are the fundamental mechanism for change. His First Law of Motion states that in the absence of a force (or acceleration) the velocity proceeds unchanged. It follows that x , $\dot{x}$, and $\ddot{x}$ are enough to generate the entire history and future for $x(t)$. Separate laws for $\dddot{x}$, $\ddddot{x}$, and higher derivatives are unnecessary. Newton had in mind that the gravitational attractive forces felt by apples and stars were proportional to the masses of the interacting bodies and inversely proportional to the inverse square of their separation. It is interesting that this inverse-square "law" is specific to three-dimensional space. In two dimensions the corresponding force is $-(m_1m_2/r_{12})$ rather than $-(m_1m_2/r_{12}^2)$.

For instance, a two-dimensional particle with coordinates ($x, \; y$) and unit mass, attracted to the origin by an attractive force $(-1/r)$, satisfies conservation of (kinetic plus potential) energy, $\dot{E} = \dot{K} + \dot{\Phi} \equiv 0$:

$$K = (1/2)(\dot{x}^2 + \dot{y}^2) \; ; \; \Phi = \ln(|r|) = (1/2)\ln(x^2 + y^2) \longrightarrow$$

$$\dot{K} = \dot{x}\ddot{x} + \dot{y}\ddot{y} \ = \dot{x}(-x/r^2) + \dot{y}(-y/r^2) \ ; \ \dot{\Phi} = (1/r^2)(x\dot{x} + y\dot{y}) \ .$$

Because the x and y terms separately cancel a linear combination (corresponding to an ellipse) also satisfies the conservation of energy. In a "conservative" system, with constant total energy $E = K(v) + \Phi(r)$, the change of kinetic energy with time compensates that due to the changing potential, $\dot{K} + \dot{\Phi} = 0$.

"Generalized coordinates" $\{ q \}$ (angles are the most common case) and their conjugate momenta $\{ p \}$, can be treated with *Lagrangian* mechanics where the Lagrangian is the difference, $\mathcal{L}(q, \dot{q}) = K - \Phi$, between the kinetic and potential energies. Lagrange's equations of motion define the momenta and their time-rates-of-change :

$$\{ \ p \equiv (\partial \mathcal{L}/\partial \dot{q}) \ ; \ (d/dt)(\partial \mathcal{L}/\partial \dot{q}) = \dot{p} = (\partial \mathcal{L}/\partial q) \ \} \ .$$

In the Cartesian case with $K(\dot{q})$ and $\Phi(q)$ Lagrange's motion equations reproduce Newton's. Lagrange's equations *generalize* Newton's approach to systems with curvilinear coordinates and also facilitate the inclusion of constraints (fixed bond lengths, fixed kinetic energies, ...) .

Hamilton's equations of motion are a particularly useful additional generalization of Newton's approach. In Newtonian and Lagrangian mechanics accelerations depend upon the second derivatives of the coodinates. In *Hamiltonian* mechanics the coordinates $\{ \ q \ \}$ and momenta $\{ \ p \ \}$ are independent variables. Their time development is governed by Hamilton's first-order equations of motion ,

$$\{ \ \dot{q} = +(\partial \mathcal{H}/\partial p)_q \ ; \ \dot{p} = -(\partial \mathcal{H}/\partial q)_p \ \} \ .$$

The underlying Hamiltonian is typically the sum of the kinetic and potential energies, $\mathcal{H}(q, p) = K(p) + \Phi(q)$. The Hamiltonian is also basic to quantum mechanics.

For us the most important consequence of Hamiltonian mechanics is Liouville's Theorem. In classical mechanics the Theorem states that the comoving "phase volume" is unchanged by the motion equations :

$$\otimes(q, p, t) = \prod^{\#} dqdp \ \longrightarrow \dot{\otimes} \equiv 0 \ .$$

Here # is the number of "degrees of freedom". Each degree of freedom q and its corresponding momentum p together represent two independent phase-space coordinates. The theorem is easy to prove. We will go through all of the details in Section 2.3 , and show that flows in phase space, described by Hamilton's equations of motion, obey a many-dimensional analog of the continuum continuity equation for an incompressible fluid :

$$\dot{\rho} \equiv (\partial \rho / \partial t) + u \cdot \nabla \rho \equiv 0 \ .$$

In quantum mechanics, the momentum in the classical Hamiltonian is replaced by a differential operator $p \rightarrow i\hbar(\partial/\partial q) = i(h/2\pi)(\partial/\partial q)$, in Schrödinger's stationary-state equation $\mathcal{H}\psi = E\psi$ for the wave function ψ corresponding to the energy E . h is Planck's constant.

The classical motion equations are either first-order or second-order ordinary differential equations and can be solved with a variety of numerical methods. Despite this simple structure, applications of the equations can produce complicated results, even for a one-body problem, as we show in the following Section. Around 1900 Poincaré recognized what is now called chaos, or the (exponential) sensitivity of results to initial conditions. Chaos can be present even in the one-body problem, as we shall soon see.

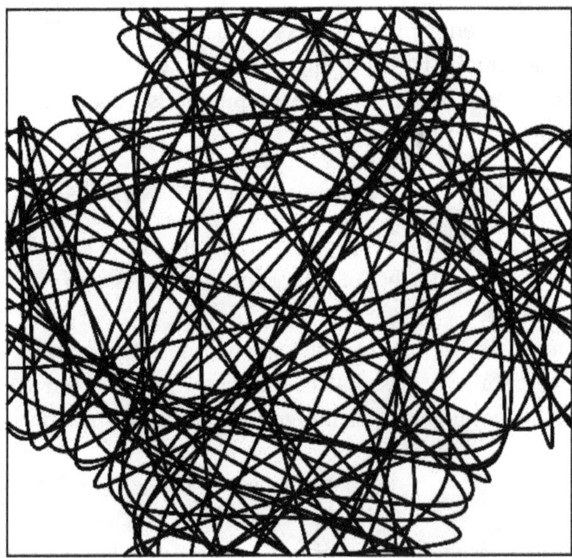

Fig. 1.1 Cell model dynamics. A single particle is accelerated by four fixed "scatterers".

1.2 Controlling Mechanical Boundaries

Most applications of mechanics take place within a fixed region in space. The one-dimensional harmonic oscillator has a periodic solution near the coordinate origin , $x \propto \cos(\omega t)$, where the frequency $\omega = 2\pi\nu$ depends on the force constant and the mass of the oscillator. A zero-pressure solid or fluid with fixed center of mass has no tendency to explore its surroundings, instead just vibrating and/or rotating as time goes on. Many-body systems can be confined in a rigid container but show much less number dependence in their properties if *periodic* boundaries are used.

Figure 1.1 illustrates this concept. In the Figure a single moving particle (a mass point) moves at constant energy in a "cell" defined by four fixed "boundary" particles. Periodic boundaries can be implemented by imagining that these fixed scatterers define the unit cell of an infinite periodic lattice, with the same dynamics going on in each of the cells. Whenever the moving particle leaves its cell, another just like it enters at the opposite side, with unchanged velocity. Then the motion continues. Single-particle models of this kind were once used to estimate the dependence of the energy and pressure on density in condensed phases (liquids, solids, and pressurized gases). The need for such "cell models" disappeared as accurate computer simulations of the many-body problem became commonplace.

Figure 1.2 shows a snapshot of two shear flows implemented by two square and *oppositely-moving* boundary regions. Their motion drives periodic shear flows in two separated fully-Newtonian periodic regions. In the two moving regions which drive the flow a particle at r is tethered to its moving square-lattice site at r_o with a quartic attractive potential ,

$$\phi_{\rm tether}(\delta r) \equiv (1/4)(\delta r)^4 \equiv (1/4)(r - r_o)^4 \ .$$

Additionally, in both the driving and the driven regions *all* pairs of particles interact with a short-ranged and very smooth pair potential chosen to minimize numerical integration errors :

$$\phi_{pair}(r < 1) \equiv 100[\ 1 - r^2\ ]^4 \ .$$

Finally, in addition to the tether and pair forces, the moving boundary regions are "ergostated" so as to keep the total energy of the system constant. The ergostat forces extract the irreversible heat created by the shear flow process.

This shear-flow simulation technique makes it possible to measure not only the shear viscosity but also the *nonlinear* contributions of shear flow

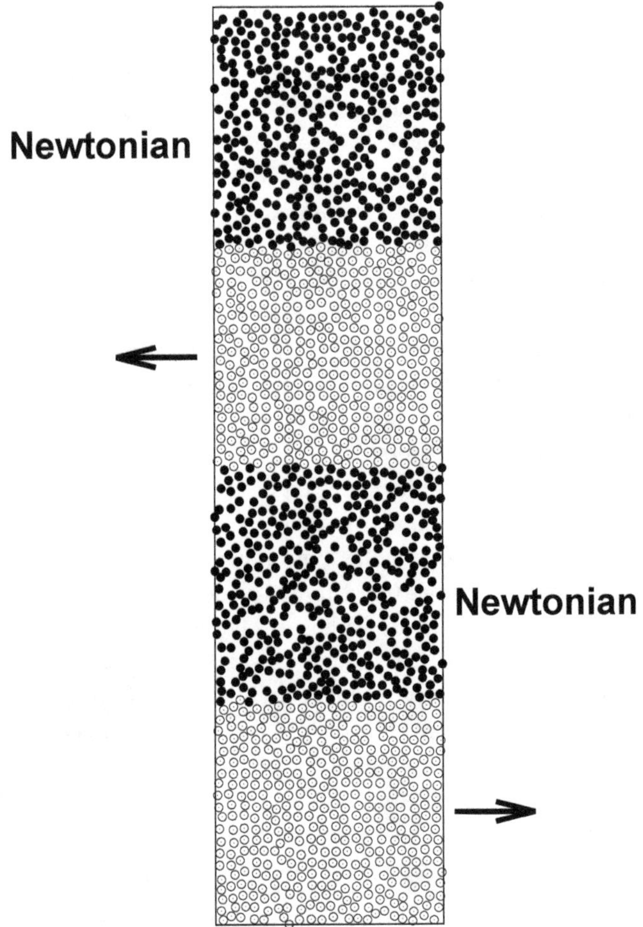

Fig. 1.2 Shear Flow. Two oppositely moving square regions cause Newtonian shear flows in the unconstrained fluid. All particles in the two moving regions are tethered to moving square-lattice sites. The horizontal and vertical boundaries of the four-region system are periodic. The smooth soft-disk pair potential is $\phi(r < 1) = 100(1 - r^2)^4$.

to the pressure tensor and the kinetic temperature. For instance, if the shearing motion is in the x direction and varies with y the boundary-driven results for two three-dimensional cubes sheared by two moving cubic-lattice regions show that $T_{xx} > T_{zz} > T_{yy}$.

There are two oversimplified *fully-periodic* models of shear flow in which the driving is homogeneous, the "Doll's" and "S'llod" algorithms. Neither of these models accounts for the correct ordering of the three kinetic temperatures and both of them predict nonlinear pressure effects which are far too large.

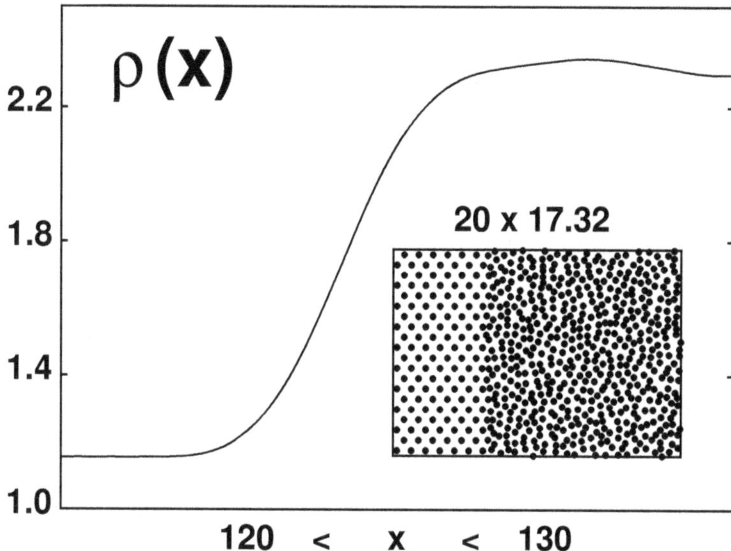

Fig. 1.3 A stationary shockwave. Cold triangular-lattice particles flow in at the left and hot particles exit on the right. The density of the incoming solid doubles at the shockwave while the velocity is cut in half. The mass flux ρu , momentum flux, $P_{xx} + \rho u^2$, and energy flux, $(\rho u)[\ e + (P_{xx}/\rho) + (1/2)u^2\ ] + Q_x$, are all constant throughout the flow.

Figure 1.3 and **Figure 1.4** illustrate two more kinds of steady flow, but now with compression or expansion driving the flows, rather than simple shear. For both these compressible flow types "moving" piston-like boundaries are supplemented at the left and right by introducing "new" right-moving particles (on the left) and by extracting or discarding "old" particles (on the right). In both flow types it is relatively easy to find conditions giving far-from-equilibrium steady flows. Thus the computed results represent the properties of nominally steady left-to-right flows.

If the material on the left is a simple low-density fluid the velocities of the two boundary regions can be adjusted to maintain a relatively-narrow stationary "shockwave" transition region near the center of the system. The shockwave is illustrated in **Figure 1.3**. The shocked fluid is hotter, denser, and has higher entropy than does the cooler, less-dense material being introduced on the left. If this cool $\longrightarrow$ hot situation is reversed, so

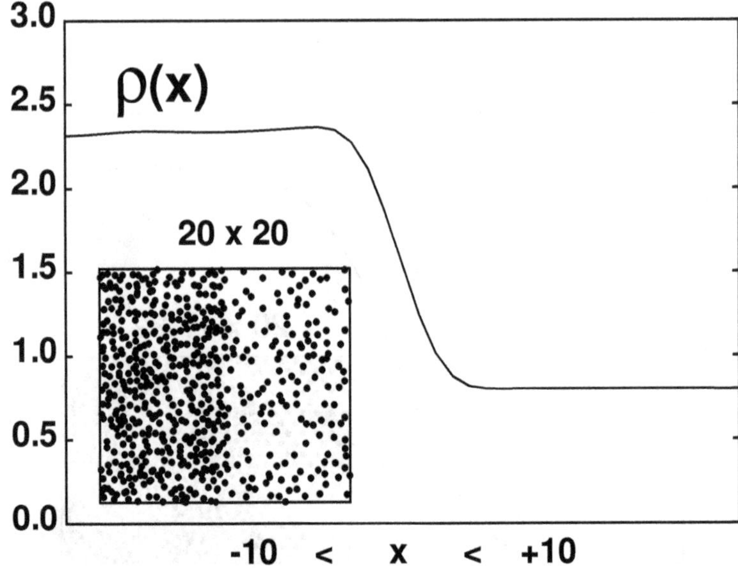

Fig. 1.4 A simulation of Joule-Thomson expansion done with Karl Travis. Fluid enters from the left, heated by the collapse of particle columns from an unstable square lattice. A repulsive potential barrier in the center of the flow plays the rôle of a porous plug, slowing and cooling the fluid. The mass flux and energy flux are constant in this flow too, but the momentum flux drops at the plug potential, which exerts a momentum flux counter to the flow. Although the exiting fluid has been cooled by the throttling process the entropy is increased, primarily due to the increase in specific volume.

that hot material entering at the left becomes a cooler less-dense material exiting to the right then the boundary velocities can again be adjusted to model the classic "Joule-Kelvin (or Joule-Thomson) experiment". See the throttling illustration in **Figure 1.4**. In the laboratory this "throttling flow" can be implemented by pushing hot dense fluid through a porous plug (for instance fiberglass) without any work being done or any heat being transferred at the interface between the left hand (slow and hot) and right hand (fast and cool) flows.

In the simulation snapshot of **Figure 1.4** all pairs of particles interact with the soft-disk potential :

$$\phi(r < 1) = (1 - r^2)^4 \ .$$

The throttling is accomplished by a potential step in the middle of the coordinate range spanned by the calculation. Choosing the x coordinate equal to zero there the step potential is

$$\phi_{\rm step}(\ -1 < x < +1\ ) = (1/4)(1 - x^2)^4 \ .$$

All three of the steady-state simulation types just described—shear, shockwave, and throttling—are *irreversible* from the thermodynamic standpoint. The irreversible heat is extracted by special ergostat forces in the shear flows and by discarding higher-entropy material at the right boundary in the shock and throttling problems.

Both these latter two simulation types are relatively simple to analyze because there is no external heat transfer and all the thermodynamic work is done in the boundary regions at the two ends of the experiment where the fluid is in an equilibrium state. Both the left and right work contributions are simple equilibrium PdV work, where the pressure responsible is nominally constant, as is also the rate of volume change. Next we consider *thermal*, rather than mechanical, boundary conditions, for use in problems with conductive heat transfer at the boundaries.

1.3 Controlling Thermal Boundaries

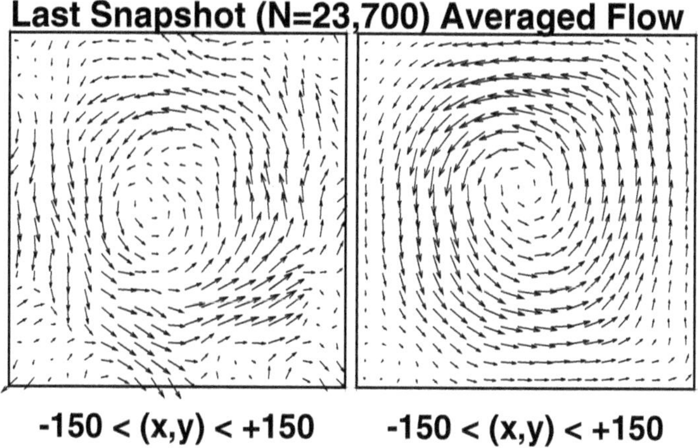

Fig. 1.5 Snapshot and a time-averaged view of a Rayleigh-Bénard flow's velocity field. The flow is driven by thermal expansion and gravity, with the lower boundary hot and the upper boundary cold. The details of the simulation are described in Chapter 4 of our 2012 book, *Time Reversibility, Computer Simulation, Algorithms, Chaos.*

Thermal boundaries can be used to impose temperature on selected degrees of freedom. **Figure 1.5** shows a "Rayleigh-Bénard" cell containing a viscous heat-conducting compressible fluid in a gravitational field *and* a temperature gradient. If the gradient is sufficiently small there is no net fluid flow of the type shown in the Figure. In that case this geometry can be used to determine the fluid's thermal conductivity, $\kappa = -Q_y/(dT/dy)$, where Q_y is the vertical heat flux responding to the vertical temperature gradient (dT/dy) . Larger gradients can stimulate convection. Fluid at the bottom of the cell contacts the hot boundary there, causing the fluid to expand and move upward. Fluid at the top of the cell contacts the upper cold boundary, causing it to compress and descend. If the buoyant forces are sufficiently large then a circulating flow, either stationary or chaotic, can be excited within the cell. Relative to Fourier heat conduction the resulting convective conduction of heat by mass motion can be more efficient, by *orders of magnitude*, in moving heat from the bottom of the cell to the top. Rayleigh-Bénard convection is particularly interesting in that the observed flow (two convection rolls *versus* four, for instance) is often neither steady nor uniquely determined by the boundary conditions.

In many-body simulations we will see that temperature is best defined by the mean value of the kinetic energy (relative to the local macroscopic flow velocity). For nominally fixed boundaries (imposed, for example, by damped harmonic springs, designed to absorb sound waves rather than to reflect them) the kinetic, or potential, or total energy can be controlled so as to reproduce the desired kinetic temperature. Velocity scaling, that is, multiplying velocities by a correction factor to control the average $\langle v^2 \rangle$, is one of the earliest methods of temperature control. Alternatively velocities can be selected from an equilibrium [Maxwell-Boltzmann] distribution from time to time.

Nosé discovered a particularly elegant and more useful thermostating method based on integral feedback, described later in this Chapter. His ideas were particularly useful in focusing computational research on chaos (sensitive dependence on initial conditions) and ergodicity (exploring all accessible states consistent with the specified macroscopic variables, such as energy). To introduce Nosé's ideas it is first necessary to describe Gibbs' understanding of temperature on the basis of his statistical-mechanical canonical ensemble.

1.4 Gibbs' Statistical Mechanics

Josiah Willard Gibbs formulated statistical mechanics at about the same time as did Ludwig Boltzmann, 1883 . Both Gibbs and Boltzmann identified the most likely macrostate of a many-body system as that which could be realized in the greatest number of microscopic "ways". We will presently quantify the notion of "ways" or "states". To illustrate this idea for a system with a discrete number of "states", consider N particles in a one-dimensional box. Let us count the number of ways that half the particles are in the left half, and half in the right half—we suppose that both sides are equally likely. With 4 particles the number of ways to have them equally divided is 6, out of the 16 different possibilities. With 8 particles the number of equally-divided ways is 70 out of 256 .

The probability of finding *exactly* half the particles on each side decreases slowly, being a bit more than 1/11 for 76 particles. Evidently the probability of finding the left and right numbers between 71 and 81 is roughly one half. We can use Stirling's large-N approximation to $N!$, $[\ N! \simeq (N/e)^N\sqrt{2\pi N}\ ]$ to estimate factorials which occur in the binomial distribution of the probabilities :

$$prob = ways/2^N = N!/(N_{left}!N_{right}! \times 2^N) \simeq$$

$$\frac{(N/e)^N\sqrt{2\pi N}}{(N/2e)^N(\pi N)2^N} = \sqrt{2/\pi N}\ ,$$

The error for $N = 76$ is about one third of a percent. Evidently the fluctuations around the most likely division, $N_{left} = N_{right} = (N/2)$, are of order $\sqrt{N}$, already negligibly small relative to N for $N = 10^4$.

In applying Gibbs' statistical mechanics to many-body systems with *continuous* particle trajectories, it is useful to generalize the notion of discrete states' probabilities to a continuous probability density $f(q,p)$ in a "phase space" spanned by all the particle coordinates $\{\ q\ \}$ and their momenta $\{\ p \equiv (\partial\mathcal{L}(q,\dot{q})/\partial\dot{q})\ \}$. This choice is the usual and natural one to make in describing the motion of # Hamiltonian degrees of freedom. "Natural" because Liouville's Theorem shows that the hypervolume, evidently proportional to the "number of states", is conserved by Hamilton's equations of motion. The detailed microstate of such a system corresponds to a single point in the 2#-dimensional $\{\ q,p\ \}$ phase space.

Gibbs and Boltzmann both made the connection that the phase volume associated with a macrostate is the exponential of the thermodynamic entropy. Let us consider macrostates characterized by their energy and allowed to interact through some weak coupling. By maximizing the phase volume of two systems by allowing energy transfer between them, the phase volume of these interacting states (with fixed total energy) is evidently maximized when the two phase volumes vary equally with energy :

$$(\partial \ln \Omega_1/\partial E_1) + (\partial \ln \Omega_2/\partial E_1) = 0 \rightarrow (\partial S/\partial E)_1 = (\partial S/\partial E)_2 \rightarrow T_1 = T_2\ ,$$

the condition of thermal equilibrium. By choosing one of the systems to be an ideal gas (for which the phase volume can be calculated analytically) one finds that $kT = m\langle\ v_x^2\ \rangle$ and that the distribution of velocities approaches the Maxwell-Boltzmann distribution, $\propto \exp(-mv^2/2kT)$.

Nosé had the clever idea of finding equations of motion consistent with the Maxwell-Boltzmann distribution function. His approach was relatively complicated and based on Hamiltonian mechanics. We consider here an interesting example, a one-dimensional "Nosé-Hoover" oscillator. If we choose the mass, force constant, temperature, and relaxation time all equal to unity, the differential equations to be solved are

$$\dot{q} = p\ ;\ \dot{p} = -q - \zeta p\ ;\ \dot{\zeta} = p^2 - 1\ .$$

It is easy to verify that the Gaussian distribution,

$$f(q,p,\zeta) = \exp[\ -(q^2 + p^2 + \zeta^2)/2\ ]/(2\pi)^{3/2}\ ,$$

is "stationary" (independent of time) by considering the flows into and out of a small cube in the three-dimensional (q,p,ζ) space :

$$(\partial f/\partial t) = -(\partial(f\dot{q})/\partial q) - (\partial(f\dot{p})/\partial p) - (\partial(f\dot{\zeta})/\partial\zeta) =$$

$$f[\ qp\ ] + f[\ p(-q - \zeta p) + \zeta\ ] + f[\ \zeta(p^2 - 1)\ ] \equiv 0\ .$$

The apparent simplicity of a Gaussian distribution conceals within it a chaotic solution, shown in cross section in **Figure 1.6**, as well as two of the infinitely many nonchaotic solutions.

A representative *chaotic* solution was obtained by starting out with $\{\ q,\ p,\ \zeta\ \} = \{\ 0,\ 5,\ 0\ \}$. It is evident that it occupies only a fraction of the oscillator phase space. The solution looks "chaotic", and in fact it is – the largest Lyapunov exponent is positive. For a *chaotic* many-body example see **Figure 2.3** on page 31 .

Giancarlo Benettin suggested measuring this largest Lyapunov exponent by following two different trajectories in (q,p) *phase space*, with the second constrained (by rescaling the separation after each timestep) to stay close to the first. Spotswood Stoddard and Joseph Ford had used the same idea separately in coordinate space and in momentum space. "Chaos" refers to the situation in which the future is *exponentially sensitive* to changes made in the present.

For the data shown in **Figure 1.7** the separation of 0.000001 was added initially to q and then rescaled at each timestep :

$$\delta \equiv \sqrt{(q_2 - q_1)^2 + (p_2 - p_1)^2 + (\zeta_2 - \zeta_1)^2} \longrightarrow 10^{-6}\ .$$

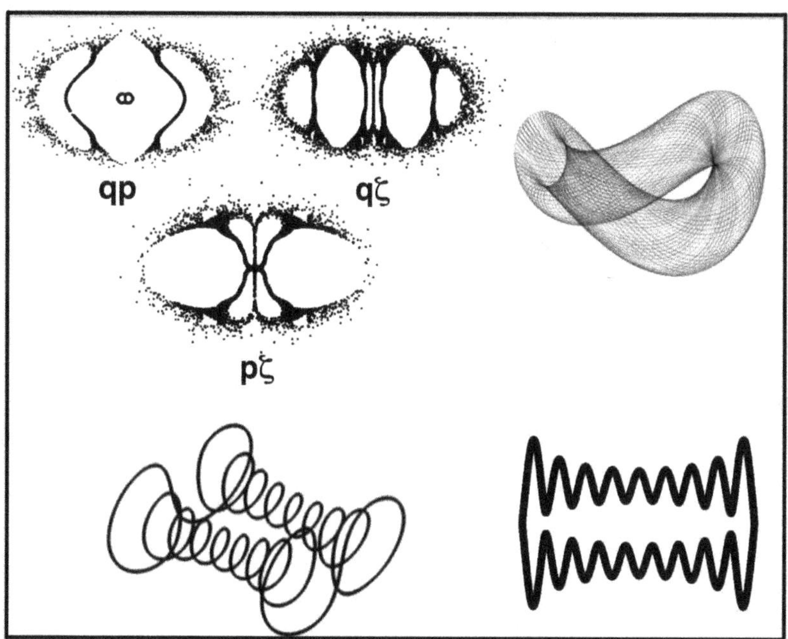

Fig. 1.6 Three "Poincaré" cross sections from a chaotic Nosé-Hoover oscillator trajectory are shown at the upper left. The initial conditions are $\{ q, p, \zeta \} = \{ 0, 5, 0 \}$. The "holes" in the sections are occupied by "regular" toroidal regions enclosing nonchaotic periodic orbits like those indicated at the upper right and lower left. The $\{ q, p \}$ projection of the lower left $\{ q, p, \zeta \}$ solution, with nonchaotic initial conditions $\{ q, p, \zeta \} = \{ 0,\ 1.43,\ 0 \}$, appears at the bottom right. The initial conditions for the torus at the upper right are $\{ q, p, \zeta \} = \{ 0, 2, 0 \}$.

The longtime average of the scale factors' logarithms is simply related to the Lyapunov exponent λ :

$$\lambda = (-1/dt)\langle \ln[\ 0.000001/\delta\ ]\ \rangle \ .$$

Rescaling the separation between the two solutions of the Nosé-Hoover oscillator equations makes it possible to average the rate of exponential divergence over a relatively long trajectory. Without this rescaling only a relatively short trajectory could be explored. The sum total of 10^7 Lyapunov-exponent estimates is about 140,000 , giving an average $\lambda \simeq 0.014$.

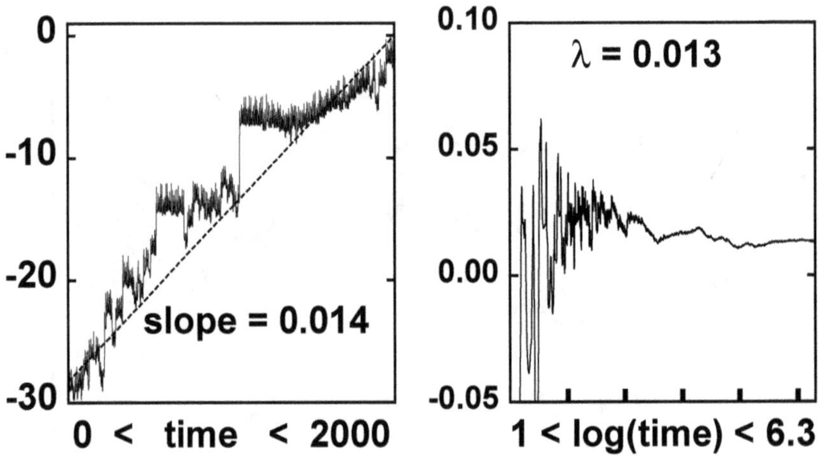

Fig. 1.7 Time-dependence of the exponential rate at which two Nosé-Hoover oscillators separate. The ordinates are $\ln \sqrt{\delta_q^2 + \delta_p^2 + \delta_\zeta^2}$ and the time-averaged slope for a longer simulation in which the phase-space separation is rescaled at the end of every timestep.

1.5 Nosé-Hoover Control of Kinetic Temperature

We have just illustrated Nosé's idea for simulating canonical ensemble dynamics for an oscillator. But that is only the simplest example (and actually fails to generate the ensemble) so it hides the generality and the usual applicability of his ideas. It is relatively easy to implement Nosé's monumental contribution to statistical mechanics and simulation. We impose what the engineers call "integral feedback control" on particle velocities. Feedback, based on the difference between the current temperature $\langle\ (p^2/mk) = (mv^2/k)\ \rangle$ and the target temperature T, forces the velocities to satisfy a time-averaged condition corresponding to a unique kinetic temperature. If the dynamics is sufficiently mixing to approach *all* velocity states then the resulting velocity distribution is guaranteed to be the Maxwell-Boltzmann distribution.

To take advantage of Nosé's idea, notice that adding a frictional force $-\zeta p$ to the differential equations of motion of the thermostated particles :

$$\{\ \dot{p} = F - \zeta p\ \}\ ;\ \dot{\zeta} = (1/\#)\sum^{\#}[\ (p^2/mkT) - 1\ ]/\tau^2\ ,$$

necessarily has a stationary distribution with $\langle\ p^2\ \rangle = mkT$. To see this just consider the time average of the $\dot{\zeta}$ equation. It is certainly less than obvious that the resulting stationary distribution is a simple Gaussian :

$$f(q,p,\zeta) \propto e^{-(q^2+p^2+\zeta^2)/2}\ ,$$

including Gibbs' canonical distribution for the oscillator variables (q, p) . We use the phase-space continuity equation to prove this result in Section 5.7.1 . This surprise, and the motion equations leading to it, are not unique.

Rather than choosing a *linear* force to control the second moment of momentum, *cubic* or *quintic* frictional forces can be used to control the fourth or sixth moments :

$$\dot{p} = F - \zeta p^3 \ ; \ \dot{\zeta} \propto [\ p^4 - 3p^2 mkT\ ] \longrightarrow \langle\ p^4\ \rangle \simeq 3(mkT)^2 \ ;$$

$$\dot{p} = F - \zeta p^5 \ ; \ \dot{\zeta} \propto [\ p^6 - 5p^4 mkT\ ] \longrightarrow \langle\ p^6\ \rangle \simeq 15(mkT)^3 \ .$$

When applied to the simplest demonstration problem, a single harmonic oscillator (so that $F = -\kappa q$), it turns out that the momentum distribution is far from ergodic. The motion has a single chaotic solution as well as an infinite number of nearly periodic torus-type solutions. It is only the sum total of all of these many solutions which is Gaussian.

In the not-too-many-body case it is usual to imagine an "ergodic" solution without separation into disjoint parts. In that same case it is highly *unusual* to remark that the N-body phase space likely contains $N!$ similar but disjoint parts corresponding to the permutations of the particle positions along with their momenta.

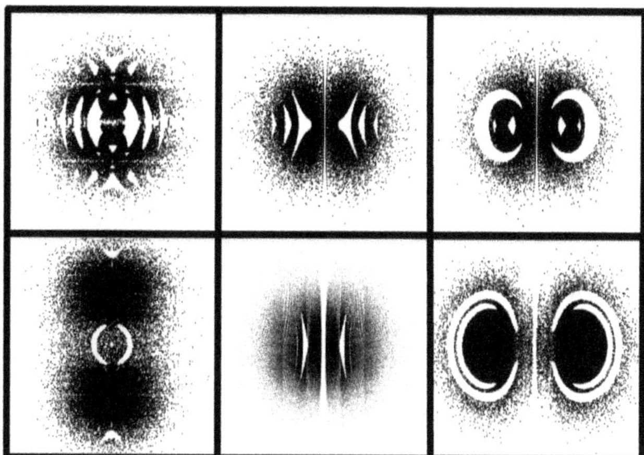

Fig. 1.8 Phase-space cross sections like those of Figure 1.6 for dynamics in which $\langle\ p^4\ \rangle$ and $\langle\ p^6\ \rangle$ are constrained. $(\ q, p\ )$, $(\ q, \zeta\ )$, and $(\ p, \zeta\ )$ are shown, from left to right, with fourth-moment-control sections above and sixth-moment-control sections below.

Figure 1.8 shows the "Poincaré surfaces" for $(\zeta = 0, p = 0, q = 0)$ using a large initial kinetic energy ($p = 5$). Control of the second, fourth, and sixth moments of the momentum distribution produces three quite different chaotic phase-space densities. The momentum distribution functions are quite different too, as is shown **Figure 1.9** .

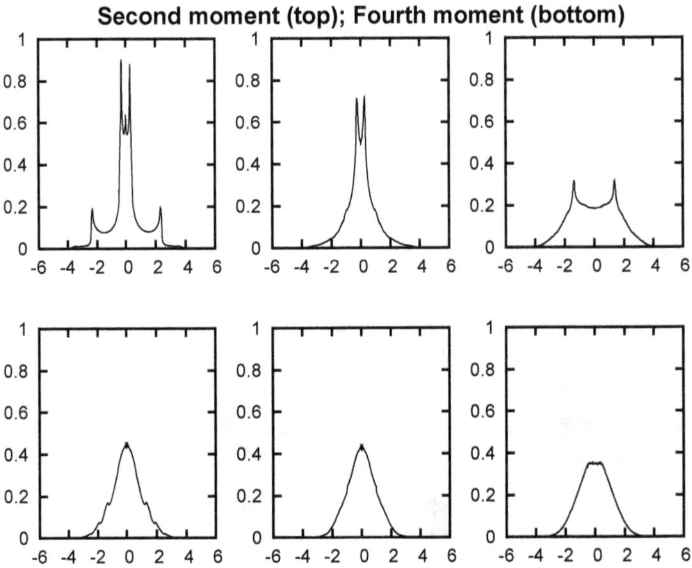

Fig. 1.9 Probability densities for q, p, and ζ for a chaotic harmonic oscillator with $\langle p^2 \rangle = 1$ in the top row and $\langle p^4 \rangle \simeq 3$ in the bottom row. The corresponding friction-coefficient equations are $\dot{\zeta} = p^2 - 1$ and $\dot{\zeta} = p^4 - 3p^2$. A useful chaotic initial condition is $(q, p, \zeta) = (0, 5, 0)$.

Although the addition of a frictional force to the equation of motion appears suspicious or arbitrary we will see later that this addition can be made "natural" by an application of Hamiltonian mechanics. These modified equations of motion are said to be "thermostated" in the sense that any solution of them will obey the same chosen relation linking the thermostated moments as do the corresponding Maxwell-Boltzmann equilibrium moments.

1.6 Nonequilibrium Multifractal Distributions

Probability densities can be smooth and continuous, like the Maxwell-Boltzmann velocity distribution. They can also be singular, but integrable, like the probability density $f(x)$ for the harmonic oscillator's spatial distribution. With the mass, force constant, and energy all equal to unity, that distribution is proportional to the time spent in dx during half the oscillator's period :

$$f(x)dx = dt/\pi = (dx/\dot{x}\pi) \rightarrow f(x) = 1/(\pi\sqrt{2-x^2}) \ ,$$

with integrable singularities at the turning points, $x = \pm\sqrt{2}$. In the vicinity of these singular points the *fraction* of the probability within the range dx varies as $dx^{(1/2)}$. If the fractional powerlaw, here $(1/2)$, varies throughout space the distribution is said to be "multifractal".

A very interesting consequence of Nosé's thermostating idea occurs for stationary *nonequilibrium* states. Such a state is so rare that its phase-space distribution is *typically* multifractal. The probability density is singular *everywhere.* The distribution is necessarily integrable because it is normalized, but the probability of finding a nonequilibrium state "by accident" is zero. The probability of finding a time-reversed state, with negative entropy production, is likewise zero.

The simplest nonequilibrium steady state is a version of the cell model with a vertical gravitational field. Despite the field, the speed of the falling particle, $\sqrt{v_x^2 + v_y^2}$, is kept constant, "thermostated", by using a friction coefficient. If the scatterers are hard disks this model is the isokinetic "Galton Board", named after Sir Francis Galton's classroom generator of the binomial distribution. If the scatterers are soft the cell model problem, viewed from a coordinate sytem with the center of mass located *between* the two particles, becomes the "color conductivity" problem in which each particle develops a current, roughly proportional to the field strength and generates fractal states like those of the Galton Board.

Notice that the entropy $-k\langle \ \ln(f) \ \rangle$ computed from a fractal distribution is *divergent.* This extremely interesting result, which we will consider in more detail later demonstrates that the concept of a nonequilibrium entropy is fatally flawed. In fact the divergence of the Gibbs entropy is more closely related to entropy production than to entropy itself. The rate of nonequilibrium entropy production is roughly proportional to the *squares* of the gradients which distinguish nonequilibrium states from their equilibrium relatives.

Despite the clearcut evidence from this Galton Board problem (all the details are given in Section 8.5, page 236) and *many* other thermostated nonequilibrium examples there is literature suggesting that the nonequilibrium entropy is finite rather than divergent. This mistaken claim is simply one of George Stell's *Setbacks in Physics* examples, a misinterpretation of a problem previously solved correctly in the literature.

If we consider the sensitivity of the scattering event to small configurational perturbations (See **Figure 1.10**) we see a spreading, with perturbations undergoing *exponential* growth in both directions of time. This growth is responsible for the Hamiltonian version of the Second Law of Thermodynamics, and shows that the width of regular distributions, pursued either forward or backward in time, has a tendency to spread.

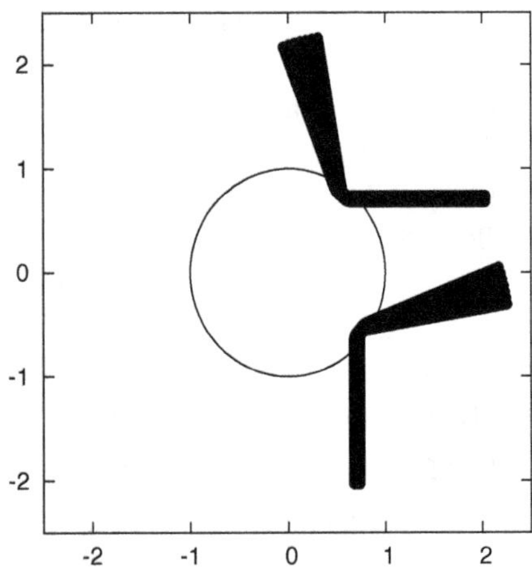

Fig. 1.10 Parallel trajectories moving up and moving to the left spread as a consequence of scattering from a fixed scatterer. The resulting exponential instability of the moving particle's trajectory is the same whether the motion is followed forward or backward in time.

1.7 Nonlinear Transport

Simulations of nonequilibrium states make it possible to explore nonlinear effects. The coupling of shear and normal stresses can be seen in the tendency of paint to climb a rotating stirring rod and in the tendency of paint flowing down a trough (and so again under shear) to rise or to fall in the center of the trough. For simple atomistic particles the magnitude of these nonlinear effects is quite small unless the gradients are sensible on the same length scale as the spacing between neighboring particles. This *is* the case in a strong shockwave.

In **Figure 1.11** we show the dependence of the differences in the normal components of the pressure tensor, P_{xx} , P_{yy} , and P_{zz} , for a periodic shear flow with P_{xy} nonzero. There is no significant change in the shear stress

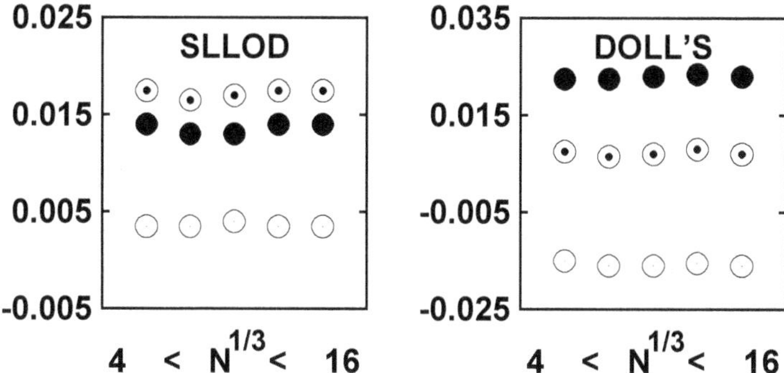

Fig. 1.11 Number-dependence of the normal stresses in steady shear as a function of system size. Open circles indicate $(1/2)(P_{xx} - P_{yy})$, circled dots indicate $(1/2)(P_{xx} - P_{zz})$ and filled circles indicate $(1/2)(P_{yy} - P_{zz})$. These normal stresses are much smaller than the shear stress $-P_{xy} = 0.344$ induced by the variation of u_x with y , $(du_x/dy) = 0.50$. The pair potential is $100(1 - r^2)^4$. The density and the energy per particle are unity. The data are taken from our 2008 work with Janka Petravic.

$-P_{xy}$, as the system size is increased from $6 \times 6 \times 6$ to $14 \times 14 \times 14$, an advantage of periodic boundaries. It is noteworthy that neither of these periodic algorithms, Doll's and Sllod, gets the correct (boundary-driven) ordering of the tensor temperature components : $T_{xx} > T_{zz} > T_{yy}$.

1.8 Time-Reversible Thermostats and Thermometers

Gibbs' statistical mechanics is most simply developed with a basis in the pressure and entropy of a classical ideal gas. Such a gas is so dilute that its potential energy can be ignored. A variational calculation shows that the most likely distribution of velocities is the familiar Maxwell-Boltzmann distribution ,

$$f(p_x, p_y) = e^{-p_x^2/2mkT} \times e^{-p_y^2/2mkT}/(2\pi mkT) \ .$$

A sufficiently large scale ideal-gas "thermometer" can eventually impose its (constant) temperature on any otherwise isolated system with which it is in thermal contact. The Nosé-Hoover motion equations ,

$$m\ddot{x} = F - \zeta m\dot{x} \ ; \ \dot{\zeta} = [\ (m\dot{x}^2/kT) - 1\ ]/\tau^2 \ ,$$

are an example of "integral feedback". If the kinetic energy is less/more than its target value the friction coefficient ζ becomes more negative/more positive. The longtime average value of kinetic energy is necessarily $kT/2$

for each degree of freedom. The Nosé-Hoover equations are a simple thermostat, a model for the interaction of a system with a heat reservoir. Rather than modelling the *many* degrees of freedom making up a macroscopic thermostat Nosé's ideas make it possible to replace all that structure with a single variable, the friction coefficient ζ . The ideal-gas thermometers shown

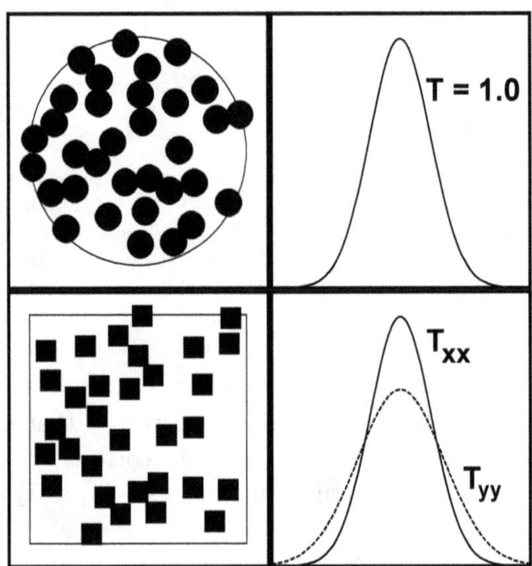

Fig. 1.12 The ideal-gas thermometer is an *equilibrium* collection of many very small disks, spheres, parallel squares, or parallel cubes. Used as a diagnostic of temperature the thermometer measures $kT_{(xx \text{ or } yy)} \equiv \langle\ mv^2\ \rangle_{(x \text{ or } y)}$.

in **Figure 1.12** , if sufficiently complex and macroscopic, can cause degrees of freedom with which it interacts to approach its own Maxwell-Boltzmann temperature T . Baranyai has carried out this idea explicitly. He considered a thermometer composed of a few hundred particles linked together with short-ranged forces. The thermometer was about the same size and mass as the system particles. The temperature was then estimated from the thermometer's comoving kinetic energy. Because this thermometer has fluctuations of its own (and because the rotational and translational degrees of freedom complicate its interpretation) it is simpler and more clearcut to *define* local temperatures in terms of averaged kinetic energies. A two-dimensional version of Baranyai's thermometer is shown in **Figure 1.13** . Neville Temperley suggested a relatively-simple mechanical model for defining nonequilibrium states, a massive piston exposed to a steady flow. His idea was to use the measured stress on the piston, and its fluctuations, to define the nonequilibrium pressure tensor and the temperature.

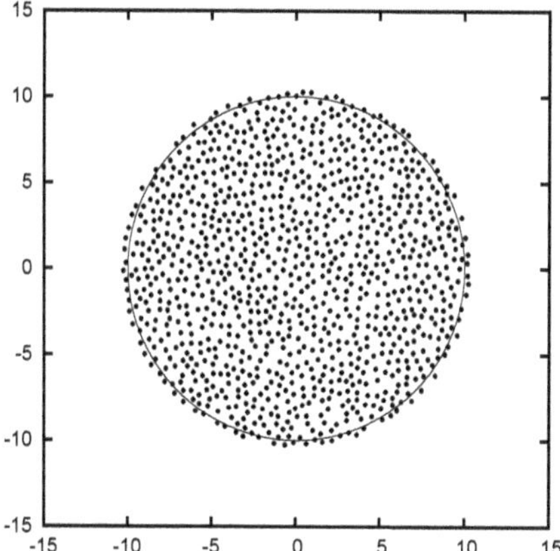

Fig. 1.13 A Baranyai thermometer is made up of many strongly interacting particles held together by attractive forces. The "thermometer" responds to collisions with nearby system particles.

1.9 Upcoming Background for Our Numerical Examples

Let us pause in this introduction to describe the organization of the remainder of this book. We began with a description of particle mechanics, as it was developed by Newton, Lagrange, Hamilton, and Nosé. The connection between particle mechanics and macroscopic physics is most easily made by applying Gibbs' and Boltzmann's statistical mechanics. Because temperature is the new variable distinguishing mechanics from thermodynamics we have developed motivation for defining temperature in kinetic terms. Thermometers are a natural outgrowth of this approach.

With temperature and thermometers available we can go on to explore the correspondence of atomistic simulations with continuum mechanics. The link between point and continuum properties is best made with the "weight functions" borrowed from smooth-particle applied mechanics. The smooth-particle technique provides twice-differentiable (in space) continuum fields as a description of particle properties defined at points. This approach makes it possible to explore in detail the differences in time-symmetry of particle mechanics and continuum mechanics.

The remainder of the book contains the explanation and application of all those numerical methods which are useful in connecting the microscopic and macroscopic approaches. After more than fifty years of computer simulation there is no possibility of a comprehensive coverage. On the other hand we are confident that nothing important need be left out. Our goal is to provide sufficient background that the reader can construct simple computer programs to apply all of these techniques. It is possible, and even commonplace, for students to find and use packaged software for simulation projects. We don't endorse that approach, believing that home-grown codes directed toward interesting problems are the most reliable route to the understanding, retaining, and refining of significant ideas, making it possible to transfer this knowledge to others through the various avenues of electronic and conventional publication and face-to-face meetings.

1.10 Summary

Although classical mechanics, formulated by Newton, Lagrange, and Hamilton, is an all-important means to understanding physical systems, it is not enough for understanding nonequilibrium systems. Steady states are the simplest nonequilibrium systems, shear flows, heat flows, shockwaves, and so on. Control variables with a relationship to Gibbs' statistical mechanics are valuable simulation tools. They have the fringe benefit of a better understanding of nonequilibrium steady states in the form of multifractal phase-space distributions. This introductory Chapter sketches out the regions to be explored in depth in the remainder of the book.

1.11 References

There is no shortage of references detailing Newtonian, Lagrangian, and Hamiltonian mechanics. Pars' *Treatise on Analytic Dynamics* is specially useful. Our paper with Janka Petravic on shear flows, "Simulation of Two-and-Three-Dimensional Dense-Fluid Shear Flows *via* Nonequilibrium Molecular Dynamics : Comparison of Time-and-Space-Averaged Stresses from Homogeneous Doll's and Sllod Shear Algorithms with Those from Boundary-Driven Shear", Physical Review E **78**, 046701 (2008), is a useful place to start an exploration of stationary shear flows. Unless one defines temperature in terms of a mechanical measurement, that concept can be elusive. For recent theoretical attempts to confront the concept headon see Peter Daivis' "Thermodynamic Relationships for Shearing Linear Viscoelastic Fluids", Journal of NonNewtonian Fluid Mechanics **152**, 120-128 (2008) . A wide variety of thermostated heat flows is considered in our paper with Ken Aoki and Stephanie De Groot, "Time-Reversible Deterministic Thermostats", Physica D **187**, 253-267 (2004) .

For the connection of Nosé-Hoover mechanics to Hamiltonian mechanics see Dettmann and Morriss' "Hamiltonian Reformulation and Pairing of Lyapunov Exponents for Nosé-Hoover Dynamics", Physical Review E **55**, 3693-3696 (1997) . Benettin, Galgani, Giorgilli, and Strelcyn's paper describing the computation of the Lyapunov exponents can be found in Meccanica (1980) . See also Spotswood Stoddard's work with Joe Ford in the 1973 Physical Review A .

For some of the perils of classical Lagrangian and Hamiltonian mechanics see our arχiv 1303.6190 : "Hamiltonian Thermostats Fail to Promote Heat Flow" in Communications in Nonlinear Science and Numerical Simulation **12**, 3365-3372 (2013) . This paper contains a fairly comprehensive set of computational thermostat references, from the early days of molecular dynamics up to 2013 . For additional reading about computational thermostats we specially recommend a look at some of Arek Brańka's work with Kris Wojciechowski, for instance "Nosé-Hoover Chain Method for Nonequilibrium Molecular Dynamics Simulation", Physical Review E **61**, 4769-4773 (2000) ; "Generalization of Nosé and Nosé-Hoover Isothermal Dynamics", Physical Review E **62**, 3281-3292 (2000) .

Clint Sprott's clear and profusely illustrated *Chaos and Time-Series Analysis* (Oxford University Press, 2003) is a rewarding introduction to chaotic geometry.

The Joule-Thomson adiabatic expansion experiment and the history of its interpretation have been reviewed by Sir John Rowlinson's "James Joule, William Thomson, and the Concept of a Perfect Gas", Notes and Records of the Royal Society **64**, 43-57 (2010) . The simulation in **Figure 1.4** is described in Wm. G. Hoover, Carol G. Hoover, and Karl P. Travis' "Shock-Wave Compression and Joule-Thomson Expansion", Physical Review Letters **112**, 144504 (2014) .

For more details of the smooth-particle averaging technique as applied to atomistic simulations see Bill's 2006 book, *Smooth Particle Applied Mechanics; the State of the Art* .

There are eight more arχiv contributions, two with Francisco Uribe, describing the simulation of stationary shockwaves with molecular dynamics and continuum mechanics. Two of these, along with two older papers (1979 and 1980) with Brad Holian, Bill Moran, and Galen Straub can be found with a search on the American Physical Society's website [prola.aps.org] . Gibbs' *Elementary Principles in Statistical Mechanics* has been reprinted by Oxbow Press. We also recommend Bill's *Molecular Dynamics* and *Computational Statistical Mechanics*, both available free at our website [http://williamhoover.info], and Denis Evans' and Gary Morriss' *Statistical Mechanics of Nonequilibrium Liquids*, likewise free at those authors' websites.

1.12 Problems

1. Sum up the vertical forces between a uniformly charged line occupying the x axis and a charge located at $(0, y)$ for $\{ y \} = \{ 1, 2, 4, 8 \}$. What is the functional form of the interaction energy from which the force follows by differentiation? Use a charge density of unity.

2. Sum the interactions between a uniformly charged xy plane and a charge located at $(0, 0, z)$ for $\{ z \} = \{ 1, 2, 4, 8 \}$ to establish the functional form of the interaction energy.

3. Find the stationary distributions $f(q, p, \zeta)$ corresponding to two sets of equations of motion : first, $\{ \dot{q} = p \ ; \ \dot{p} = -q - \zeta p^3 \ ; \ \dot{\zeta} = p^4 - 3p^2 \}$, and second, $\{ \dot{q} = p \ ; \ \dot{p} = -q - \zeta^3 p \ ; \ \dot{\zeta} = p^2 - 1 \}$.

4. For a one-dimensional momentum distribution with $\langle \ p^2 \ \rangle = 1$ what are the maximum and minimum values of $\langle \ p^4 \ \rangle$ and $\langle \ p^6 \ \rangle$?

5. Explain why two-dimensional *flows* cannot exhibit chaos while two-dimensional *maps* can.

Chapter 2

Formulating Atomistic Simulations

Topics

Newtonian Mechanics of Particles / Lagrangian Mechanics of Particles with Constraints / Hamiltonian Mechanics and Liouville's Theorem / Time-Reversible Leapfrog Algorithm / Energy Control Using Gauss' or Lagrange's or Dettmann-Morriss' Ideas / Boundary Conditions / Smooth Particle Averages / Constraints and Gibbs' Entropy /

2.1 Newton's Mechanics for Many-Body Particle Systems

Our goal here is to develop a simple but flexible microscopic description of the particle dynamics underlying macroscopic nonequilibrium flows of mass, momentum, and energy. Our starting point is atomistic point-mass classical mechanics. We picture N-body systems of interacting particles. We expect to analyze their reaction to heat and to work, as described by the laws of thermodynamics. We will consider situations in which atomistic analogs of density, velocity, energy, heat flux, and stress evolve with time. We will develop *smooth-particle techniques* which optimize the correspondence of underlying microscopic dynamics to a macroscopic continuum description.

Because the atomistic description is time-reversible while the continuum one is not, we expect the comparison to yield an improved understanding of both the small-scale and the large-scale approaches. We start with the interacting particle descriptions given to us by Newton's, Lagrange's, and Hamilton's mechanics and relate that mechanics to the continuum description given to us by fluid mechanics. We will need to include boundary conditions and constraint forces in the particles' motion equations in order to model work, heat transfer, and irreversible processes. We begin with a brief sketch of computational particle mechanics.

To start out, Newton's second-order motion equations, $\{ F = m\ddot{r} \}$ require initial conditions, $\{ r, \dot{r} \}$. A regular lattice, perhaps with random displacements and/or Maxwellian velocities, is the usual choice. The FORTRAN fragment :

```
index = 0
do ix = 1,4
do iy = 1,4
index = index + 1
x(index) = ix - 2.5
y(index) = iy - 2.5
enddo
enddo
```

generates a 4×4 array of $\{ x, y \}$ coordinates, centered on the origin. Because Newton's equations are *second*-order in the time, the *first* time derivatives, $\{ \dot{x} = v_x,\ \dot{y} = v_y \}$, need to be specified too. These velocities can either develop naturally, responding to chosen boundary conditions, constraints, or driving fields, or they can be selected from an appropriate Maxwell-Boltzmann distribution. Selecting random velocity components in the range :

$$\{ -8\sqrt{kT/m} < v < +8\sqrt{kT/m} \} ,$$

and accepting those with a probability $e^{-mv^2/2kT}$ produces an approximate thermal distribution. These velocities can be further improved by first removing the average velocity ,

$$v \longrightarrow v - \langle v \rangle = v - (1/N)\sum v ,$$

and then *scaling* the velocities in order to match the desired average value of the kinetic temperature T :

$$v \longrightarrow v\sqrt{kT/m\langle v^2 \rangle} .$$

Most often *boundary conditions* are required too, with the simplest choice *periodic*. With the coordinate origin at the center of the system any particle coordinate x which comes to exceed the maximum allowed, $+(L_x/2)$, is reïntroduced at $x - L_x$ with its velocity unchanged. Likewise any coordinate less than minimum, $-(L_x/2)$, is reïntroduced by adding

L_x . Provided the forces in Newton's equations come from a potential, $\{ F_x = m\dot{v}_x = -(d\Phi/dx) \}$, the total energy $E = \Phi + K$ is conserved :

$$\dot{E} = \dot{\Phi} + \dot{K} = \sum[\ (d\Phi/dx)\dot{x} + mv_x\dot{v}_x\ ] = \sum[\ -F_xv_x + mv_x\dot{v}_x\ ] \equiv 0\ .$$

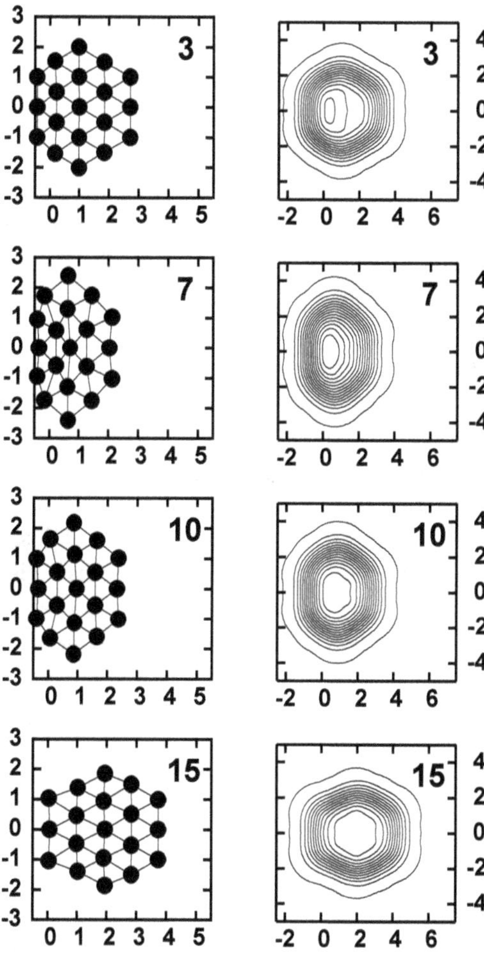

Fig. 2.1 The hexagonal solid bounces "inelastically" from the left wall at $x = 0$ according to a quartic wall potential. The loss in speed defines the "coefficient of restitution", $|\ v_{\text{after}}\ |/|\ v_{\text{before}}\ |$. The four snapshots of the bouncing hexagon correspond to the discrete times 3 , 7 , 10 , and 15 . The corresponding density contours, at equally-spaced values 0.1 , 0.2 , . . . , use Lucy's weight function with a range of 2.0 . Density is calculated as is outlined in Section 2.7 , $\rho(r) = \sum w(|r - r_i|)$.

The description of two- or three-dimensional many-body systems requires nothing more than adding on the terms involving the velocities and forces in the y and z directions.

An alternative to periodic boundaries is a smooth short-ranged repulsive boundary force, for instance :

$$\mid x \mid > (L/2) \longrightarrow \phi_{\rm B} = (1/4)[\mid x \mid - (L/2)\,]^4 \ .$$

Likewise, for the particle-particle interactions, purely-repulsive smooth and short-ranged forces can be generated with soft pair potentials such as $\phi(r < 1) = (1 - r^2)^4$ or $(1 - r)^3$. To minimize numerical integration errors it is desirable that the potential approaches zero *very smoothly* at the cutoff, here unity, with ϕ' and ϕ'' both vanishing there (rather than *jumping* discontinuously to zero) . Integrating the motion equations over a timestep dt leads to energy cutoff errors of order $\phi(1 - vdt)$ where v is a typical particle speed.

To fix ideas we consider the Newtonian collision problem illustrated in **Figures 2.1-2.2** . **Figure 2.1** shows four snapshots of a simple "many-body" problem, designed to determine the "coefficient of restitution" for a cold $(3 + 4 + 5 + 4 + 3 = 19)$-body hexagonal crystallite bouncing from a smooth repulsive barrier at $x = 0$. The coefficient, the ratio of the speeds after and before bouncing, characterizes the inelastic loss of energy in the collision process. At sufficiently low speeds the collision is nearly elastic so that the restitution coefficient approaches unity. In the snapshots we use a harmonic Hooke's-Law potential for all the nearest-neighbor particle pairs, $\phi(r) = (\kappa/2)(\mid r \mid - 1)^2$ and an initial speed for the stress-free hexagon of 0.25 . For convenience the force constant κ and the particle masses are both set equal to unity.

There is a smooth cubic repulsive boundary force from the one-sided boundary potential, $\phi(x < 0) = x^4 \longrightarrow F(x < 0) = -4x^3$, for any negative values of the horizontal coordinates $x < 0$. **Figure 2.2** shows the time dependence of the center-of-mass speed, $(1/19)\mid \sum \dot{x}_i \mid$ of the hexagon for an initial speed of 0.25 .

At low velocities the hexagon behaves like a rigid body, bouncing back with nearly the same speed as the original approach speed. In the collision process a part of the kinetic energy is first converted into soundwaves and then later into heat, as evidenced by the loss of the kinetic energy associated with the center-of-mass motion.

For this hexagonal-solid problem, where the interacting pairs don't change with time, it is convenient to start out with a list of nearest-neighbor

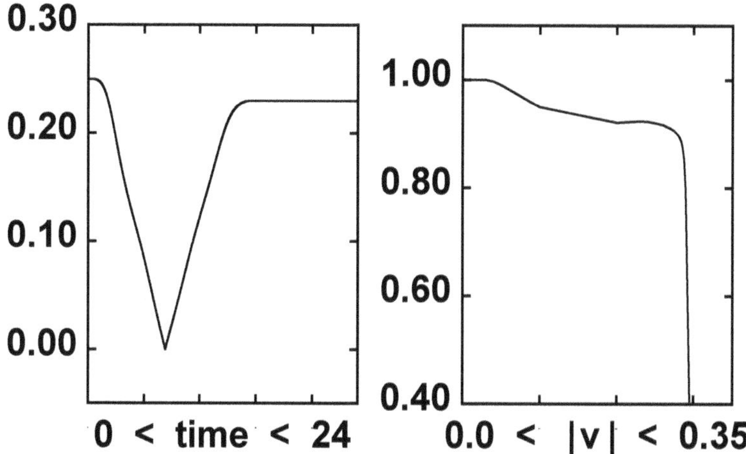

Fig. 2.2 The time dependence of the center-of-mass speed $(1/19)|\sum \dot{x}|$ of the bouncing hexagon, with an initial speed of 0.25 , is shown at the left. The coefficient of restitution, at the right, decreases to 0.40 as the initial speed is increased from 0 to 0.295 .

particle pairs. For more elaborate simulations the computational space can be divided up into a gridwork of cells, so that only particle pairs in the same or neighboring cells need be considered.

Chapter 5, "Numerical Molecular Dynamics and Chaos", is devoted to the details of the numerical methods for solving such problems. In Section 2.4 we preview the simplest of the many integrators for Newton's motion equations (the "Leapfrog" algorithm) :

$$\{ x_{t+dt} \equiv 2x_t - x_{t-dt} + (dt)^2 \ddot{x}_t \ ; \ y_{t+dt} \equiv 2y_t - y_{t-dt} + (dt)^2 \ddot{y}_t \} \ .$$

Armed with such an integrator, and given a formulation of the forces, including boundary forces, initial coordinates, and velocities, it is easy to solve Newton's Cartesian motion equations. Many-body Cartesian problems are simple to solve but can be complex to analyze. A physical *understanding* of the solution, the most rewarding goal of such work, requires the formulation and evaluation of a number of macroscopic concepts, including density, temperature, pressure, and heat flux, all of them to be described later.

Let us turn next to a modified, more general form of mechanics, *Lagrangian* mechanics, which is particularly well-suited to the treatment of problems with *generalized* coordinates, such as angles, or with *constraints*, such as rigid bond lengths or fixed kinetic energy.

2.2 Lagrange's Mechanics for Systems with Constraints

In addition to boundary conditions it is often desirable to specify *constraints*, fixed relationships among the coordinates, velocities, and accelerations. Perhaps the simplest constraint is that defining a rigid diatomic molecule by requiring that the two point masses making it up maintain a fixed separation, $\mid r_{12} \mid = \mid r_1 - r_2 \mid$.

More complicated constraints can impose fixed values of macroscopic variables, such as stress or heat flux. Such constraints can be "global", applying to whole systems, or "local", affecting only a few degrees of freedom. Both instantaneous and time-averaged constraints can be developed. **Figure 2.3** shows trajectories for a 25-atom crystalline solid with a single "cold" particle and a single "hot" one. The time-averaged temperatures of both "thermostated" particles are constrained with integral feedback, effective over a time of order τ :

$$\{ \dot{p} = F - \zeta p \; ; \; \dot{\zeta} = [\ (p^2/mkT) - 1\ ]/\tau^2 \ \} \ .$$

Lagrange's mechanics is a useful basis for problems with constraints. Rather than the energy sum $E(x, v_x) = K + \Phi$ of Newtonian mechanics, Lagrange's mechanics is based on an energy *difference*, the Lagrangian $\mathcal{L}(\dot{q}, q)$:

$$\mathcal{L}(\ q, \dot{q}\ ) \equiv K(q, \dot{q}) - \Phi(q) \ .$$

Here the "generalized" coordinates $\{\ q\ \}$ can include angles or other nonlinear combinations of the Cartesian coordinates. The Lagrangian equations of motion describing the evolution of the coordinates and velocities follow from Hamilton's Least-Action Principle. For Cartesian coordinates the Least-Action Principle shows that Lagrange's motion equations are equivalent to Newton's. That Principle states that the observed trajectory linking two coordinate sets at two times, $\pm dt$, is the one for which the "action integral", $\int_{-dt}^{+dt} \mathcal{L} dt'$, is an extremum (usually a minimum, so as to give the *least* action) . Suppose that the time interval dt is small. If we consider the effect of a variation $\delta q(0)$ on the action integral there are two contributions, coming from the dependence of the Lagrangian on the two not-so-independent variables q and $\dot{q}$:

$$\delta \int_{-dt}^{+dt} \mathcal{L}(q, \dot{q}) dt' = \int_{-dt}^{+dt} [\ (\partial \mathcal{L}/\partial q)\delta q + (\partial \mathcal{L}/\partial \dot{q})\delta \dot{q}\ ] dt' \ .$$

Integrating the last term by parts, and noting/assuming that δq vanishes at the endpoints, gives the form :

$$\delta \int_{-dt}^{+dt} \mathcal{L} dt' = \int_{-dt}^{+dt} [\ (\partial \mathcal{L}/\partial q)\delta q - (d/dt)(\partial \mathcal{L}/\partial \dot{q})\delta q\ ] dt' \ .$$

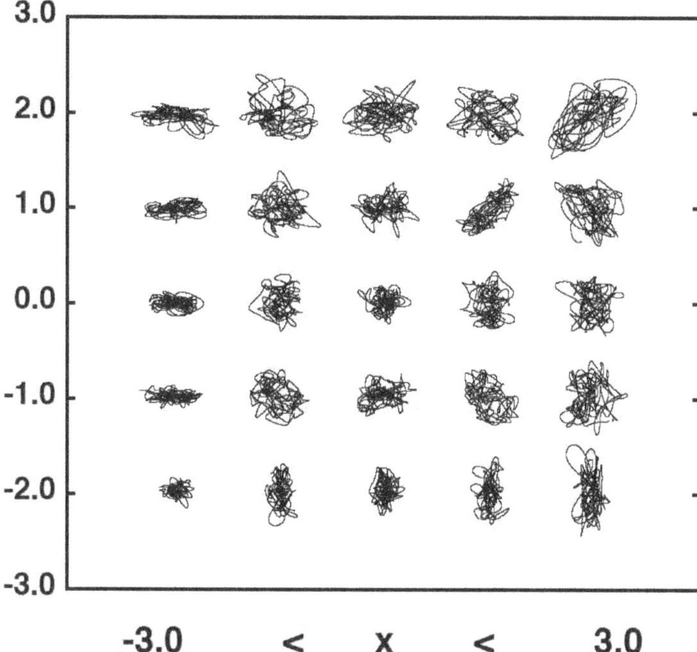

Fig. 2.3 Trajectories of a square-lattice ϕ^4 crystal with a cold particle at the lower left and a hot one at the upper right. Each particle is tethered to its lattice site with a quartic potential and linked to its nearest neighbors with a Hooke's-law potential.

Then, because δq is arbitrary (so that it could be concentrated at any arbitrary time inside the time interval $-dt < t' < +dt$) its coefficient must vanish. The consequence is the set of Lagrange's evolution equations, one for each $\{ \ q, \dot{q} \ \}$ pair :

$$\delta \int_{-dt}^{+dt} \mathcal{L} dt' = 0 \longrightarrow \{ \ (d/dt)(\partial \mathcal{L}/\partial \dot{q}) = (\partial \mathcal{L}/\partial q) \ \} \ .$$

These are Lagrange's equations of motion, written in terms of generalized coordinates and their time derivatives $\{ \ q, \dot{q} \ \}$. In the Cartesian case, with $\mathcal{L}(x, \dot{x}) = \sum (m\dot{x}^2/2) - \Phi(x)$, Lagrange's equations are the same as Newton's :

$$\{ \ (d/dt)(\partial (m\dot{x}^2/2)/\partial \dot{x}) = m\ddot{x} = -(\partial \Phi/\partial x) = F(x) \ \}.$$

In generalized coordinates the only extra work involved is formulating the kinetic energy in terms of the coordinates and their time derivatives.

To illustrate Lagrange's mechanics with a generalized angular coordinate, consider a freely rotating and vibrating two-dimensional diatomic molecule, with two identical masses, $m_1 = m_2 = m$, and with the molecule's center of mass at the origin. The two masses are separated by a variable distance $2r(t)$, and the orientation of the molecule is described by the polar angle $\theta(t)$. In the center-of-mass frame each mass has half the kinetic energy :

$$(K/2) = (m/2)(\dot{r})^2 + (m/2)(r\dot{\theta})^2 \ .$$

Adding a Hooke's-law spring constant κ , with potential energy $\phi \equiv (\kappa/2)(2r - 2r_0)^2$ to govern the vibrational motion, the molecule's Lagrangian is :

$$\mathcal{L} = m(\dot{r}^2) + m(r^2\dot{\theta}^2) - (2\kappa)(r - r_0)^2 \ ,$$

so that the equations of motion are :

$$(d/dt)(\partial\mathcal{L}/\partial\dot{r}) = (2m\ddot{r}) = (2mr\dot{\theta}^2) - 4\kappa(r - r_0) \ ;$$

$$(d/dt)(\partial\mathcal{L}/\partial\dot{\theta}) = (d/dt)(2mr^2\dot{\theta}) = 4mr\dot{r}\dot{\theta} + (2mr^2)\ddot{\theta} = (\partial\mathcal{L}/\partial\theta) = 0 \ .$$

The last equation's transparent conservation of the "angular momentum" $\sum[\ 2mr^2\dot{\theta}\ ]$ is a fringe benefit of using the polar angle θ as a generalized coordinate. Then the *second* time derivatives of r and θ follow easily from Lagrange' s motion equations.

Let us first take this opportunity to simplify the notation by choosing the particle mass m , the Hooke's law spring constant κ , and half the spring's restlength r_0 all equal to unity. Then Lagrange's motion equations are :

$$\ddot{r} = r\dot{\theta}^2 - 2(r - 1) \ ; \ \ddot{\theta} = -(2\dot{r}\dot{\theta}/r) \ .$$

The best checks of the solution are conservation of the energy $K + \Phi$ and the angular momentum $2r^2\dot{\theta}$. Additionally, because both sides of Lagrange's motion equations are left unchanged by time reversal, $+t \longleftrightarrow -t$, any solution generated for increasing time could be played backward and would still obey the same motion equations.

In addition to the convenience of generalized coordinates Lagrangian mechanics simplifies the implementation of "constraints" on the motion. The motion of a *rigid* diatomic molecule, with the separation of the two unit masses set equal to 2 , could be described and enforced by choosing and using a Lagrange multiplier λ . Again setting the spring's force constant to unity, the constrained Lagrangian is :

$$\mathcal{L} = K - \Phi + \lambda(r^2 - 1) = (\dot{r}^2) + (r^2\dot{\theta}^2) - 2(r - 1)^2 + \lambda(r^2 - 1) \ .$$

The motion equations for the constrained oscillator become :

$$\ddot{r} = r\dot{\theta}^2 - 2(r-1) + \lambda r \ ; \ \ddot{\theta} = -(2\dot{r}\dot{\theta}/r) \ .$$

The multiplier λ will constrain the separation $2r \equiv 2$ provided that the *second* derivative of the constraint vanishes :

$$\ddot{r} \equiv 0 \longrightarrow \lambda = 2[\ 1 - (1/r)\ ] - \dot{\theta}^2$$

Of course the *initial conditions* must also obey the desired constraint and its *first* time derivative :

$$r \text{ constant } \longrightarrow r(t=0) \equiv 1 \text{ and } \dot{r}(t=0) \equiv 0 \ .$$

The diatomic-molecule problems, both with and without the radial constraint, are excellent pedagogical exercises for assimilating Lagrangian mechanics. It is specially noteworthy that once the constraint(s) are successfully implemented the resulting mechanics again conserves energy and angular momentum.

Constraints can also be applied to measure the sensitivity of the dynamics to small perturbations. To illustrate, we will consider a two-body generalization of the one-dimensional harmonic oscillator problem with its Newtonian motion equation $\ddot{x} = -x$.

Why is sensitivity to perturbations interesting? A recurring theme in many-body dynamics is the *exponential* sensitivity to small changes in the initial conditions. This sensitivity is called "chaos". Two nearby *chaotic* solutions have a (time-averaged) tendency to separate exponentially fast, $\simeq e^{\lambda t}$, from one another. This chaotic property is quantified by the perturbation's growth rate λ , the "Lyapunov exponent" . It can be measured with Lagrangian mechanics.

Let us illustrate the approach here for a nonchaotic harmonic-oscillator problem. We choose *two* coupled oscillators separated in coordinate space by 0.5 :

$$q_1 = -1.0 \ ; \ \dot{q}_1 = 0.0 \ ; \ q_2 = -1.5 \ ; \ \dot{q}_2 = 0.0 \ ,$$

and formulate their joint Lagrangian so as to maintain this separation :

$$L = (1/2)[\ \dot{q}_1^2 + \dot{q}_2^2 - q_1^2 - q_2^2\ ] + (\lambda/2)[\ (q_1 - q_2)^2 - 0.25\ ] \ .$$

The equations of motion,

$$\ddot{q}_1 = -q_1 + \lambda(q_1 - q_2) \ ; \ \ddot{q}_2 = -q_2 + \lambda(q_2 - q_1) \ ,$$

can be combined with the second derivative of the constraint equation ,

$$(\dot{q}_1 - \dot{q}_2)^2 + (\ddot{q}_1 - \ddot{q}_2)(q_1 - q_2) = 0 \ ,$$

giving

$$(\dot{q}_1 - \dot{q}_2)^2 - (q_1 - q_2)^2 + 2\lambda(q_1 - q_2)^2 = 0 \longrightarrow$$

$$\lambda = (1/2)\left[1 - \frac{(\dot{q}_1 - \dot{q}_2)^2}{(q_1 - q_2)^2} \right]$$

Figure 2.4 shows the two oscillators' trajectories with the constant-separation constraint included. The usual method of determining the tendency toward instability is to rescale one of the trajectories toward the other. The present approach instead treats the two trajectories symmetrically.

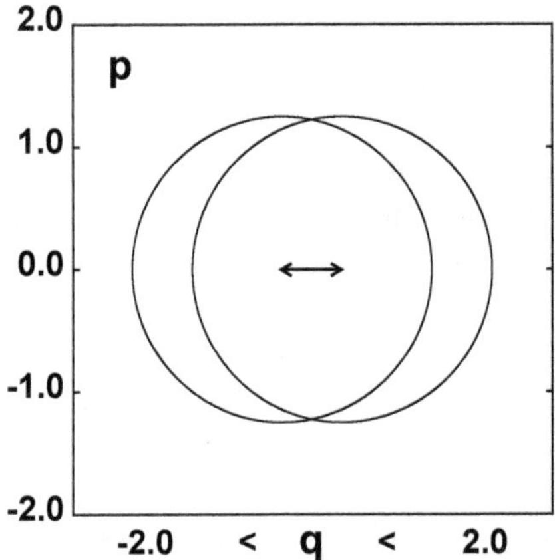

Fig. 2.4 $(q,p) = (q,\dot{q})$ orbits of two one-dimensional harmonic oscillators with unit masses and force constants together with a constrained separation, $|q_1 - q_2| = 0.5$.

2.3 Hamilton's Mechanics and Liouville's Theorem

Hamiltonian mechanics is specially important because it makes a crucial connection with Gibbs' *statistical* mechanics. Hamiltonian mechanics is also the basis for Schrödinger's version of quantum mechanics. The underlying Hamiltonian function $\mathcal{H} = \sum q\dot{p} - \mathcal{L}$, is in most cases the total energy, $\mathcal{H}(q,p) = K(p) + \Phi(q)$. But, just as in Lagrangian mechanics, *any* convenient generalized coordinates can be used to describe the system. Rather than $\{ q, \dot{q} \}$ the Hamiltonian is based on $\{ q, p \}$ with the momenta $\{ p \equiv (\partial\mathcal{L}/\partial\dot{q}) \}$ conjugate to the $\{ q \}$ replacing the not-so-independent

$\{ \dot{q} \}$ with the fully-independent variables $\{ p \}$. Accordingly, the Hamiltonian equations of motion are first-order in the time :

$$\{ p \equiv (\partial \mathcal{L}/\partial \dot{q}) \} \longrightarrow \{ \dot{q} = +(\partial \mathcal{H}/\partial p) \ ; \ \dot{p} = -(\partial \mathcal{H}/\partial q) \} \ .$$

In the event that the generalized coordinates are Cartesian, Hamilton's first-order motion equations reproduce Newton's second-order ones :

$$\mathcal{H}(q,p) = K(p) + \Phi(q) \longrightarrow \{ m\ddot{q} = (d/dt)p = F(q) \} \ .$$

The fundamental consequence of Hamilton's Mechanics is Liouville's Theorem. This is particularly clear in the Cartesian case where $\dot{q} = (p/m)$ is independent of q and $\dot{p} = F(q)$ is likewise independent of p. This means that the product $dq \times dp$ is unchanged by the motion, for all the coordinate-momentum pairs. A comoving element of phase volume is sheared, but neither grows nor shrinks, so that the comoving phase volume is conserved.

Liouville's Theorem is just this same result but for generalized coordinates and their momenta. The Theorem states that Hamilton's mechanics conserves comoving phase volume, the (hyper-) volume in phase space, $\prod dqdp$, where the product includes all # "degrees of freedom" (the number of coordinate-momentum pairs required to define the system's configurational and dynamical state).

Let us follow Gibbs in terming a small phase-space volume the "extension in phase". We adopt the symbol $\otimes$ for it. Any such extension in phase could be constructed of infinitesimal hypercubes. Consider the comoving deformation of a typical hypercube following Hamilton's motion equations. The motion distorts any plane of conjugate variables $dq \times dp$.

During a time interval dt all the pairs $\{ \dot{q}, \dot{p} \}$ obey Hamilton's motion equations. To compute the volume change during dt it is enough to keep all terms linear in dt , which vanish when summed up :

$$dq \times dp \xrightarrow{dt} [\ dq \ + \ dtdq(\partial \dot{q}/\partial q) \ ] \times [\ dp \ + \ dtdp(\partial \dot{p}/\partial p) \ ] =$$

$$[\ dq \ + \ dtdq(\partial/\partial q)(+\partial \mathcal{H}/\partial p) \ ] \times [\ dp \ + \ dtdp(\partial/\partial p)(-\partial \mathcal{H}/\partial q) \ ]$$

$$\longrightarrow$$

$$dq \times dp \left[\ 1 + dt[\ (\partial^2 \mathcal{H}/\partial q \partial p) - (\partial^2 \mathcal{H}/\partial p \partial q) \ ] \ \right] \equiv dq \times dp \ [\ ! \ ] \ .$$

Every plane described by a generalized coordinate-momentum pair is unchanged in area by the motion, just as was the case with Cartesian coordinates where $(\partial \dot{q}/\partial q)$ and $(\partial \dot{p}/\partial p)$ *both* vanish. Evidently an infinitesimal phase-space hypercube $\otimes$ which follows Hamilton's motion equations stretches or shrinks in each of the pairs of orthogonal q and p directions according to the signs of $(\partial\dot{q}/\partial q)$ and $(\partial\dot{p}/\partial p)$. The sum of all these $2\#$ changes vanishes :

$$\dot{\otimes} = (d/dt)\prod[\ dq \times dp\ ] \equiv 0\ ;\ [\ \text{Liouville's Theorem}\ ]\ .$$

Liouville's theorem is fundamental to Gibbs' connection of statistical mechanics with thermodynamics as is detailed in Chapter 3 .

Although Liouville's Theorem suggests that the geometry of Hamiltonian flows is "simple", like the flow of an incompressible fluid in phase space, the flow is typically highly complex. A small portion of the cell-model flow of **Figure 1.1** is shown as the motion of a grid defined by the 25 particles shown in **Figure 2.5** .

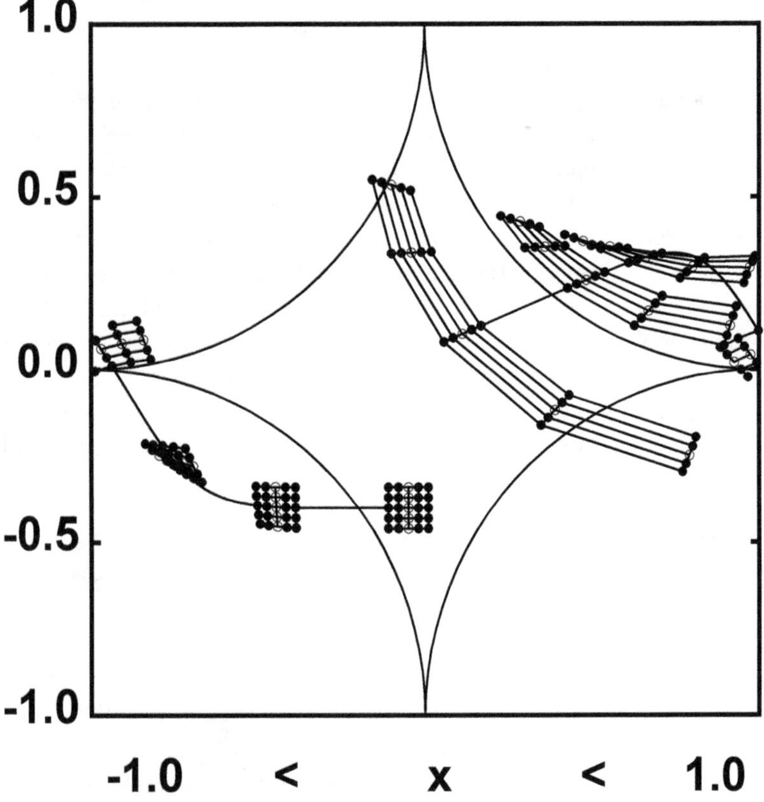

Fig. 2.5 Illustration of the exponential spreading of trajectories in the one-particle cell model. Snapshots of the deforming initially-square (x, y) grid defined by 25 particles are shown at six equally-spaced times. The initial particle velocities $(\dot{x}, 0)$ are all identical.

Inasmuch as dynamics underlies *all* simulations we turn next to an algorithm useful in solving Newtonian Cartesian dynamical problems numerically, the "Leapfrog Algorithm".

2.4 Time-Reversible Leapfrog Algorithm

In order that the reader can experiment with some simple dynamical simulations (see also the energy-based Problems at the end of this Chapter) we take the time here to describe the simplest of the many methods for solving Newton's Cartesian-coordinate equations of motion. The "Leapfrog" or "centered-second-difference" algorithm for solving $\{ F = ma = m\ddot{r} \}$ is based on Taylor's series expansions of the coordinates' time dependence. For simplicity we consider a simple one-variable $x(t)$ problem where the force $m\ddot{x}$ depends upon the coordinate. Newton's motion equation can be approximated by differencing two two-term Taylor's series for velocity :

$$F(x) = ma = m\ddot{x} \simeq (m/dt)[\ v_{(+dt/2)} - v_{(-dt/2)}\ ] \simeq$$

$$(m/dt^2)[\ x(t+dt) - 2x(t) + x(t-dt)\ ] = m[\ \ddot{x} + \tfrac{dt^2}{12}\ \ddddot{x} + \ldots\ ]\ ,$$

leading to the "Leapfrog Algorithm" :

$$x(t+dt) \equiv 2x(t) - x(t-dt) + F(t)(dt^2/m)\ .$$

This antique appears in Feynman's *Lectures* and is attributed in Milne's *Numerical Calculus* to Störmer's 1907 work. It has four very interesting features :

[i] It is an explicit and self-starting algorithm. Two successive coordinate sets give all the rest by simple summations, resembling those generating a Fibonacci series :

$$x(t \pm dt) = [\ (F(t)dt^2/m)\ ] + 2x(t) - x(t \mp dt)\ .$$

[ii] Like Newton's motion equations, this finite-difference algorithm is automatically time-reversible. Exchanging the two initial coordinate sets required to "start" the algorithm allows one to generate, equally well, the "past" as well as the "future".

[iii] It conserves (q,p) phase volume, provided that the momenta $\{ p \}$ are *defined* in such a way that reversing the direction of the sequence of

coordinate sets also changes the signs of the momenta. One could use, for instance $\{ \ p \equiv [\ q(t+dt) - q(t-dt) \ ]/2dt \ \}$.

[iiii] Levesque and Verlet considered a useful *integer* version of the algorithm, with the $\{ \ F(t)dt^2/m \ \}$ rounded off to integers. With a little care this algorithm can be made precisely reversible, "bit-reversible" for as long as desired, so that in principle such a solution, in a bounded coordinate space, is periodic.

To illustrate the periodic character, consider a harmonic oscillator with mass and force constant and timestep all equal to unity. For convenience we start out at a "turning point" $x_0 = 1$ at time $t = 0$. We find $x_{\pm dt} = x_{\mp dt} = \frac{1}{2}$, directly from the equations of motion :

$$x_{+dt} = 2x_0 - x_{-dt} - x_0 dt^2 = x_0 - x_{-dt} \longrightarrow x_{+dt} = x_{-dt} = \tfrac{1}{2}x_0 = \tfrac{1}{2} \ .$$

The "future" coordinates in this case are identical to the "past" because permuting the initial values of $x(t-dt)$ and $x(t+dt)$ leaves both of them unchanged. Extend the trajectory "forward" or "backward" and the result is the same repeating set of six coordinates :

$$\ldots, +1, +\tfrac{1}{2}, -\tfrac{1}{2}, -1, -\tfrac{1}{2}, +\tfrac{1}{2}, +1, +\tfrac{1}{2}, -\tfrac{1}{2}, -1, -\tfrac{1}{2}, +\tfrac{1}{2}, +1, \ldots$$

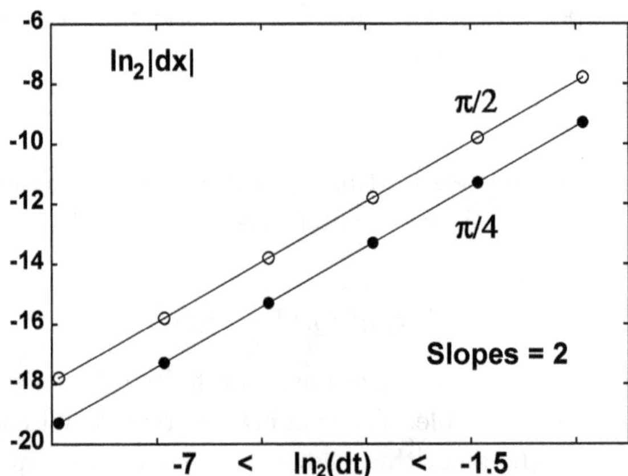

Fig. 2.6 Logarithmic plot (base 2) of $x(t)$ error *versus* dt at times $(\pi/2)$ and $(\pi/4)$. The timestep data vary from $(\pi/384) \leftarrow (\pi/12)$ along the abscissa. The errors are accurately proportional to dt^2 . The single-step leapfrog error is proportional to dt^4 . After two time integrations ($\ddot{x} \rightarrow \dot{x} \rightarrow x$) the error is that of a "second-order" integrator.

Here the period is $6dt = 6$, so that the error in period length (a "phase error" as opposed to an amplitude error) is $|dx| = 2\pi - 6 = 0.2832$. In **Figure 2.6** we plot the coordinate error for six different timesteps ranging downward from $(\pi/12)$ to $(\pi/384)$ and two different choices of maximum time, $(\pi/4)$ and $(\pi/2)$. These errors, relative to $\cos(t)$, vary accurately as dt^2 . There is no exponential divergence of changes to the initial conditions here. To implement a large-integer bit-reversible Leapfrog algorithm it is advisable first to read and understand Problem 4 at the end of this Chapter. In Chapter 5 we describe the Leapfrog algorithm in more detail and consider also algorithms with single-step errors of order dt^5 or dt^6 .

2.5 Energy Control by Gauss or Lagrange or Hamilton

Thermodynamics and its correspondence to equilibrium simulations have been well understood since about 1970. Our primary goal in this book is to show how to simulate and understand *Nonequilibrium* systems, the simplest of which are nonequilibrium steady states. Shear flows or heat flows, or the more complex shockwave and Joule-Kelvin flows [the same as "Joule-Thomson" flows] illustrated in **Figures 1.3-1.4** , all have the common feature of generating "irreversible" heating. Unless a mechanism for extracting this heat can be found the dynamics cannot simulate steady states. The first computational "thermostats" or "ergostats" focused on constraining the kinetic energy or the total energy by velocity scaling. If at time t the kinetic temperature $T(t)$ exceeded the target temperature $T_{\rm target}$ the velocities were simply rescaled ;

$$v_{\rm after} = v_{\rm before}\sqrt{T_{\rm target}/T(t)} \ .$$

Similarly, if the total energy exceeds the target energy, velocity rescaling can extract the excess :

$$v_{\rm after} = v_{\rm before}\sqrt{[\ E_{\rm target} - \Phi(t)\ ]/[\ E(t) - \Phi(t)\ ]} \ .$$

2.5.1 *Gauss' Principle of Least-Constraint* $\longrightarrow K = K_0$

Before 1980 neither of these *ad hoc* discontinuous velocity-scaling approaches appeared to be related to Newton's or Lagrange's or Hamilton's versions of mechanics. But in 1980 author Bill and Denis Evans noticed that Gauss' Principle of Least Constraint could be applied to such problems and furnished a differential equation of motion equivalent to isokinetic velocity scaling with $K(t) = K_0 = \sum m\dot{r}^2/2$:

$$\{ \ m\ddot{r} = F - \zeta m\dot{r} \ \} \ ; \ \zeta = \sum^{\#} F \cdot \dot{r} / \sum^{\#} m\dot{r}^2 \longrightarrow$$

$$\sum^{\#} m\ddot{r} \cdot \dot{r} = \sum^{\#} F \cdot \dot{r} - \left[\sum^{\#} F \cdot \dot{r} / \sum^{\#} m\dot{r}^2 \right] \sum^{\#} m\dot{r} \cdot \dot{r} \equiv 0 \ .$$

Gauss' Principle declares that the summed-up squares of the additional forces $\{ \ -\zeta m\dot{r} \ \}$ used to impose a *constraint* (here kinetic energy is held constant) should be the smallest possible. Because to linear order in dt forces perpendicular to velocity provide no change in kinetic energy it is clear that Gauss' constraint forces must act *parallel* to the velocities. Gauss' Principle goes beyond the Newton/Lagrange/Hamilton approaches in that it can be applied to *any* constraint. Let us next apply a conventional Lagrange-multiplier approach to the same problem.

2.5.2 *Applying a Lagrange Multiplier* $\longrightarrow K = K_0$

Author Bill and Tom Leete collaborated on this Lagrange-Multiplier idea in 1982 . It was in the relatively early days of thermostated molecular dynamics. Tom was finishing his Master's Thesis at West Virginia University. They worked out the algorithm, which Bill described in a "Nonlinear Fluid Phenomena" conference talk in Boulder that June. The conference proceedings are Volume **118A** of the journal Physica.

The idea follows a straightforward, though tedious, path. Consider the constrained Lagrangian, designed to keep the kinetic energy $K(t)$ at its initial value, $K_0 \equiv K(t=0) = \sum m\dot{q}^2/2$:

$$\mathcal{L} \equiv (1/2) \sum^{\#} m\dot{q}^2 - \Phi + \lambda(K - K_0) \ .$$

The constraint, $K - K_0 = 0$, is to be imposed by choosing $\lambda(t)$ correctly. Differentiating the constraint with respect to time gives a restriction on the accelerations :

$$\lambda \longrightarrow \dot{K} = \sum^{\#} m\dot{q} \cdot \ddot{q} = 0 \ .$$

Lagrange's equations of motion, including the contribution from the Lagrange Multiplier $\lambda(t)$ are :

$$\{ \ (d/dt)(\partial\mathcal{L}/\partial\dot{q}) = \dot{p} = (\partial\mathcal{L}/\partial q) \ ; \ (d/dt)m\dot{q}[\ 1 + \lambda(t) \ ] = F(q) \ \} \longrightarrow$$

$$\{ \; m\ddot{q}(1+\lambda) + \dot{\lambda} m\dot{q} = F(q) \; \} \; .$$

This differential equation for λ can be solved by applying the constant-kinetic-energy constraint. Multiply each of the # equations of motion by $\dot{q}$ and sum, eliminating the summation which includes all of the products, $\{ \; m\dot{q}\ddot{q} \; \}$, as that sum vanishes. The result is :

$$\dot{\lambda} \times 2K_0 = \sum^{\#} \dot{q}F(q) = -\dot{\Phi} = -\dot{E} \longrightarrow \lambda = [\; E_0 - E \;]/(2K_0) \; .$$

The general relationship between Lagrange's and Hamilton's mechanics then provides an explicit expression for the *isokinetic* Hamiltonian :

$$\mathcal{H}(q,p) = \sum^{\#} \dot{q}p - \mathcal{L}(q,\dot{q}) = 2\sqrt{K(p)K_0(\dot{q})} - K_0 + \Phi \; .$$

The foregoing algebra is unnecessary. To simplify the notation we define

$$K_p \equiv K(p) \; ; \; K_0 \equiv K(\dot{q}) \; .$$

Then simply consider the equations of motion that follow from the foregoing Hamiltonian :

$$\mathcal{H}(q,p) = 2\sqrt{K_p K_0(\dot{q})} - K_0 + \Phi \;\; \longrightarrow$$

$$\{ \; m\dot{q} = p\sqrt{K_0/K_p} \; ; \; \dot{p} = F \; \} \longrightarrow$$

$$(d/dt)\sum(m\dot{q}^2/2) = (d/dt)\sum(p^2/2m)(K_0/K_p) = (d/dt)K_0 \equiv 0 \; .$$

The vanishing time derivative of the kinetic energy $K_0 = \sum(m\dot{q}^2/2)$ confirms that the Hamiltonian approach replicates the considerably more complex Lagrangian one. Two simple example problems which demonstrate the equivalence are illustrated in **Figure 2.7** . The numerical solutions for either the Lagrangian or the Hamiltonian description are easily obtained using the Runge-Kutta solution methods discussed in Chapter 5 . The coordinates plotted in the Figure correspond to solutions of the isokinetic oscillator Hamiltonian :

$$\mathcal{H}_{\rm oscillator} = \sqrt{p_x^2 + p_y^2} - (1/2) + (1/2)(x^2 + y^2) \;\; \longrightarrow$$

$$\dot{x} = p_x/|\; p \;| \; ; \; \dot{y} = p_y/|\; p \;| \; ; \; \dot{p}_x = -x \; ; \; \dot{p}_y = -y \; .$$

It is interesting that the isokinetic oscillator equations for $\ddot{p}_x$ and $\ddot{p}_y$,

$$\ddot{p}_x = -p_x/|\; p \;| \; ; \; \ddot{p}_y = -p_y/|\; p \;| \; ,$$

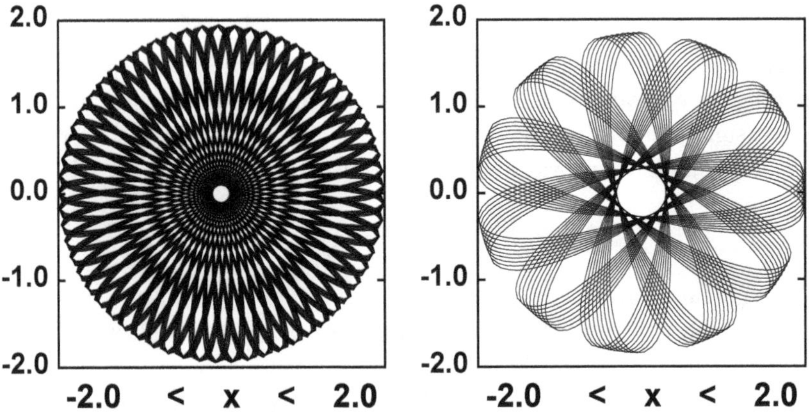

Fig. 2.7 (x, y) plots for isokinetic harmonic oscillators with initial coordinates $(1, 1)$. Inital velocities are $(3/5, 4/5)$ and $(5/13, 12/13)$ at left and right respectively. The mass and force constant are both unity with elapsed times of 2000 and 400 at the left and right. Lagrangian and Hamiltonian simulations of each problem give identical results.

have exactly the same form as do Newton's motion equations for $\ddot{x}$ and $\ddot{y}$ in the constant radial field, $\phi = \sqrt{x^2 + y^2}$. This is no coincidence. Apart from a constant the isokinetic oscillator Hamiltonian is quadratic in the coordinates and "radial" in the momenta :

$$\mathcal{H}_{\text{oscillator}} = \sqrt{p_x^2 + p_y^2} + (1/2)(x^2 + y^2) \ .$$

The Cartesian Hamiltonian for a radial potential just permutes the rôles of the coordinates and momenta :

$$\mathcal{H}_{\text{radial}} = \sqrt{x^2 + y^2} + (1/2)(p_x^2 + p_y^2) \ .$$

It is also possible to show that Lagrange's equations of motion are simply related to Gauss' for the isokinetic problem :

$$\{ \ (m\ddot{q})_{\text{Lagrange}} = (m\ddot{q})_{\text{Gauss}}/[\ 1 + \lambda \ ] \ \} \ .$$

In the equilibrium case, where fluctuations in the energy are small, Lagrange's equations behave perfectly well, keeping the kinetic energy constant.

Nonequilibrium steady states are a different matter. In such cases E grows steadily while λ decreases, soon causing the accelerations to diverge. We conclude that Lagrangian mechanics isn't useful for the study of nonequilibrium steady states. We will eventually come to this same verdict for Hamilton's equations of motion. Let us turn next to a relatively general approach to equations of motion, useful at equilibrium and (sometimes) away from equilibrium, Hamilton's *Principle of Least Action.*

2.5.3 *Using Hamilton's Least-Action Principle* $\rightarrow K = K_0$

Hamilton's Principle of Least Action is a fundamental basis for both classical and quantum mechanics, as Feynman often emphasized. Let us apply it here to obtain an isokinetic extension of the Leapfrog algorithm. Hamilton's Principle itself states that the time integral of the Lagrangian, $\mathcal{L}$, over "all possible paths" is a minimum for the path actually chosen (by Nature). Let us consider, from a finite-difference point of view, the motion of a selected degree of freedom from time $t - dt$ to time $t + dt$. We assume that the coordinates $q(t-dt)$ and $q(t+dt)$ are both known, and fixed, while the intermediate coordinate $q(t)$ can be chosen freely. Evidently the action integral is the sum of a kinetic contribution :

$$dt \times m \left([\, (q(t) - q(t - dt) \,]^2 + [\, q(t + dt) - q(t) \,]^2 \, \right) / 2dt^2 \ ,$$

and a potential one :

$$-(dt/2) \times [\, \phi(t - dt) + 2\phi(t) + \phi(t + dt) \,] \ .$$

Let us set the variation with respect to $q(t)$ equal to zero :

$$(m/dt)[\, 2q(t) - q(t - dt) - q(t + dt) \,] + dtF(t) = 0 \ . \ [\text{ Variation 1 }]$$

As usual, Feynman was right. The Least-Action result is just a rearranged version of our basic Newtonian finite-difference leapfrog algorithm :

$$q(t + dt) = 2q(t) - q(t - dt) = dt^2F(t)/m \ . \ [\text{ Leapfrog Algorithm }]$$

So far so good. Let us now additionally impose a *constant kinetic energy constraint* on the Lagrangian through the Least Action Principle. If the Lagrangian contains the additional term $\zeta(K - K_0)$ with ζ chosen so that the finite-difference representations of $K(t - \frac{dt}{2})$ and $K(t + \frac{dt}{2})$ are equal :

$$\sum^{\#}(m/2)[\, (q(t) - q(t - dt) \,]^2 - \sum^{\#}(m/2)[\, q(t + dt) - q(t) \,]^2 = 0 \ ,$$

we can proceed to subtract half the $q(t)$ variation of this kinetic-energy constraint,

$$\sum^{\#}\zeta(m/2)[\, q(t - dt) - q(t + dt) \,] \ , \ [\text{ Variation 2 }]$$

from the summed-up (over #) version of the Least-Action variation, "Variation 1" above. Taking the limit of small dt , with the combined variations vanishing, then provides the familiar motion equations generated by Gauss' Principle :

$$\{ \ 0 \equiv -\zeta m dt\dot{q} - m dt\ddot{q} + dtF \ \} \longrightarrow \{ \ m\ddot{q} = F - \zeta m\dot{q} \ \} \ .$$

Just as in the application of Gauss' Principle the Lagrange multiplier ζ needs to be chosen to satisfy the constant-kinetic-energy constraint :

$$\zeta = \sum^{\#}[\ F \cdot \dot{q}\ ]/\sum^{\#} m\dot{q}^2 \longrightarrow [\ K \equiv K_0\ ] \ .$$

We recognize these Least-Action equations of motion as identical to those coming from Gauss' Principle, and so have shown that the Hamilton's Least-Action Principle and Gauss' Least-Constraint Principle agree as to how the isokinetic constraint is to be implemented. Of course both these approaches are equivalent to "velocity scaling" in the small-timestep limit.

2.6 Boundary Conditions—Virial and Heat Theorems

An "understanding" of many-body behavior links raw simulation data, $\{\ q,p\ \}$, to macroscopic thermodynamic and hydrodynamic descriptions through spatial and temporal averaging. The formal structure of these links is best based on "thought experiments" analogous to laboratory experiments, but with their basis in simulations. The simplest realization considers a system of N mass-point particles interacting with a finite-range pair potential and obeying Hamiltonian mechanics :

$$\mathcal{H}(\ q,p\ ) = K + \Phi \ ; \ K = \sum^{N}(p_i^2/2m) \ ; \ \Phi = \sum_{i<j}^{N}\phi(r_{ij}) \ .$$

Imagine that in addition to pair forces exerted on each particle i , $-\sum_j \nabla_i\phi_{ij}$, short-ranged boundary forces $\{\ F^b\ \}$ are exerted by two parallel walls perpendicular to the x axis.

Multiplying each particle's force in the x direction, perpendicular to the walls, by the wall "area" (just a length in two dimensions) L_y and its x coordinate while combining the pair terms into local contributions gives :

$$\sum^{N} mx_i\ddot{x}_i = \sum_{i<j}(x_{ij}^2/r_{ij})F_{ij} - P_{xx}^{\rm right}L_y(L_x/2) - P_{xx}^{\rm left}L_y(L_x/2) \ .$$

If we imagine a large-N many-body system confined by reflecting boundary "walls" at $x = \pm(L_x/2)$ the time average of the wildly fluctuating forces on the walls satisfies the *Virial Theorem* :

$$\langle\ P_{xx}V = \sum_{i<j}(x_{ij}^2/r_{ij})F_{ij} - (d/dt)\sum^{N} mx_i\dot{x}_i + \sum^{N} m\dot{x}_i\dot{x}_i\ \rangle \ .$$

The mean value of the second sum vanishes while the last sum defines the kinetic temperature T_{xx} :

$$P_{xx}V = \sum_{i<j}(x^2F/r)_{ij} + NkT_{xx} \ .$$

For a small volume this same expression is the local momentum flux component P_{xx} . To derive it in three-dimensional space one can consider the flow of x momentum across sampling planes $dydz$ during a small time interval dt . A pair of interacting particles contributes to P_{xx} whenever the sampling plane intercepts a straight line joining the two particles. Similarly we can derive the virial theorem for shear stress :

$$P_{xy}V = \sum_{i<j}(xyF/r)_{ij} + NkT_{xy} \ ,$$

which can be used for measuring solid's shear moduli (with NkT_{xy} vanishing) .

Heat flux can be treated analogously. Now imagine that the walls exchange heat with the confined system. Next *define* individual particle energies by including in each of them half of each pair interaction and then compute the time-rate-of-change of the $\{ \ e_i \ \}$:

$$\{ \ \dot{e}_i = m\dot{v} \cdot v - (1/2)\sum_j F_{ij} \cdot v_{ij} + \dot{e}_{\rm walls} \ \} \ .$$

Just as before, multiply by x_i , sum, and time average, introducing the heat fluxes [the energy transfer per unit area and time] Q_x at the walls. Use the identity $\langle \ (d/dt)(x_ie_i) \ \rangle = \langle \ \dot{x}_ie_i + x_i\dot{e}_i \ \rangle \equiv 0$ to get

$$\langle \ \sum_i -\dot{x}_ie_i = (1/2)\sum_{i<j} x_{ij} \ F_{ij} \cdot (v_i + v_j) - Q_xV \ \rangle \longrightarrow$$

$$\langle \ Q_xV \equiv \sum_i (\dot{x}e)_i + (1/2)\sum_{\rm pairs} x_{ij} \ F_{ij} \cdot (v_i + v_j) \ \rangle \ .$$

The last sum includes terms of the form $x_{ij}F_{ij} \cdot (v_i + v_j)$. Both the virial theorem and the heat theorem can be written differently for solids, by introducing the mean value of x_i rather than the instantaneous one. Also, by considering the momentum and energy balance within a small volume element $dxdy$ a local dynamic representation of the pressure and heat flux results, and has exactly the same form as the global expressions above. For the details see Bill's *Molecular Dynamics*, or *Computational Statistical Mechanics*, both of them available free at our website [http://williamhoover.info] .

There is an interesting difference between the pressure-tensor and heat-flux-vector expressions. At equilibrium the pressure tensor can be expressed as a power series in the density,

$$PV/NkT = 1 + B_2(T)\rho + B_3(T)\rho^2 + B_4(T)\rho^3 + B_5(T)\rho^4 + \dots .$$

The nth coefficient B_n in the series, the "nth virial coefficient", was expressed in terms of n-body integrals by Joseph and Maria Mayer. Improvements in the Mayers' approach have made it feasible to compute as many as a dozen terms in the series for simple potentials, like hard spheres. Although the heat flux has a similar appearance it vanishes at equilibrium so that there is no known analog of the virial series for heat flux.

In the next Section we consider in detail a useful technique for incorporating *atomistic* data in local definitions of *continuum* "field" variables. Thus atomistic point values of temperature, energy, pressure tensor, and heat-flux vector provide smoothly differentiable continuum values of these same variables as well as density and flow velocity.

2.7 Smooth-Particle Averages of Atomistic Properties

Smooth-particle averaging provides a method for converting the continuum evolution equations, *partial* differential equations for the momentum and energy, into a finite set of *ordinary* differential equations for the evolving motion and energy of "particles". The motion and energy of the continuum at points other than those particles is easily evaluated by "smooth-particle" interpolation. The algorithm for solving continuum problems in this way is explained in Section 4.3, page 85 . The averaging procedure underlying the algorithm is itself a useful interpolation tool for the continuous twice-differentiable representation of values available at a finite set of moving points, just the information available from molecular dynamics.

Any detailed visualization and interpretation of such particle computer experiments, "simulations", requires *defining* values locally, at specific points in space. Given individual *particle* values at the time t of interest it is straightforward to define average manybody *continuum* properties at the point r by averaging the properties of nearby particles. If *continuous differentiable* definitions are desired then a "weight function" w , with at least *two* vanishing spatial derivatives, is ideal for this purpose.

For simplicity we assume that the adopted weight function $w(r)$ is normalized to unity and has a range h , beyond which it vanishes :

$\int_{-h}^{+h} w_{\rm 1D}(|x|)dx \equiv 1$; $\int_0^h 2\pi r w_{\rm 2D}(r)dr \equiv 1$; $\int_0^h 4\pi r^2 w_{\rm 3D}(r)dr \equiv 1$.

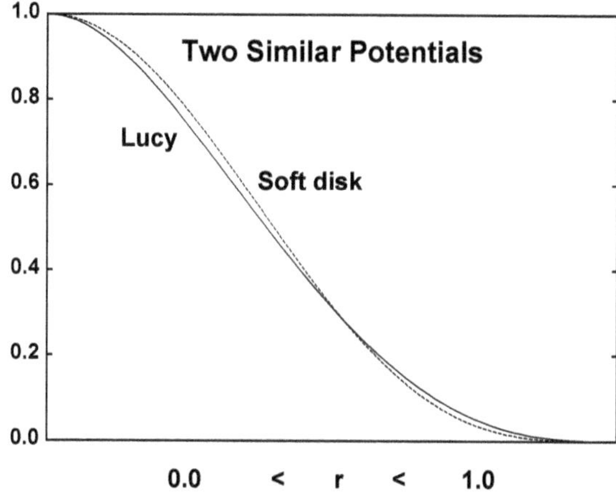

Fig. 2.8 Lucy's unnormalized weight function, $(1 - 6r^2 + 8r^3 - 3r^4)$, compared to the soft sphere (or soft disk) potential, $(1 - r^2)^4$.

From the physical point of view w should have a smooth maximum, be normalized to unity, and have its function value, and two spatial derivatives vanish at the cutoff distance h. The simplest polynomial satisfying these five conditions is Lucy's quartic weight function :

$$w_{\rm Lucy}(r < h) = C[\ 1 + 3(r/h)\ ][\ 1 - (r/h)\ ]^3 \ .$$

See **Figure 2.8** . The normalization coefficient C depends upon the dimensionality of the averages :

$$C_{\rm 1D} = (5/4h)\ ;\ C_{\rm 2D} = (5/\pi h^2)\ ;\ C_{\rm 3D} = (105/16\pi h^3)\ .$$

In applying this averaging idea the formal integrals are replaced by particle sums. For instance, in an N-particle system in D-dimensional space the average density and velocity at a grid point r_g are :

$$\rho(r_g) = \sum^N m w_{\rm D}(|r_i - r_g|)\ ;\ u(r_g) = \sum^N m\dot{r}_i w_{\rm D}(|r_i - r_g|)/\rho(r_g)\ ,$$

where the sums include all particles which are within the range h of the grid point r_g .

To illustrate this smooth-particle averaging technique we consider again the 19-particle bouncing hexagon of **Figure 2.1** . In **Figure 2.9** we compare two density-contour plots to the corresponding density-gradient plots.

In both cases we use Lucy's weight function with a range twice the stress-free nearest-neighbor spacing to compute contours during the collision process.

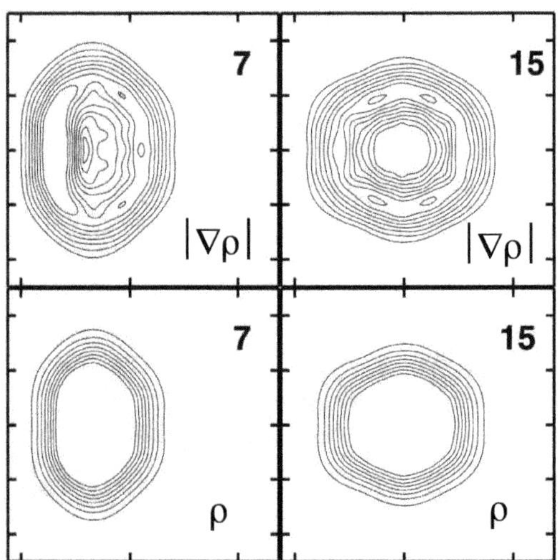

Fig. 2.9 Lucy-function densities and density gradients for the deformed hexagonal crystallite of **Figure 2.1** . The *magnitude* of the gradient, $\sqrt{(\partial\rho/\partial x)^2 + (\partial\rho/\partial y)^2}$, is most easily computed from smooth particle representations of $(\partial\rho/\partial x)$ and $(\partial\rho/\partial y)$.

The weight-function definition of averages automatically provides corresponding definitions of spatial gradients of these averages. This is handy, as gradients are basic to continuum definitions of the fluxes and time derivatives of mass, momentum, and energy. Newton's viscous stress is proportional to the momentum gradient while Fourier's heat flux is proportional to the temperature gradient. The time derivatives of momentum and energy are likewise proportional to the divergences of the stress tensor and the heat-flux vector.

With Lucy's weight function the local definitions of density, momentum, and energy can be differentiated with respect to the spatial coordinates. For instance, the density gradient, plotted in **Figure 2.9** for a bouncing hexagon, is :

$$\nabla\rho \equiv \sum m\nabla w(|r_i - r_g|) \ .$$

The two vanishing derivatives of $w(r < h)$ at $r = h$ guarantee the continuity of the corresponding gradients, $\{ \ \nabla\rho, \ \nabla u, \ \dots \ \nabla e \ \}$. Because these same spatial gradients govern the continuum evolution equations detailed comparisons of particle simulations with the predictions of continuum mechanics can be carried out.

2.8 Constraints and Gibbs' Entropy

Constraints in classical mechanics can reduce the number of "degrees of freedom" required to describe system states. The description of a two-dimensional diatomic molecule, with four degrees of freedom, can be reduced to two degrees of freedom by constraining its center of mass to a fixed location. The remaining two degrees of freedom, r and θ, can further be reduced to one, by constraining the separation between the two masses to be constant :

$$\{ x_1, y_1, x_2, y_2 \} \longrightarrow \{ r, \theta \} \longrightarrow \theta \ .$$

Such coordinate-dependent constraints are easily implemented by reducing the number of differential equations of motion and the corresponding contributions to the total energy.

Constraints in *nonequilibrium* situations typically involve coordinates and momenta, as well as the friction coefficients governing the motion or even the size and shape of the system. For instance, consider again isokinetic mechanics, where the equations of motion include a friction coefficient ζ chosen to keep the kinetic energy of # degrees of freedom constant :

$$\{ \dot{p} = F(q) - \zeta p \} \ ; \ \zeta \equiv \sum^{\#} F \cdot (p/m) / \sum^{\#} (p^2/m) \longrightarrow \sum^{\#} (p^2/2m) \text{ constant} \ .$$

When frictional constraints are incorporated in the motion, Liouville's Theorem holds in a modified form, taking into account any nonzero contributions to the flow of phase-space density :

$$(d \ln f/dt) = (\dot{f}/f) \equiv - \sum [\ (\partial \dot{q}/\partial q) + (\partial \dot{p}/\partial p) + (\partial \dot{\zeta}/\partial \zeta) \ ] \ .$$

Then the conclusions drawn from Liouville's Theorem are qualitatively different. No longer does the probability density flow through phase space like an incompressible fluid. With frictional forces the density can build up, in regions where heat has been extracted. Likewise it can decrease in regions where heat has been supplied by external forces through negative values of the friction coefficient ζ .

$$\dot{f}(q,p)/f(q,p) = \sum^{\#} \zeta \; .$$

Away from equilibrium the phase-space probability density can quickly diverge (while remaining integrable). Such cases are completely different to the smooth Gibbsian distributions familiar from classical statistical mechanics. The changing phase-space density is intimately related to the Second Law of Thermodynamics, as is explained in the next Chapter.

2.9 Summary

Algorithms are all-important in the solution of nonlinear problems in atomistic mechanics. One of the characteristic properties of an atomistic formulation of mechanics is time-reversibility (the movie is described by the same motion equations whether played forward or backward). Another property, less universal, is recurrence, with the system returning close to its initial state. A third property, rarer still, is ergodicity, coming close to *all* phase-space states consistent with appropriate conserved quantities. The simple Leapfrog algorithm, implemented in an integer configuration space, can exhibit all of these properties. More complex situations, involving geometric or thermodynamic constraints, can be added to mechanics through Lagrangian ideas or as an application of Liouville's phase-space continuity relation. Coupled with smooth-particle methods for representing atomistic data as twice-differentiable continuum fields, atomistic simulations are a powerful means for understanding and influencing macroscopic behavior.

2.10 References

The 1982 Boulder Conference "Nonlinear Fluid Phenomena" was organized by Howard Hanley and published as Physica **118A** in 1983. The lively nearly *verbatim* discussions of the topics discussed there still make interesting reading today. Experimentalists, theoreticians, and simulators all contributed to the success of that meeting. For the Levesque-Verlet "Bit-Reversible" Leapfrog algorithm see "Molecular Dynamics and Time

Reversibility", Journal of Statistical Physics **72** , 519-537 (1993). For the details of smooth-particle averaging see Bill's *Smooth Particle Applied Mechanics – The State of the Art* (World Scientific, Singapore, 2006).

For more details concerning the use of Least Action to derive motion equations see R. E. Gillilan and K. R. Wilson, "Shadowing, Rare Events, and Rubber Bands. A Variational Verlet Algorithm for Molecular Dynamics", Journal of Chemical Physics **97** , 1757-1772 (1992) as well as Bill's "Temperature, Least Action, and Lagrangian Mechanics", Physics Letters A **204** , 133-135 (1995).

2.11 Problems

1. Write a computer program that sums up the potential energy of a periodic Hooke's-Law crystal by adding the $3N$ nearest-neighbor bond energies. Use a triangular lattice with a nearest-neighbor separation of unity and a density of $\sqrt{(4/3)}$. Compute numerical values of the moduli B and C_{11} :

$$VB = -V^2(dP/dV) = V^2(d^2E/dV^2) \; ; \; VC_{11} = (d^2E/d\epsilon_{xx}^2) \; .$$

ϵ_{xx} is the infinitesimal dimensionless "longitudinal strain", $(\delta L_x/L_x)$.

2. Determine whether or not a powerlaw dependence of the coefficient of restitution on speed is useful for the bouncing hexagon example shown in **Figures 2.1 and 2.2** . Use the Leapfrog integrator and calculate the center-of-mass speed from $[\ |x(t+dt) - x(t-dt)|\ ]/(2dt)$.

3. By generalizing the deformation in Problem 1 to include shear, show that the longitudinal and transverse sound speeds of the two-dimensional Hooke's-Law lattice are $\sqrt{9/8}$ and $\sqrt{3/8}$ where the stress-free nearest-neighbor spacing, particle mass, and force constant are all equal to unity.

4. Suppose two successive Leapfrog integers are `I = -1` and `J = +1` with `F(J)*dt*dt = 0.5` , so that the next integer in the approximate trajectory is `K = J + J - I + F(J)*dt*dt` so that `K = 3` . Show that the attempt to reverse the step from `K` to `I` fails. Explain why and provide two remedies.

5. Reproduce the trajectories shown in **Figure 2.7** using the isokinetic Hamiltonian equations of motion given on page 41 :

$$\dot{x} = p_x/|\,p\,| \; ; \; \dot{y} = p_y/|\,p\,| \; ; \; \dot{p}_x = -x \; ; \; \dot{p}_y = -y \; .$$

Solve the equations with a "predictor-corrector" technique. That is, first *predict* the new values of $\{ x, y, p_x, p_y \}$ by adding on dt multiplied by the derivatives evaluated at the current values :

$$\texttt{yynew(j)} = \texttt{yynow(j)} + \texttt{yypnow(j)} * \texttt{dt} \longrightarrow \texttt{yypnew(j)} \ .$$

Then, *correct* these four estimates with four expressions of the form :

$$\texttt{yynew(j)} = \texttt{yynow(j)} + \texttt{0.5d00} * \texttt{dt} * (\texttt{yypnew(j)} + \texttt{yypnow(j)}) \ .$$

Apply the corrector a few times before proceeding to the next step.

6. Why is it that the time-averaged time derivative of a bounded quantity vanishes, so that $\langle \ (d/dt)(x_i e_i) \ \rangle \equiv 0$?

Chapter 3

Thermodynamics, Statistical Mechanics, and Temperature

Topics

Thermodynamics and Ideal-Gas Temperature / Thermodynamics of the Classical Ideal Gas / Laws of Thermodynamics / Zeroth Law / The First Law / The Paradoxical Second Law / Temperature from Entropy / Jaynes' Idea / Nosé's Idea / Gibbs' Canonical Ensemble / Configurational Temperature / Temperature from Kinetic Theory, the Ideal-Gas Thermometer / Tensor Temperature with Hard Parallel Cubes /

3.1 Thermodynamics and Ideal-Gas Temperature

Mechanics (or dynamics) describes the displacements and the motions of matter responding to forces. The models considered range from atomistic to astrophysical in size. *Thermo*dynamics is quite different. Thermodynamics is a macroscopic study of the the mechanical and *thermal* state changes undergone by matter. It is sometimes called thermostatics to distinguish its subject matter from fluid dynamics, solid mechanics, and hydrodynamics. "Thermostatics" emphasizes the slow speed of thermodynamic motions, so slow relative to the sound speed that the states considered are equilibrium states. In addition to mechanical forces, which can do work and change energy by compression, expansion, or shear, thermodynamics also includes temperature variations, along with the flow of heat in response to these temperature differences. Thus the concepts necessary to undertaking thermodynamic "thought experiments" and analyses include temperature and heat reservoirs in addition to mechanical tools like frictionless pistons and smooth indentors.

Statistical mechanics furnishes a bridge between atomistic mechanics and macroscopic thermodynamics. "Statistical" indicates averaging over the many microscopic (q,p) coordinate-momentum states consistent with a single macroscopic (P,V,E) pressure-volume-energy state. A new temperature-based state function, *Entropy*, is specially useful in interconnecting macroscopic thermodynamics with microscopic statistical mechanics. Thermodynamics is based on Laws more comprehensive than those of mechanics. These Laws are the fruits of experience, and can be applied to a wide variety of models of matter. The models need to have well-defined reproducible "states". These states are characterized by the macroscopic properties, pressure, density, energy, temperature, $\dots$. Describing solids can require additional mechanical descriptors such as dislocation density, locked-in stresses, and void fractions. Because thermodynamics includes temperature among the state variables, a means of measurement is necessary, a *thermometer.* And because temperature is a universal equilibrium property an unlimited variety of thermometers can be imagined or defined in terms of their construction.

In all of our work we will adopt the ideal-gas thermometer and temperature scale, based on the simplest possible microscopic model of a thermodynamic system, a monatomic classical collection of very many identical point masses, homogeneous, isotropic, and always at equilibrium so that its properties are unchanging. We imagine that the ideal-gas particles are very light, $m \simeq 0$, in order that the thermometer can react quickly when called upon to make a measurement. If the system whose temperature is to be measured is in motion, then the ideal-gas thermometer moves with it. Temperature, like pressure, is defined and measured in a *comoving* frame.

The properties of an equilibrium ideal gas are very well understood. They are limiting properties, for sufficiently many particles at sufficiently low density and at sufficiently long times that fluctuations and past history can both be ignored. In this limit the distribution of velocities is Gaussian, with a second moment defining the kinetic temperature :

$$f(v_x,v_y) \propto \exp[\ -(mv_x^2+mv_y^2)/2kT\ ]\ ;\ \langle\ mv_x^2\ \rangle = \ \langle\ mv_y^2\ \rangle = kT\ .$$

Further, simple probabilistic kinetic-theory calculations show that the mean velocity of a massive particle (mass $M >> m$) colliding elastically with such a gas is damped, at a rate proportional to its velocity, and so approaches zero exponentially. At the same time the mean-*squared* velocity of such a massive particle in thermal contact with an ideal gas approaches kT in each of the D space directions :

$$\langle\ (d/dt)Mv^2\ \rangle \propto [\ DkT - Mv^2\ ]\ .$$

The relaxation times for the massive particle's mean velocity and the mean-squared velocity are of the order of (M/m) collision times. They depend upon the interparticle force law and the dimensionality. This ability of an ideal gas to impose its own temperature on a more massive particle is *the* fundamental thermal property of a heat reservoir. Likewise the fact that an ideal-gas thermometer measures the kinetic temperature of degrees of freeedom with which it interacts supports the *kinetic* definition of temperature. The kinetic-theory analysis of collisions is necessarily statistical, and predicts the flow of heat from hotter to colder despite the time reversibility of the underlying collisional laws.

3.2 Thermodynamics of the Classical Ideal Gas

An understanding of thermodynamics is necessarily based on models, some of them mathematical and some physical. The simplest understanding is that based on the conceptual model of an "ideal" gas, an isotropic gas composed of particles so numerous and so small that its properties are easy to calculate and understand. The fundamental mechanical property of such a gas is its pressure, the force per unit area exerted by the gas on its container. Let us carry out a thought-experiment on an isotropic D-dimensional ideal gas at equilibrium. We evaluate its pressure component, $P = P_{xx}$, the force per unit area parallel to the x axis. To do this we consider collisions at a boundary perpendicular to that axis. Evidently an elastic wall collision of a gas particle with mass m and x velocity component v_x contributes an impulsive force (with an instantaneous momentum transfer) to the wall. The momentum transfer from gas particle to wall is $+2mv_x$. The compensating momentum transfer from wall to particle is $-2mv_x$. The force on the wall (momentum transfer per unit area and time) is the pressure-tensor component P_{xx} . This force reflects all the particles' momentum changes $\{+mv_x \rightarrow -mv_x\}$ on unit area of the wall in unit time .

Suppose that the container of our D-dimensional ideal gas is a cube with length L , area $L \times L$, or volume $L \times L \times L$ corresponding to the choices $D = 1, 2,$ or 3 . Averaged over the time between right-wall collisions, $\tau_x = (2L/v_x)$, and summed up over all particles, the mean pressure is simply related to the (kinetic) energy of the gas (because the potential energy is assumed negligible) :

$$E = K = (mv^2/2) \; ; \; N\langle \, mv_x^2/L^D \, \rangle = P = (2E/DV) = (NkT/V) \, .$$

The ideal-gas temperature scale resulting from this mechanical thought experiment *defines* the (kinetic) temperature T :

$$T \equiv (PV/Nk)_{\text{ideal}} \ .$$

Temperature is defined, and can be measured, in terms of the PV product for N gas particles in a sufficiently large volume V . Boltzmann's constant k is the conventional choice made to relate mechanical and thermal properties.

In addition to its simplicity this ideal gas-model for temperature has the advantage of realism. Real gases, when sufficiently dilute, all approach the ideal-gas model for pressure and temperature as a limiting low-density case. Thermodynamics is based on the notion that all the materials it treats have stationary equilibrium states. For the equilibrium ideal gas the pressure is isotropic, with $P_{xx} = P_{yy} = P_{zz} = P = (NkT/V)$ in three dimensions and $P_{xx} = P_{yy} = P = (NkT/V)$ in two.

3.3 The Three Laws of Thermodynamics

The scope of Thermodynamics is broad. Any System which can be characterized by its equilibrium States is included. Thermodynamics describes two modes of interaction of such a System with its Surroundings. *Work* changes the System energy through the variation of a *mechanical* coordinate, such as the length of the box. In addition to energy changes incurred by work, *heat* changes the System energy *without* any corresponding coordinate change. Leaving chemistry aside, performing work and transferring heat are the two modes of energy change described by thermodynamics. Energy is traditionally defined as "the power to do Work". It could equally well be defined as "the power to generate Heat".

All of the foregoing unconventionally-capitalized words are thermodynamic idealizations which can be represented by computational models and which approximate properties of portions of the real world of which we are a part. For simplicity we typically imagine that our systems are homogeneous fluids composed of identical particles interacting with pairwise-additive potential functions and obeying classical mechanics. These simplistic ideas allow us to ignore the complexity associated with molecular interactions, solid-phase defects like dislocations and vacancies, and the fluid-phase complexity associated with bubbles, drops, and interfaces. The State variables of typical homogeneous fluids include the volume occupied V, the number of particles N, and the temperature T and energy E and pressure P . The Notion of State results from the observation that an Isolated System, where Work and Heat are absent, eventually reaches mechanical and thermal Equilibrium, at which the time-rates-of-change in local values of the State variables are absent. Fluctuations, assumed small, are ignored.

3.3.1 *The Zeroth Law of Thermodynamics*

The basis of Thermodynamics is observational. Two assumptions (or observations) vital to the subject shape the thermodynamic description of states :
[1] For a given number (or mass) of particles a homogeneous fluid's states can be described by *two* independent state variables. Common choices for the two are volume and temperature or pressure and energy. With these choices the thermodynamic state of a homogeneous simple fluid can be represented by a unique point in the (V,T) plane or the (P,E) plane. Other pairs of state variables, not necessarily single-valued, could be chosen from a long list.
[2] Temperature and pressure are particularly useful state variables because two systems, or even three (at the gas-liquid-solid "triple point"), in mechanical and thermal contact, can come to an equilibrium state in which the coexisting systems (or phases) share the same pressure and the same temperature.

The observation that two systems having the same temperature (as measured by an ideal-gas thermometer) would remain in thermal equilibrium were each exposed to the temperature of the other (perhaps with a thin conducting wire) is termed the "Zeroth Law of Thermodynamics". The parallel mechanical observation for pressure (shared through a movable piston), though equally important, has no special name. The Zeroth Law of Thermodynamics, like most of thermodynamics, applies in the large-system limit, when fluctuations and surface effects can be ignored.

Small systems behave differently. To see this consider a simple one-dimensional illustration, the equilibration of two four-particle systems with the *same* initial kinetic temperature but with two *different* sets of particle velocities , $kT = \langle\ (mv^2/k)\ \rangle$:

$$\{\ -0.8, -0.6, +0.6, +0.8\ \} \text{ and } \{\ -1, 0, 0, +1\ \}\ .$$

What happens when these two systems interact? Suppose that the result of a collision between Particles i and j is the exchange of velocities, $v_i \to v_j$ and $v_j \to v_i$. A statistical average, weighting each of the $4 \times 4 = 16$ possible collision types with $|v_i - v_j|$, and summing up each pair of particles' contribution to the energy transfer between the two systems, shows that at first the system with two unit-speed particles has an overall tendency to lose energy to the system with the more uniform velocities, showing that no instantaneous Zeroth Law of Thermodynamics holds for systems with only a few degrees of freedom. Thermodynamics makes definite predictions to the extent that the systems contemplated are large enough that fluctuations can be ignored.

The Zeroth Law states that thermometry is possible for equilibrium systems so that *any* equilibrium system can serve as a thermometer. We will see that both statistical mechanics and kinetic theory suggest choosing the ideal gas thermometer. Because the ideal gas has a temperature proportional to its pressure the ideal-gas thermometer provides a *mechanical* basis for thermodynamics.

3.3.2 *The First Law of Thermodynamics*

Conservation of energy is a familiar idea and can be motivated by the observation that without it perpetual motion machines (capable of creating energy) would exist. Another basis for energy conservation is microscopic mechanics (either classical or quantum) which implies an unchanging energy for any isolated system. Let us look into energy conservation in more detail for a System coupled to its Surroundings. It is imagined that these Surroundings can change the energy of our System by [1] doing work on it or [2] exchanging heat with it. We imagine that both types of changes are sufficiently slow and small as to be thermodynamically "Reversible", not to be confused with mechanically time-reversible. [*All* of the mechanical processes we consider will be officially time-reversible, though Lyapunov instability can make accomplishing that reversibility difficult, or even impossible, in practice.]

In thermodynamics, "reversible" changes proceed *through a series of equilibrium states.* In practice this limiting case depends upon the neglect of second-order deviations when the rates of doing work and transferring heat are small. Nonzero rates imply velocity and temperature gradients which correspond to irreversible processes, again thermodynamically, but not mechanically.

Where *solids* are involved work and heat can be ambiguous when irreversible processes like the motion of dislocations in plastic flow are present, directly converting some of the "work" directly to "heat". The system energy changes according to the rate $\dot{Q}$ at which heating increases the system energy less the rate $\dot{W}$ at which the system performs work, decreasing its energy :

$$\dot{E} = \dot{Q} - \dot{W} \ . \ [\text{ First Law of Thermodynamics }]$$

Evidently in any *cyclic* process governed by the First Law the work done and the heat taken in must be equal ,

$$\oint dQ = \oint dW \ \longleftrightarrow \ E \text{ is constant} \ .$$

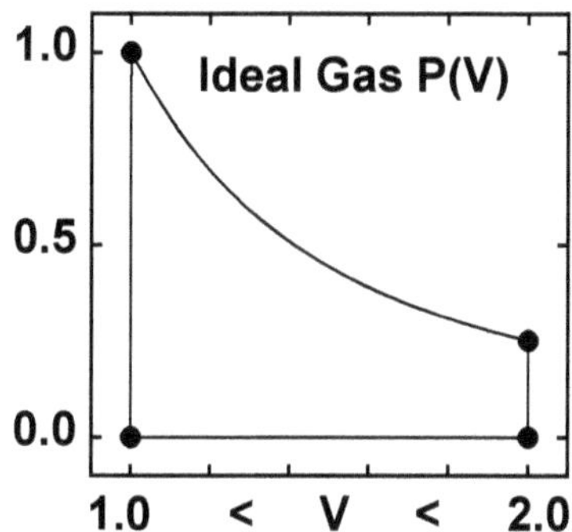

Fig. 3.1 A thermodynamic ideal-gas cycle with the temperature on the ($Q = 0$) adiabat $P = 1/V^2$ varying from 1 at $V = 1$ to 0.5 at $V = 2$. The net work done in one clockwise cycle is 0.5 : $\int_1^2 (1/V^2)dV = (1/2)$, agreeing with the *net* heat taken in, summing up the heat in, 1 at $V = 1$, less the heat out, (1/2) at $V = 2$.

To visualize such processes it is useful to picture a specific example cycle. Consider, for instance, the ideal-gas pressure-volume cycle shown in the **Figure 3.1**. A cold gas, with $(P, V) = (0, 1)$, is [i] heated at constant volume to unit pressure, and [ii] is then allowed to expand, to twice its original volume, *adiabatically*, and is next [iii] cooled again to zero pressure and temperature, and is finally [iiii] compressed at zero pressure back to its initial state :

$$(P, V) = (0, 1) \stackrel{i}{\longleftrightarrow} (1, 1) \stackrel{ii}{\longleftrightarrow} (0.25, 2) \stackrel{iii}{\longleftrightarrow} (0, 2) \stackrel{iiii}{\longleftrightarrow} (0, 1) \ .$$

An "adiabatic" process is one without heat transfer, including the compression step [iiii] at zero pressure and the two-fold expansion [ii] . Each of the four steps of which the cycle is composed is "reversible", proceeding through a series of equilibrium states. So far as the First Law is concerned, the cycle can be carried out as the text describes or in the reversed direction. The First Law declares that during this four-step cycle energy is conserved : the heat taken in at $V = 1$ is unity, while that given off at $V = 2$ is 0.5 ; the difference, 0.5 , is equal to the work done in the adiabatic expansion [ii] :

$$\oint dQ = 1.00 - 0.50 = \oint dW \equiv \int_1^2 dV/V^2 = 0.50 \ .$$

Evidently the system undergoing the state changes of **Figure 3.1** is able to do work as a result of the heat supplied to it. The First Law of Thermodynamics declares that the work done is equal to the (net) heat taken in. The thermodynamic efficiency of this cycle is 0.50 , as only half the heat taken in is converted to work. To convert *all* the heat to work would require expansion to an infinite volume and cooling to zero temperature, neither of which is possible.

There are other limits on the possible. We could imagine reversing the entire four-step process. In the reversed cycle the work done *on* the system would be equal to the net heat extracted from it. The First Law of Thermodynamics is a conservation law, simply stating that all changes to the total energy, when summed up, give zero. The First Law has nothing to say about the feasibility of reversing cycles such as the one illustrated in the figure. Reversibility is treated by the Second Law, to which we now turn.

3.3.3 *The Paradoxical Second Law of Thermodynamics*

The *Second* Law of Thermodynamics is very different to the First. It is *not* a conservation law. The Second Law of Thermodynamics states that the summed-up *Entropy* changes, for *any real* process, are positive. Evidently thermodynamic processes are not really reversible, but can only be approximately so.

Any formulation of the Second Law needs to include a definition of the entropy. In an informal sense entropy is a kind of waste, or lost information. In a more formal setting we will show that entropy is proportional to the logarithm of the number of available (q,p) phase-space energy "states" , expressed in terms of a bounded integral over the classical (q,p) phase space :

$$S = k \ln \Omega(N,V,E) \equiv (k/N!) \ln \prod^{\#} [\ \textstyle\int dq \int dp/h \ ]_{\mathcal{H}<E} \ ,$$

where k is Boltzmann's constant and h is Planck's. Planck's constant and the $1/N!$ [for N identical particles] assure the agreement of the classical and quantum approaches in the low-density, high-temperature ideal-gas limit. There are two paradoxes associated with the one-way nature of the Second Law. We will state and consider them next.

Liouville's Theorem states that probability $f dq^{\#} dp^{\#}$ flows through phase space like an incompressible fluid, with $\dot{f} \equiv 0$, as a consequence of Hamilton's equations of motion. If the available phase-space is bounded a system governed by Hamilton's motion equations and Liouville's phase-volume incompressibility theorem must eventually return near its initial point. This inevitable return is called "Poincaré recurrence". But because the time required (a "Poincaré-cycle" time) to access all such states exceeds the Age of the Universe once the number of degrees of freedom is a few dozen, this relationship involving the number of states is more than a little paradoxical. How can a system property—the entropy—depend on something unobservable—the number of states in an (unobservably-long) Poincaré cycle? The irreversibility of the Second Law, an overall entropy increase, must somehow be unrelated to the inevitability of recurrence, for together the two concepts are a contradiction, Zermélo's Paradox.

The Second Law is paradoxical in another way too ("Loschmidt's Paradox"). It is conceptually hard to square the one-way nature of the Second Law with the two-way time-reversible nature of mechanics. Evidently a Poincaré cycle could be reversed by the simple expedient of changing the signs of the momenta and integrating the (time-reversible) equations of motion backwards rather than forwards.

We will see that the price of understanding these apparent one-way-*versus*-two-way mechanical contradictions of the Second Law lies in generalizing our mechanical models to apply to *nonequilibrium* situations through the imposition of constraints. A macroscopic nonmechanical understanding of the *Second* Law of Thermodynamics rests instead on an energy-based conceptual background including ideal-gas thermometry and the analysis of thermodynamic efficiency for cyclic processes converting (partially) heat to work. This approach rests on a macroscopic thermodynamic definition of (the changes in) a new state function, the Entropy S :

$$\Delta S \equiv \int dQ/T \text{ [for reversible processes] .}$$

Changes in thermodynamic entropy can be computed or measured by summing up *reversible* heat inputs divided by the temperature at which the heat enters or leaves :

$$\dot{S} = \dot{Q}_{\text{reversible}}/T \ .$$

Though Q is itself not a state function, entropy is. To show this as simply as possible it is necessary to compute ΔS for an ideal gas and then to argue that the entropy change for any system reversibly interacting with such a gas is $-\Delta S$. Once this is done it becomes apparent that Entropy is a state function associated with Heat transfer in the same way that Energy is a state function associated with performing mechanical Work.

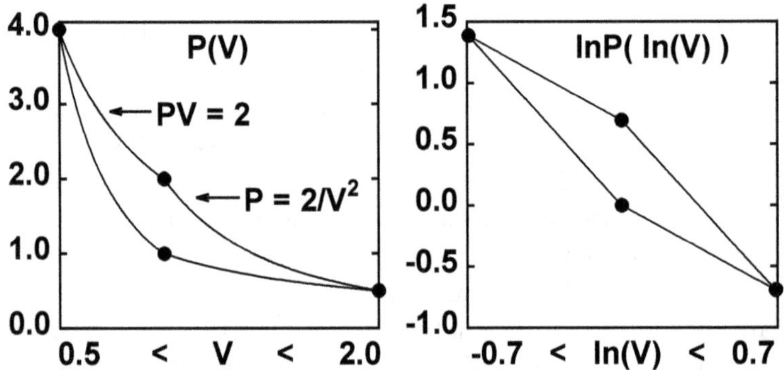

Fig. 3.2 An ideal-gas Carnot Cycle with upper temperature 2 and lower temperature 1 . The work done and heat taken in on the upper isotherm are equal to $2\ln 2$. Half this work is regained on the lower isotherm so that the efficiency of the cycle is $1/2$. In the general case the ratio of the net work done to the heat taken in at the higher temperature is $(T_H - T_C)/T_H$, here equal to 0.5 .

The usefulness and generality of the entropy concept can be seen by considering heat transfer between two temperatures $[\ T_{\rm hot}, T_{\rm cold}\ ]$ through the medium of an ideal gas. Two views of an ideal-gas thermodynamic cycle are given in **Figure 3.2** (a Carnot cycle). There is isothermal expansion doing work at the higher temperature and a compensating compression of the gas (negative work) at the lower. For definiteness we consider a two-dimensional ideal gas with initial values of $(P,V) = (1,1)$. The "adiabat" (short for "adiabatic equation of state") for the gas follows from the First Law of Thermodynamics for a process without any heat transfer (an "adiabatic" process) :

$$PV = E \rightarrow \dot{P}V + P\dot{V} = \dot{E} = -\dot{W} = -P\dot{V} \rightarrow$$

$$V\dot{P} = -2P\dot{V} \rightarrow (P/P_0) = (V_0/V)^2 \ .$$

The logarithmic scales show that $V_2/V_1 = V_3/V_4$. Evidently the two adiabats, along which Q vanishes, make no contribution to the entropy because no heat is added or extracted.

On the other hand the two *isothermal* (and therefore isoenergetic) sections of the cycle do involve heat transfer, with the heat transfer proportional to the temperature :

$$\Delta T = 0 \longrightarrow \Delta E_{12} = \Delta E_{34} \equiv 0 \longrightarrow$$

$$\Delta Q_{12} \equiv \Delta W_{12} = NkT_{\rm hot} \ln(V_2/V_1) \ ;$$

$$\Delta Q_{34} \equiv \Delta W_{34} = NkT_{\rm cold} \ln(V_4/V_3) \ .$$

As a consequence, the entropy change in the Carnot cycle is $Nk\ln(V_2V_4/V_1V_3)$, which vanishes, from $V_2/V_1 = V_3/V_4$, as it must for a cyclic reversible process.

So let us imagine a cyclic reversible process in which the work done by the ideal gas is $\oint PdV$. For such a gas, with pressure proportional to energy density, the work done in an adiabatic expansion is necessarily equal to the gas' energy change :

$$\dot{E} = -P\dot{V} \propto -(E/V)\dot{V} \longrightarrow EV \text{ constant } \longrightarrow P \propto (1/V)^2 \ .$$

Now introduce temperature and imagine that the ideal-gas system and its surroundings always share the same temperature and pressure so that the total energy is constant. In the ideal gas $P \propto (1/V)$:

$$\dot{E} \equiv 0 \longrightarrow \dot{Q} = \dot{W} = -(NkT/V)\dot{V} \longrightarrow \dot{Q}/T = -Nk\dot{V}/V \ .$$

In the Carnot cycle illustrated in **Figure 3.2** the net work done in the two isothermal segments is $\ln 2$ while that in the adiabatic segments cancels. Because the heat taken in during the hot expansion is $2\ln 2$ the efficiency of this cycle is given by the ratio $[\ T_{\rm hot} - T_{\rm cold}\ ]/T_{\rm cold}$. In fact this last result is perfectly general, for any Carnot cycle. Because any reversible cycle whatsoever can be composed of a sum of Carnot cycles (with all the work and heat associated with the internal isotherms and adiabats canceling, as each is traversed twice, in opposite directions) it is *generally* true that entropy *is* a state function. The Second Law statement that entropy can only increase is a reminder that work can easily be converted to heat (as in Sir Benjamin Thompson's [= Count Rumford's] boring cannon with dull drill bits) while only a part of the heat taken in can be converted to work. So a cyclic form of the Second Law of Thermodynamics is :

$$\oint dQ/T \geq 0 \ .$$

Equality holds only at equilibrium, for which the heat transfer is (nearly) reversible.

3.4 Temperature from Gibbs' Entropy and Jaynes' Failure

Statistical mechanics was developed by Gibbs and Boltzmann with the goal of connecting the classical mechanics of Chapter 2 to the thermodynamics of this Chapter. Their subject is termed "statistical mechanics", to identify a type of microscopic mechanics relying on microscopic *averages* designed to match macroscopic properties. Hamilton's $\{ q, p \}$ version of mechanics lends itself to probabilistic approaches because it gives directly the flow equation for the probability density in phase space $f(q,p,t)$:

$$(\partial f/\partial t) = -\sum^{\#}[\ (\partial f\dot{q}/\partial q) + (\partial f\dot{p}/\partial p)\ ] \longrightarrow$$

$$\dot{f} = (\partial f/\partial t) + \sum^{\#}[\ \dot{q}(\partial f/\partial q) + \dot{p}(\partial f/\partial p)\ ] =$$

$$-f\sum^{\#}[\ (\partial\dot{q}/\partial q) + (\partial\dot{p}/\partial p)\ ] \equiv 0 \stackrel{\mathcal{H}}{\longrightarrow} \dot{f} \equiv 0\ .$$

What we have here called the "flow equation" is nothing more sophisticated than the "continuity equation" of continuum mechanics which equates the change of mass in a small fixed element of space to the differences in the *flows* of mass (due to differences in the velocity u) at the boundaries of the element. For an infinitesimal element :

$$(\partial\rho/\partial t) = -\nabla\cdot(\rho u)\ .$$

Square or cubic elements, from which any other shapes could be constructed, are the simplest choice.

Because Liouville's Theorem shows that probability density $f(q,p,t)$ flows unchanged through phase space, like an incompressible fluid, there is no possibility for volume or density to change as the flow progresses. Evidently filling the accessible part of phase space at *constant* density gives a distribution which doesn't change with time, the microsopic statistical equivalent of macroscopic equilibrium.

Boltzmann and Gibbs showed that a consistent link between phase-space microstates and macroscopic thermodynamic states can be based on the microscopic definition for entropy :

$$S(N,V,E) = k\ln[\ \prod^{\#}\int dq \int dp\ \delta(\mathcal{H}-E)\ ]\ .$$

As is usual in thermodynamic considerations, the ideal gas is the simplest case to consider. The coordinate integration gives $V^N/N! \simeq (Ve/N)^N$

for N identical ideal-gas particles in a D-dimensional volume V , where the dimensionality D can be 1, 2, or 3. The momentum integration (if we ignore the vanishing center of mass velocity) gives the (hyper) volume of a DN-dimensional sphere of radius $\sqrt{\sum p^2}$, $\propto \sqrt{(2mE)}^{DN}$. Apart from an additive constant the D-dimensional ideal-gas entropy is :

$$S_{\text{ideal}} = NK \ln[\ (V/N)(E/N)^{(D/2)}\ ] \rightarrow PV = NkT = (2E/D)\ .$$

where the mechanical and thermal equations of state follow from the Second Law of Thermodynamics for reversible processes : $TdS = dE + PdV$.

Let us imagine that we have a sample of ideal gas so large and massive that it can be considered a heat *reservoir*, with an unchanging temperature T . By imagining the weak coupling of such an ideal-gas reservoir (indicated by $_r$) to a much smaller system (indicated by $_s$) we have the additivity of entropy (because the states which can be occupied by the reservoir and system are independent of one another and only depend upon their respective energies and volumes) :

$$\Omega_{\text{total}} = \Omega_s(E_s, V_s)\Omega_r(E_r, V_r) \longrightarrow$$

$$S_{\text{total}} = S_s(N_s, E_s, V_s) + S_r(N_r, E_r, V_r)\ .$$

It is evident that maximizing the total entropy (corresponding to maximizing the number of phase-space microstates) with respect to the energies at fixed volume gives

$$(\partial S/\partial E)_s = (\partial S/\partial E)_r = (1/T)\ ,$$

for the maximum, corresponding to *thermal* equilibrium. If we consider the additional possibility that the reservoir and system share a total volume V the entropy must additionally be maximized with respect to the division of the volume :

$$(\partial S/\partial V)_s = (\partial S/\partial V)_r = (P/T)\ ,$$

corresponding to *mechanical* equilibrium at the common pressure P . Apart from an additive constant, the ideal-gas reservoir entropy is

$$S = Nk[\ \ln(V/N) + (D/2)\ln(E/N)\ ]$$

so that the maximum-entropy conditions correspond to thermal equilibrium, with $T_r = T_s$, as well as mechanical equilibrium with $(P/T)_r = (P/T)_s$:

$$\{\ (\partial S/\partial E) = (DNk/2E) = (1/T)\ ;\ (\partial S/\partial V) = (Nk/V) = (P/T)\ \}\ .$$

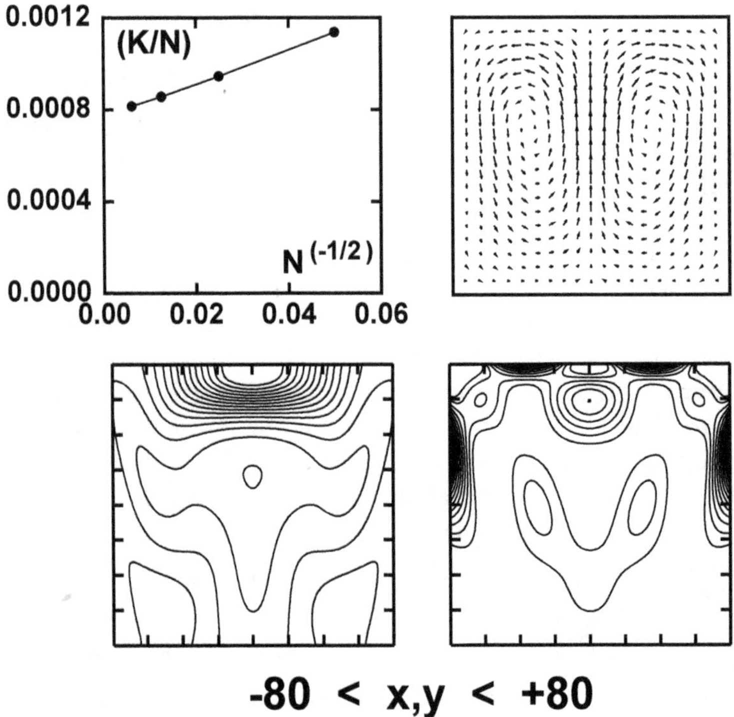

Fig. 3.3 Two-roll convection for a two-dimensional ideal gas with $e = T$ at a Rayleigh number $\mathcal{R} = \Delta T g L^3/(\nu D) = 18,000$. Here $L = (1/g) = 160$ and the mean mass density is unity. The lower hot boundary temperature is 1.5 and the upper one is 0.5 so that ΔT is unity. Both the kinematic viscosity $\nu = \eta/\rho$ and the thermal diffusivity $D = \kappa/(\rho C_V)$ are equal to $L/\sqrt{\mathcal{R}} = \sqrt{64/45}$. The contour plots show the second invariant of the stress tensor [$\sigma_{xy}^2 + (1/4)(\sigma_{xx} - \sigma_{yy})^2$] (to the right) and the square of the temperature gradient (to the left), which are both of the same order as $(\Delta T/L)^2$. The kinetic energy, for this 160×160 grid, is compared to kinetic energies for 80×80, 40×40, and 20×20 grids in the upper left part of the figure.

Evidently Gibbs microcanonical (fixed-energy) approach provides a basis for thermodynamics in terms of maximizing the entropy function $S(E, V)$ and the ideal-gas definition of temperature.

Edwin Jaynes suggested that an extended version of this same sort of entropy maximization approach could be applied to *nonequilibrium* systems. For example, one could constrain the mass or energy current in the presence of an external field, computing the corresponding maximum-entropy distribution function variationally. There are two reasons why this approach is a failure :

[i] For systems with a fixed current not just that current, but also its first, second, . . . time derivatives must *all* be fixed (which is impractical) ;
[ii] For macroscopic systems there are often two or more *different* solutions for the *same* boundary conditions (as in Rayleigh-Bénard flow) but with very different entropies and very different entropy production rates. See **Figures 3.3-3.4** which show two qualitatively different solutions for exactly the same temperature and velocity boundary conditions.

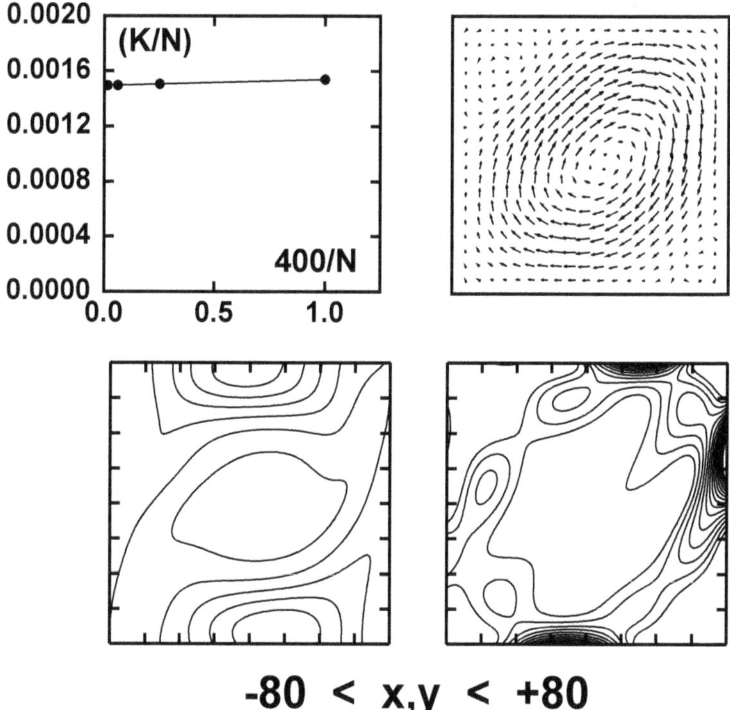

Fig. 3.4 One-roll convection for the same boundary and constitutive conditions as in Figure 3.3 . The contour intervals are $0.25L^{-2}$ for the second invariant of the stress tensor and $0.50L^{-2}$ for the squared temperature gradient in both figures. The number dependence of the kinetic energy here is much smaller than in the two-roll solutions of **Figure 3.3** .

Let us next take up an alternative way to link phase-space calculations to thermodynamics, through Gibbs' canonical ensemble, where temperature replaces energy as an independent variable.

3.5 Nosé's Picture of Gibbs' Canonical Ensemble

Gibbs' "canonical" [in the sense of "simple" or "fundamental"] ensemble is the familiar Boltzmann-factor distribution in phase space, $\propto e^{-\mathcal{H}(q,p)/kT}$. Nosé's 1984 work was motivated by, and displays well, the simplicity of the canonical ensemble relative to the microcanonical one. Gibbs pointed out that the phase-space distribution of a subset of the degrees of freedom [such as those of a single atom or molecule] depends on the number of states accessible to the "rest", where the rest (or "reservoir") are viewed as a heat bath. The number of subset states [or equivalently their extension in phase, $\otimes_s$] can then be expressed in terms of the derivative of the entropy $S_r = k\ln(\otimes_r)$ with respect to the bath's energy, $(\partial S/\partial E)_{N,V} \equiv 1/T$.

So long as the rest of the system has many degrees of freedom, and no trouble in accessing its states, the subset's probability distribution can be expressed in terms of the ideal-gas-thermometer temperature describing the rest. The "rest", if it is large enough, can be viewed as a heat reservoir with a measurable ideal-gas kinetic temperature. In this case the constant-energy $E = E_s + E_r$ distribution can be described as a constant-temperature one for the system, where the system's temperature is a bath property imposed on the system by the many other degrees of freedom.

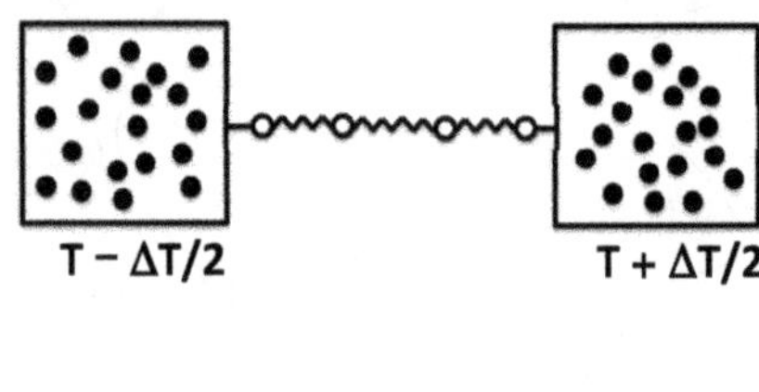

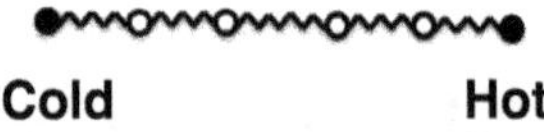

Fig. 3.5 At the top, two particulate many-body reservoirs drive a flow of heat from right to left. Below two Nosé-Hoover reservoirs, indicated by the two filled circles, play the same rôle but with a much simpler computation.

Figures 3.5-3.6 shows the conceptual links of subsystem, reservoir, and thermometer as well as Nosé's streamlining of this picture, replacing reservoir or reservoir plus thermometer with a single "time-scaling" "thermostat variable" s . In the end, as we saw in Sections 1.3-1.5 , the corresponding conjugate "momentum" p_s is nothing other than the time-reversible "friction coefficient" $p_s \longrightarrow \zeta$. A single friction coefficient is enough to represent an entire heat reservoir provided only that the subsystem in contact with the reservoir gains access to a representative sampling of its phase space $\otimes_{\text{subsystem}}$. Nosé's route to the canonical distribution revolutionized, by simplification, our conceptual picture of Gibbs' canonical distribution and its computational applications.

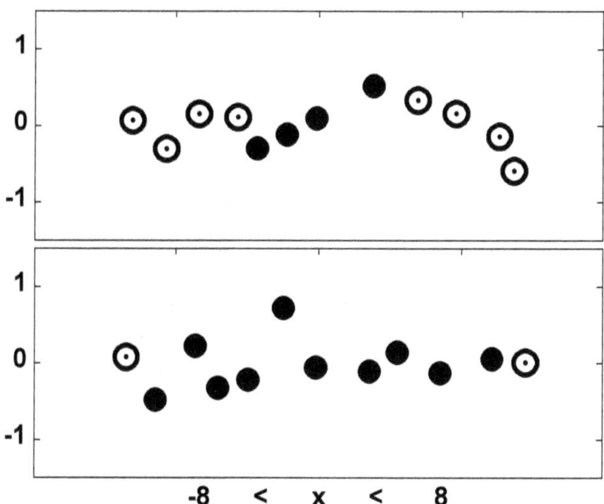

Fig. 3.6 Nosé's revolutionary representation of a heat reservoir by a single Nosé-Hoover degree of freedom. Rather than devoting many degrees of freedom to traditional thermodynamic heat reservoirs (8 reservoir particles are indicated by open circles at the top) we can control temperature by adding a single additional degree of freedom (two open-circle thermostat particles are the open circles at the bottom). The temperatures are 0.5 (on the left) and 2.0 (on the right). The trajectory traces are shown for an elapsed time of 15 . Nearest neighbors interact with Hooke's-Law springs and each particle is tethered to its lattice site by a quartic potential. See Section 7.6 page 199 for more details of this Aoki-Kusnezov ϕ^4 model.

By imagining that a large system is divided in two, subsystem plus heat reservoir, the subsystem probability can be expressed in terms of the reservoir temperature. Thus Gibbs' canonical distribution follows directly from the microcanonical one. It is only necessary that the reservoir possess an ideal-gas temperature where the ideal gas has many degrees of freedom, so that the gas' temperature remains *constant* while its energy undergoes relatively-small changes :

$$(\partial \ln f_s/\partial E) = (-1/k)(\partial S/\partial E)_V = -(1/kT) \longrightarrow f \propto e^{-E/kT} \ .$$

3.6 Two Canonical-Ensemble Temperatures, T_K and T_C

In thermodynamics constant-energy systems can include the energies of a thermal heat reservoir (a large ideal gas at a fixed temperature T) or a mechanical work reservoir (a frictionless piston exerting a fixed pressure P in a gravitational field). From the point of view of a system interacting with such reservoirs the total energy (apart from an additive constant) is the system's Helmholtz' free energy A (in the absence of the piston) or Gibbs' free energy G :

$$A(V,T) \equiv E - TS \; ; \; G(P,T) = E + PV - TS \; .$$

Consider the Helmholtz case. Heat transmitted from the reservoir to the system decreases the reservoir energy by TS, where T is the reservoir temperature and S is the system entropy. Thus $-TS$ is the depletion of the reservoir due to heat transfer. The corresponding heat reservoir's energy loss TS must be included in the total energy. Similarly, if the piston's energy gain PV is also included then the total energy is Gibbs' $G \equiv (E + PV - TS)$.

Another way to "derive" the canonical distribution comes from Gibbs' version of statistical mechanics, where S is the logarithm of the number of phase-space states (or the equivalent phase hypervolume) multiplied by Boltzmann's constant. According to that view Helmholtz' free energy is related to a sum over states :

$$A = E - TS \rightarrow e^{-A/kT} \equiv Z = e^{-E/kT} e^{S/k} = \sum^{\Omega} e^{-E/kT} \; .$$

Here, in the final equality, we have replaced the number of phase-space states in the sum, $\Omega \equiv e^{S/k}$, by an explicit sum over those states.

In evaluating the sum over phase-space states (or "Zustandsumme") it is conventional to divide by $N!h^{DN}$ in order to match the high-temperature low-density "classical-ideal-gas limit" of the quantum sum over states for N identical particles.

$$Z(N,V,T) \equiv \frac{1}{N!h^{DN}} \prod^{\#} [\int dq \int dp \,] e^{-\mathcal{H}/kT} \; .$$

If the Hamiltonian is separable into kinetic and potential parts ,

$$\mathcal{H}(q,p) = K(p) + \Phi(q) \; ,$$

two different definitions of temperature can be derived from the canonical partition function :

$$kT \langle \, \nabla_q^2 \mathcal{H} \, \rangle = \langle \, (\nabla_q \Phi)^2 \, \rangle \; = \; \langle \, F^2 \, \rangle \; \longrightarrow kT_q \equiv \langle \, F^2 \, \rangle / \langle \, \nabla_q^2 \mathcal{H} \, \rangle$$

$$kT\langle\ \nabla_p^2\mathcal{H}\ \rangle = \langle\ (\nabla_p K)^2\ \rangle\ =\ \langle\ (p^2/m)\ \rangle\ \longrightarrow kT_p \equiv \langle\ (p^2/m)\ \rangle\ .$$

Although the two possible definitions, the first "configurational" and the second "kinetic" look very similar, the kinetic definition is much more useful. We will see, in the following Section, that it follows from simple kinetic theory, though not quite so simple as in Section 3.2 . The configurational definition, on the other hand, not only has large fluctuations, but also needs to be corrected for rotation. The configurational temperature can also be negative, which is unphysical.

3.7 Kinetic Theory Temperature ; Ideal-Gas Thermometer

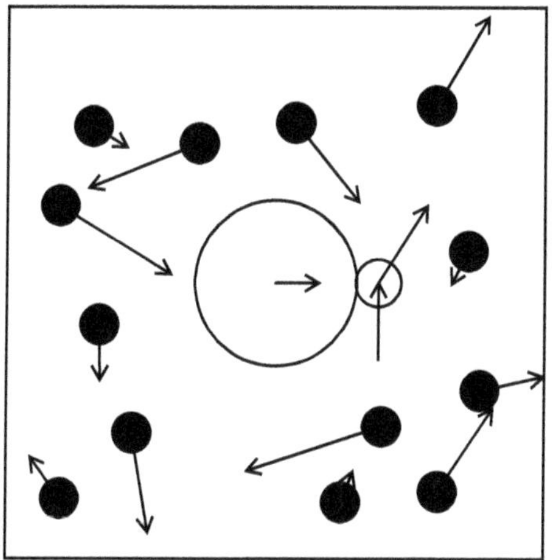

Fig. 3.7 Laboratory-frame closeup of a light heat-reservoir particle interacting with a heavy particle. The light-particle's change in radial velocity is twice the heavy particle's velocity. A simple hard-disk interaction in the center-of-mass frame corresponds to a one-dimensional momentum exchange in the radial direction. In a stationary state the large particles' velocities match the light-particle kinetic temperature, $\langle\ mv^2\ \rangle = \langle\ MV^2\ \rangle\ ;\ \langle\ V\ \rangle = 0$.

A kinetic-theory definition of the kinetic temperature can be carried out in D spatial dimensions. For simplicity we consider only the one-dimensional case here, computing the time-averaged momentum and energy transfers between an ideal-gas thermometer and a massive slow-system particle. The ideal gas has a Maxwell-Boltzmann velocity distribution, with particle mass m and velocities $\{\ \dot{x}\ \}$. The larger, slower, heavier particle with mass M and velocity $\dot{X}$ simply exchanges momentum with the smaller and more numerous ideal-gas masses. The velocities before collision and after collision (with the latter indicated by a prime $'$) conserve momentum and kinetic energy. The post-collision velocities are as follows :

$$\dot{X}' = [\ (M-m)/(M+m)\ ]\dot{X} + 2[\ m/(M+m)\ ]\dot{x} \simeq \dot{X} + 2(m/M)\dot{x}\ ;$$

$$\dot{x}' = [\ (m-M)/(M+m)\ ]\dot{x} + 2[\ M/(M+m)\ ]\dot{X} \simeq -\dot{x} + 2\dot{X}\ .$$

Consider the simple limiting-case example, illustrated in **Figure 3.7** . A light reservoir particle (the small open-circle mass m in the figure) collides [still for simplicity, on the x axis] with the massive system particle M . In the center-of-mass frame of the two colliding particles and in the large-M limit the light particle enters with $-\dot{X}$ and exits with $+\dot{X}$, for a momentum change of $2m\dot{X}$. The heavy particle appears motionless in this frame. The averaged *energy* of such heavy particles, gradually equilibrates to match that of the light particle bath, after a number of collisions of the order (M/m) . For details see the 1993 and 2008 references at the Chapter's end.

At thermal equilibrium the *averaged* effect of a light-particle thermometer on the heavy particle velocity $\dot{X}$ is the ratio of two Maxwell-Boltzmann Gaussian integrals :

$$\ddot{X} \propto \langle\ \dot{X}' - \dot{X}\ \rangle \simeq -4(m/M)\dot{X}\ ,$$

where the averaging indicated by $\langle \dots \rangle$ is weighted with the relative velocity and the Maxwell-Boltzmann thermometer's velocity distribution :

$$\langle \dots \rangle \equiv \frac{\int_{-\infty}^{+\infty} |\dot{x}-\dot{X}|\ \dots\ e^{-m\dot{x}^2/2kT}d\dot{x}}{\int_{-\infty}^{+\infty} |\dot{x}-\dot{X}|e^{-m\dot{x}^2/2kT}d\dot{x}}\ .$$

In evaluating the integrals an expansion in powers of $\sqrt{m/M}$ is used. Replacing the $\dots$ with the heavy particle's energy change, $M[\ (V')^2 - (V)^2)\ ]/2$ shows that the kinetic energy change is proportional to the difference between the heavy particle's kinetic energy and the equilibrium value, $kT/2$. Molecular dynamics simulations consistent with these results can be found in our 2008 paper in the References at this Chapter's end. Regardless of the form of the interaction energy between the heavy and light particles, the action of the thermostat is exactly the same, depending as it does on conservation of momentum and energy to convert the initial velocity $\dot{X}$ to its post-collision value $\dot{X}'$.

3.8 Anisotropic Temperature from Hard Parallel Cubes

In thermodynamics the anisotropy of temperature is seldom considered. Any difference between T_{xx} and T_{yy} would decay rapidly, on a timescale of the collision time. In processes where the anisotropicity is reinforced on this same time scale there can be a persistent difference between the longitudinal and transverse temperatures. This *is* the case for strong shockwaves in gases, liquids, or solids. The light emitted from a strongly shocked material has been shown to differ qualitatively from an equilibrium "black-body" spectrum, confirming the anisotropy of temperature on the timescale of particle-particle collisions.

To analyze such nonequilibrium temperature distributions it is perfectly feasible to imagine, to model, and to simulate three-dimensional generalizations of the one-dimensional ideal-gas thermostat with a local tensor temperature :

$$f(p) \propto \exp[\ -(p_x^2/2mkT_{xx}) - (p_y^2/2mkT_{yy}) - (p_z^2/2mkT_{zz})\ ]\ .$$

An ideal-gas thermostat composed of rotationless hard parallel squares (in two dimensions) or cubes (in three) provides a simple mechanical model for such a tensor temperature.

3.9 Summary

Thermodynamics differs from mechanics by including temperature, heat flow, and entropy in its descriptions of physical systems. All of these concepts have atomistic counterparts, making it possible to simulate thermodynamic and hydrodynamic processes. Out of the infinite variety of possible thermometers the classical ideal gas is the simplest and has a direct mechanical significance. The kinetic temperature of the ideal gas, proportional to its pressure, is the simplest temperature. This definition has a distinct advantage over entropy-based thermodynamic definitions of temperature, particularly in situations where the temperature is a tensor quantity.

3.10 References

For a more-detailed summary of the kinetic-theory calculation sketched in Section 3.7 see W. G. Hoover, B. L. Holian, and H. A. Posch, "Comment I on 'Possible Experiment to Check the Reality of a Nonequilibrium Temperature' ", Physical Review E **48**, 3196-3198 (1993). Molecular dynamics simulations confirming the kinetic theory results are given in Wm. G. Hoover and Carol G. Hoover, "Nonequilibrium Temperature and Thermometry in Heat-Conducting ϕ^4 Models", Physical Review E **77**, 041104 (2008).

A comprehensive study of the ϕ^4 model of **Figure 3.6** was carried out by Ken Aoki and Dimitri Kusnezov including collaborations with Bill, Eric Lutz, and Harald Posch. They discuss conductivity, Lyapunov instability, boundary temperature jumps, and quantum systems in a clear and readily accessible form, eleven arχiv manuscripts.

Some of the criticisms of the purely-logarithmic "thermostat" of Problem 4 are summarized in Daniel Sponseller and Estela Blaisten-Barojas' "Failure of Logarithmic Oscillators to Serve as a Thermostat for Small Atomic Clusters", Physical Review E **89**, 021301R (2014) and the references therein.

3.11 Problems

1. Show that the most likely direction for heat transfer for the four-particle systems of Section 3.3.1 is *from* the system with velocities $\{ -1,\ 0,\ 0,\ +1 \}$ *to* the system with initial velocities $\{ -0.8,\ -0.6,\ +0.6,\ +0.8 \}$. Use the kinetic-theory approximation that the probability for two particles to collide is proportional to their relative speed.

2. According to Stewart and Jacobsen's 1989 data the entropy of triple-point argon is 53.3 [Joules]/[mole kelvin] $= 6.4Nk$ so that the number of (quantum) states per atom is about $e^{6.4} = 608$. If a system composed of N argon atoms changed its N-body state randomly among $e^{6.4N}$ states, at a rate of 10^{12} changes per second, how long would it take for all the N atoms to return to their original N-body state? How large would N have to be to make this recurrence time equal to the Age of the Universe (use 14 000 000 000 years for the Age) .

3. Formulate the equations of motion for the collision of two soft parallel squares in such a way that mass, momentum, and energy are conserved without either particle's rotating.

4. Consider Campisi's "log-thermostat" Hamiltonian, $2\mathcal{H} \equiv p^2 + T\ln(r^2)$ and show that its configurational temperature is $-T$ (yes, negative) in one dimension and *infinite* in two (yes, infinite) . In two dimensions simply average the expressions for T_{xx} and T_{yy} .

5. Compare "modified" Leapfrog simulations of the ϕ^4 conductivity at temperatures 0.0001, 0.001, 0.01, 0.1, 1.0 so as to estimate the conductivity's power-law dependence on temperature. Model your calculation along the lines of **Figure 3.5** on page 68 . In order to modify the Leapfrog algorithm for the first and last particles in the chain imagine that $\zeta_{\rm cold}$ and $\zeta_{\rm hot}$ are known, for the first and last particles at time t . Calculate the coordinates of those particles at time $t + dt$ by solving a modification of the Leapfrog equation, given below, assuming that the current and past values of the coordinates, plus the current friction coefficient are known.

The time is then incremented from t to $t + dt$ in two steps : First, find the new coordinates for the first and last particles in the chain from the linear motion equations for r_{t+dt} :

$$r_{t+dt} = 2r_t - r_{t-dt} + (F_t dt^2/m) - (dt/2)\zeta_t[\ r_{t+dt} - r_{t-dt}\ ] \ .$$

Second, find the new friction coefficients by constructing centered kinetic energies $K_{1/2}$ for the cold and hot particles. $K_{1/2}$ is given by :

$$K_{1/2} \equiv (m/2)\left(\frac{r_{t+dt} - r_t}{dt}\right)^2 \ .$$

giving

$$\zeta(t + dt) = \zeta(t) + dt[\ (K_{1/2}/K_0) - 1\ ]/\tau^2 \ .$$

where $K_0 \equiv (DkT/2)$ is the target kinetic energy for the thermostated particle.

This algorithm is discussed in a 1990 Physical Review A paper by B. L. Holian, A. J. DeGroot, Wm. G. Hoover, and C. G. Hoover : "Time-Reversible Equilibrium and Nonequilibrium Isothermal-Isobaric Simulations with Centered-Difference Störmer Algorithms".

Chapter 4

Continuum Mechanics: Continuity, Stress, Heat Flux, Applications

Topics

Continuum Point of View / Mass, Momentum, and Energy Conservation Equations using with Eulerian and Lagrangian Coordinates / Heat Flux and Energy Equations / Smooth-Particle Averaging – a Link to Atomistic Mechanics / Constitutive Relations and Entropy Production / Free Expansion Example / Drop Oscillations / Shockwaves and Fluxes / Irreversibility, Time Delay, Locality /

4.1 Mechanics from the Continuum Viewpoint

It is expected that larger and larger-scale atomistic simulations will approach and come to define a macroscopic "continuum" limit. In that limit continuum concepts such as local density and flow velocity will come to describe results precisely and accurately. All the continuum's properties are "field variables", distributed *continuously* in space rather than particle properties located at mass points. The "flows" (per unit area and time) of mass, momentum, and energy should likewise approach continuous, and even smoothly differentiable, local "fluxes". The influence of the fluctuations inherent in thermal motion should fade to insignificance with increasing system size.

In favorable cases, such as an equilibrium gas with short-ranged pair interactions, it is possible to prove the existence of such a limit along with formulæ for the deviations from it, which can be as small as $(1/N)$ or as large as the surface effects of order $(1/N^{1/D})$ in D space dimensions. The *approach* to the limit is particularly interesting in cases dominated by atomistic events, where the continuum point of view is likely to be inappropriate. Fracture, plastic flow, and fragmentation, surface structure and shockwaves, and turbulent flow fluctuations all come to mind. For simple equilibrium systems the details of the number-dependence were well-investigated in the early days of computer simulation.

The simplest continuum variable is the density $\rho(r)$, the mass of particles per unit volume. To measure density we could first impose a grid on our system, dividing the system into cells. Then we could simply add up the masses for all the particles within each cell. Better, a *continuous* and differentiable density, rather than a set of discrete cell values, can be achieved by allocating the mass of each particle according to a spatially continuous and twice-differentiable weight function $w(r < h)$ with a smoothly-truncated finite range h . Then, as described in Section 2.7 , the local density becomes a continuously varying differentiable sum :

$$\rho(r) \equiv \sum_j w(r - r_j)m_j \ .$$

Note that the sum includes all particles within a distance h of the location r . Similarly, the local flow velocity $u(r)$ can be based on the ratio of two differentiable sums :

$$u(r) = \rho(r)u(r)/\rho(r) \equiv \sum_j w(r - r_j)m_j v_j / \sum_j w(r - r_j)m_j \ ,$$

where both the sums include all the particles within the range h of the location r . So long as $w(r)$ has two continuous derivatives, so too will the density and the velocity just defined. This smooth-particle approach to defining continuum properties provides, through the "range" of the weight function, h , some useful flexibility in optimizing the correspondence between particle and continuum properties.

The specific *volume*, the volume associated with a particular number of particles, seems to be less useful than density. Like mass, volume *can be* additive. The volume associated with Particle i could be defined as all that volume closer to Particle i than to any other particle. But a continuous and differentiable definition of specific volume is elusive. This is why we choose to emphasize density rather than volume in our continuum work. Once the local mass density, momentum density, and energy density are defined, we should be able to predict their evolution in time on the basis of the flow velocity and the gradients responsible for the momentum and energy fluxes. Our goal is to optimize the correspondence of particle and continuum ideas. We illustrate this goal in the following Section.

4.2 Conservation of Mass, Momentum, and Energy

Our overall goal here is to correlate the microscopic and macroscopic descriptions of the flows of fluids and solids. In what follows we introduce the partial differential equations for the conservative evolution of the field variables $\{ \rho, u, e \}$. There are two distinct finite-difference approaches depending on our choice of spatial grids, either comoving ("Lagrangian") or fixed ("Eulerian").

4.2.1 *Eulerian and Lagrangian Continuity Equations*

The continuum description of systems details the time evolution of the density, velocity, and energy density. Let us derive the flow equations describing this evolution of mass, momentum, and energy. Without any loss of generality we consider a one-, two-, or three-dimensional grid of small cells of volume dx, $dxdy$, or $dxdydz$. We can derive the "continuity equation" in the one-dimensional case by subtracting the mass leaving a cell from the mass entering, during the time interval dt . The one-dimensional derivation we outline here requires only that density and velocity vary *differentiably* in space and time. With these assumptions, the change in cell mass during dt is :

$$(\partial\rho/\partial t)dxdt = [\ (\rho u)_{-dx/2} - (\rho u)_{+dx/2}\ ]dt \rightarrow -(\partial(\rho u)/\partial x)dxdt \ .$$

For convenience we choose a cell $-(dx/2) < x < +(dx/2)$ centered on the origin. Dividing by $dxdt$ gives the one-dimensional "continuity equation" :

$$(\partial\rho/\partial t) = -(\partial[\ \rho u\ ]/\partial x) \ .$$

Exactly similar steps, with dy or $dydz$ included for two or three-dimensional flows, lead to the general continuity equation ,

$$(\partial\rho/\partial t) = -\nabla\cdot(\rho u) \text{ "Eulerian continuity equation"} \ ,$$

where ∇ is $(\partial/\partial x, \partial/\partial y)$ in two dimensions and $(\partial/\partial x, \partial/\partial y, \partial/\partial z)$ in three dimensions. The "Eulerian" adjective connotes describing flow through a fixed coordinate system.

An alternative approach follows if the grid is allowed to move with the local velocity u. Using such a "comoving" grid is often called the "Lagrangian" approach, though it has nothing to do with the Lagrangian of constrained atomistic mechanics. In the comoving case the mass in a cell doesn't change but the right side of the cell moves at a different speed to that of the left :

$$x_{\text{right}} - x_{\text{left}} \stackrel{dt}{\longrightarrow} (\partial u/\partial x)dxdt \ .$$

As a result

$$0 = (d/dt)(\rho dx) = \dot{\rho} dx + \rho(d/dt)dx = \dot{\rho} dx + \rho(\partial u/\partial x)dx \ .$$

Here we use the handy superior dot " · " to indicate a time derivative following the motion. Dividing the last expression by dx gives an alternative view of the continuity equation in one dimension :

$$\dot{\rho} = -\rho(\partial u/\partial x) \text{ “Lagrangian continuity equation in one dimension” } .$$

The calculation is about the same in the two- or three-dimensional cases where the Lagrangian continuity equation is $\dot{\rho} = -\rho\nabla \cdot u$.

Let us look at the continuity equation from the standpoint of moving point particles with masses $\{ m_j \}$. Consider now the time dependence of the smoothed-particle description of density and velocity, based on the weight function $w(r < h)$ along with the two definitions for ρ and u :

$$\rho(r,t) \equiv m\sum_j w(r - r_j) \ ; \ \rho(r,t)u(r,t) \equiv m\sum_j w(r - r_j)v_j \ .$$

Now compare the two sides of the Eulerian continuity equation using these definitions. The one-dimensional result is an *identity* :

$$(\partial\rho/\partial t) \equiv -m\sum_j w'v_j \ ; \ -\nabla(\rho u) \equiv -m\sum_j w'v_j \ ;$$

as are also the two- and three-dimensional versions :

$$(\partial\rho/\partial t) = -\nabla \cdot (\rho u) \longleftrightarrow \dot{\rho} \equiv -\rho\nabla \cdot u \ .$$

Thus the smooth-particle definitions *guarantee* that both the Eulerian *and* the (equivalent) Lagrangian forms of the continuity equation are identities.

We note in passing the important observation that (both forms of) the continuity equation are *time-reversible.* A reversed movie of any flow satisfying the continuity equation will likewise satisfy it with each term in either equation changing sign under time reversal. This exact equivalence of the smoothed-particle description to the continuum description is a very promising start on our quest of seeking a correspondence between the two points of view.

4.2.2 *Eulerian and Lagrangian Motion Equations, Pressure*

The spatial and temporal evolution of velocity according to the continuum equations of motion can be described within either an arbitrary fixed grid or a comoving grid :

$$(\partial u/\partial t) \text{ “Eulerian acceleration” or } \dot{u} \text{ “Lagrangian acceleration” } .$$

As usual the superior dot " · " is a time derivative following the motion. The Lagrangian derivation is the simpler of the two approaches. For simplicity, and with adequate generality, we consider the two-dimensional case. We focus on a comoving volume element with total momentum $\rho u dx dy$. Because the element moves *with* the flow velocity there are no convective flows carrying momentum into or out of the element. It is only through pressure forces exerted on its borders that the element's momentum can vary. Again consider an element small enough that the pressure forces shown here can be assumed to vary linearly in both time and space, throughout the element.

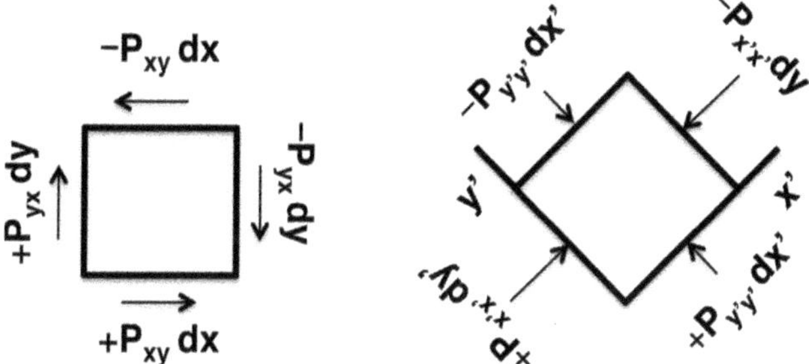

Fig. 4.1 In two dimensions the pressure tensor is the force per unit length. The shear forces shown on the left balance precisely ($P_{xy} = P_{yx}$) . In the rotated coordinates on the right, the equivalent normal stresses are parallel to the rotated axes (x', y') . The figure as drawn corresponds to negative values of (du_x/dy) or (du_y/dx) (on the left) and to a tensile stress in the y' direction (on the right).

In continuum mechanics *pressure* is defined as the force per unit area exerted by the element on its surroundings (or minus the force per unit area exerted by the surroundings on the element) , as measured in the comoving frame. See **Figure 4.1.** The pressure P is a second-rank tensor with P_{xx} the force parallel to the x axis exerted by the element across a plane perpendicular to the x axis. Likewise P_{xy} is the force per unit area across the same plane, but in the y direction. Such a force, if not balanced by an opposing rotational force [in the x direction on the plane perpendicular to the y axis] would cause the angular acceleration to diverge in the small-element limit. See Problem 2 . For this reason it is invariably *assumed* that the tensor is *symmetric*, with $P_{xy} \equiv P_{yx}$. For a linear pressure variation, valid for small volume elements, the (Lagrangian) equation of motion becomes :

$$\rho \dot{u} = \rho(\partial u/\partial t) + \rho u \cdot \nabla u = -\nabla \cdot P \ .$$

The observant reader might well notice that although we discuss and consider $(d/dt)(\rho u)$ we instead end up with $\rho\dot{u}$. In one dimension the missing step is :

$$(d/dt)(\rho u dx) = (\rho u)(d/dt)dx + dx(d/dt)(\rho u) =$$

$$(\rho u)(\partial u/\partial x)dx + (\rho dx)\dot{u} + (u dx)\dot{\rho} \equiv \rho\dot{u}dx \ .$$

where the cancellation of the first and third terms from the last line has required the use of the continuity equation in the form :

$$(u dx)\rho(\partial u/\partial x) = -(u dx)\dot{\rho} \ .$$

Finally, to make the notation of the motion equation clear, we write out all of the individual pressure-tensor terms in the two-dimensional case :

$$\rho\dot{u}_x = -(\partial P_{xx}/\partial x) - (\partial P_{yx}/\partial y) \ ;$$

$$\rho\dot{u}_y = -(\partial P_{xy}/\partial x) - (\partial P_{yy}/\partial y) \ .$$

Again the reader is reminded that it is *assumed*, on physical grounds, that the pressure tensor is symmetric, with $P_{xy} \equiv P_{yx}$.

The physics behind the continuity equation and equation of motion ,

$$\dot{\rho} = -\rho\nabla u \ ; \ \rho\dot{u} = -\nabla \cdot P \ ,$$

can be stimulating, and sometimes daunting. Although the continuity equation is time-reversible, the equation of motion may well not be. There is a tension between the time-reversibility of the underlying atomistic motion equations and the Second Law irreversibility of the macroscopic view of matter.

Typical applications of the motion equation include both reversible *and* irreversible contributions to the pressure tensor P . If pressure is a function of density and energy alone, as at equilibrium, the motion will be reversible. If pressure varies linearly with the velocity gradient, as in a Newtonian viscous fluid, the motion is irreversible. This difference suggests there will be interesting discrepancies with atomistic mechanics in most cases. And this is true. Other possible difficulties are more imaginary than real. It has, for instance, been suggested that the velocity which appears in the continuity equation is *not the same velocity* as that appearing in the equation of motion. By considering well-posed example problems such conceptual difficulties can either be avoided or explored and overcome.

4.2.3 *Heat Flux and the Energy Equations*

In thermodynamics (as opposed to the less-ambitious discipline of thermostatics) changes in the comoving energy are divided into both coordinate-dependent contributions, "work", and coordinate-independent ones, "heat". In continuum mechanics (which includes dynamics) exactly the same division holds, with the pressure tensor doing work and the "heat-flux" vector transmitting heat by *conduction* as opposed to convection (through motion with the fluid) . In continuum mechanics the First Law of Thermodynamics takes the form :

$$\rho\dot{e} = -P : \nabla u - \nabla \cdot Q \ ,$$

where Q is the heat-flux vector (energy per unit mass crossing a comoving plane of unit area in unit time) . The derivation is straightforward and almost familiar. In one dimension the simplest approach begins by considering the comoving change in energy within a comoving element ρdx . This energy change *in* the element reflects the flow of energy through the element's surfaces, including both work and heat :

$$(d/dt)(\rho(e + \tfrac{u^2}{2})dx) = \rho dx(\dot{e} + u\dot{u}) = -dx[\ (\partial/\partial x)(Pu + Q)\ ]$$

$$= -dx[\ u(\partial P/\partial x) + P(\partial u/\partial x) + (\partial Q/\partial x)\ ] \ .$$

In *one* dimension, we take advantage of the equation of motion, multiplied by dxu :

$$dxu\rho\dot{u} = -dxu(\partial P/\partial x) \ .$$

This relation can then be used to simplify the energy evolution equation :

$$\rho dx\dot{e} = -dx[\ P(\partial u/\partial x) + (\partial Q/\partial x)\ ] \ .$$

These same steps in two or three dimensions lead to the general continuum energy equation :

$$\rho\dot{e} = -P : \nabla u - \nabla \cdot Q \ .$$

The double dot notation " : " indicates a sum over all four terms in two dimensions and nine in three. Here we display all four, for the two-dimensional case :

$$P : \nabla u = P_{xx}(\partial u_x/\partial x) + P_{xy}(\partial u_y/\partial x) + P_{yx}(\partial u_x/\partial y) + P_{yy}(\partial u_y/\partial y) \ .$$

Now we can summarize the fundamental basis of fluid dynamics (or thermodynamics, as opposed to thermostatics). The continuity equation for the evolution of density, the equation of motion, for the evolution of momentum, and the energy equation, for its evolution, need to be solved together. The momentum flux (pressure tensor) and the heat flux vector need to be expressed in terms of state variables for the differential equations to be well-posed. A variety of highly-developed numerical methods exist to solve these coupled equations. These methods are typically more complex than those of molecular dynamics. This is because the continuum evolution equations are all *partial* differential equations.

The continuum's density, velocity, and energy are all of them functions of *both* space and time. The continuum evolution equations are perfectly general, applying equally to gases, fluids, solids, and mixtures, subject only to two assumptions: [i] the comoving momentum and energy fluxes, P and Q , can be related to the present and past histories of density, velocity, and energy; [ii] no additional variables (like dislocations or voids or impurities or surface forces) are required in evaluating P and Q .

In molecular dynamics the evolution equations are *ordinary* differential equations. Once the force functions and initial conditions – $\{ r, \dot{r} \}$ – are given, the corresponding accelerations, $\{ \ddot{r} \}$, can be integrated with respect to time; problem solved ! In continuum mechanics the accelerations can only be integrated if the pressure tensor and heat flux vector have specified forms. The force laws of atomistic mechanics (hard spheres, Lennard-Jones, ...) are typically much simpler than the constitutive relations of continuum mechanics.

In the continuum case, the main remaining questions are how to find and formulate the pressure tensor and the heat-flux vector. In continuum mechanics it is assumed that there are recipes, "constitutive relations", which answer these questions, giving the fluxes in terms of the assumed-known values of the density, velocity, energy, and their spatial gradients, and, possibly, their histories. The simplest constitutive relations are Newton's formulation of the shear viscosity and Fourier's formulation of the heat flux vector :

$$P_{xy} = P_{yx} = -\eta[\ (\partial u_y/\partial x) + (\partial u_x/\partial y)\ ] \ ;$$

$$Q_x = -\kappa(\partial T(\rho, e)/\partial x) \ .$$

Notice that the symmetrized form of the velocity gradient has been chosen so as to guarantee the symmetry of the pressure tensor.

Applications of the continuum equations necessarily include even such complicated situations as turbulent flows. We know these flows can be incredibly complex. Such complexity suggests (and three centuries of experience confirm) that we will find numerical difficulties, instabilities, and divergences, at least for some problems, in applying the equations. Let us consider first a numerical method that avoids some of these instabilities, while exhibiting others. It is the *smooth-particle* approach. One of its several virtues is simplicity. We turn to it in the next Section.

4.3 Smooth-Particle Simulation of Continua

The continuum equations for $\dot{r} = u$, $\dot{u}$, $\dot{\rho}$, and $\dot{e}$ are most easily "solved" by evolving *particle* values of the velocity, density, pressure tensor, and energy according to local values of the velocity and temperature gradients evaluated at each particle. The smooth-particle method approaches the need for continuous local variables by visualizing smooth particles as *very smooth*, with two continuous derivatives everywhere in space. Each particle is imagined to have its mass distributed in space according to the weight function $w(r)$. Here we will use exactly the same normalized weight function, Lucy's, as that used in Chapter 2 as an interpolation function for molecular dynamics :

$$w(|r| < h) = C_D[\,1 + 3(|r|/h)\,][\,1 - (|r|/h)\,]^3 \;;\; \int_{\text{space}}^{\text{all}} w(|r|)d\mathbf{r} \equiv 1 \;,$$

where $d\mathbf{r} = 2dr,\ 2\pi r dr,\ 4\pi r^2 dr$ in one, two, and three dimensions. The density at any location is calculated by superposing the contributions there of all of the N particles within the range h :

$$\rho(r) \equiv m \sum_{j=1}^{N} w(r - r_j) \ .$$

Particles could have different masses and different shapes, but here we assume that all of them are described by a purely-radial weight function and that all of them have the same mass, m .

This density definition automatically satisfies conservation of mass. No matter where N particles are placed, integrating their superposed densities over all space necessarily gives Nm . This smooth-particle density definition has the added bonus that it satisfies the continuum continuity equation *exactly*. We showed, in Section 4.2.1 , that the smooth-particle density and velocity obey both forms of the continuity equation :

$$(\partial\rho/\partial t) = -\nabla\cdot(\rho u) \longleftrightarrow \dot{\rho} = -\rho\nabla\cdot u \ .$$

The exact equivalence of these two expressions for the time dependence of the local density makes it unnecessary to solve the continuity equation with the smooth-particle approach. Continuity is satisfied exactly. Notice also, that the underlying equivalence of the Eulerian and Lagrangian formulations guarantees that *both* forms of the continuity equation are *identities* for smooth particles.

By choosing Lucy's weight function for evaluating the density and velocity we are guaranteed that the resulting smooth-particle density and smooth-particle velocity are not just "smooth" but "very smooth", with *two* vanishing space derivatives at $r = h$.

We have just confirmed that the smooth particles conserve mass. The smooth-particle approach can also be formulated so as to satisfy *linear momentum* conservation exactly. To show this, begin by *defining* the spatial average of some particle property f_j at the location r :

$$f(r) \equiv m \sum_j (f_j/\rho_j) w(r - r_j).$$

Here the weight function $w(r - r_j)$ is treated as a probability distribution for Particle j at r in the vicinity of r_j . This formulation for $f(r)$ provides an expression both simple and useful for the spatial derivative of $f(r)$:

$$\nabla f(r) \equiv m \sum_j (f_j/\rho_j) \nabla w_{rj} \ . \quad [\text{ Gradient Definition }] \ .$$

This derivative formulation should be regarded as a *definition* of the continuum gradient ∇f . It is useful in formulating the continuum motion and energy equations in conservative forms. To see this idea succeed for the motion equation consider the following continuum identity :

$$\dot{u} \equiv -(\nabla \cdot P)/\rho \equiv -\nabla \cdot (P/\rho) - (P/\rho^2) \cdot \nabla \rho \ .$$

Apply the ∇f "Gradient Definition" above to *both* terms on the righthand side [using both $f = (P/\rho)$ and $f = \rho$] to find the acceleration for Particle i :

$$\dot{v}_i \equiv -\sum_j [\ (P/\rho^2)_j + (P/\rho^2)_i \ ] \cdot \nabla_i w_{ij} \ ,$$

where $w_{ij} \equiv w(|r_i - r_j|)$. Because the gradients $\nabla_i w_{ij}$ and $\nabla_j w_{ij}$ sum to zero this set of motion equations conserves (linear) momentum exactly.

Analogous steps, applied to the pressure and heat-flux gradients in the energy equation, produce smooth-particle equations for the evolution of the specific energy (the energy per unit mass, e) which conserve the energy exactly :

$$\dot{e}_i \equiv -\sum_j [\ (P/\rho^2)_j + (P/\rho^2)_i \ ] : (\nabla_i w_{ij}) v_j$$

$$-\sum_j [\, (Q/\rho^2)_j + (Q/\rho^2)_i \,] \cdot (\nabla_i w_{ij}) \ .$$

Once again we use the double-dot notation , " : " , to indicate a sum over all four terms in two dimensions, and all nine in three dimensions.

What we have gained, by our definitions of smooth-particle density, averages, and gradients, is a numerical method for solving the partial differential equations of continuum mechanics in a form which strongly resembles the ordinary differential equations for molecular dynamics. Rather than integrating $\{ \dot{r},\ \dot{v} \}$ the smooth-particle approach integrates one extra equation: $\{ \dot{r},\ \dot{v},\ \dot{e} \}$. Before considering some sample solutions we need to discuss recipes for P and Q . These are required ingredients for well-posed continuum problems.

4.4 Constitutive Relations and "Entropy Production"

By far the simplest constitutive relation for continuum calculations is the hydrostatic (inviscid) ideal-gas equation of state together with an absence of heat conductivity :

$$P_{xx} = P_{yy} = P_{zz} = (\rho^2/2) \ ; \ Q = 0 \ .$$

This choice leads to very interesting motion and energy equations :

$$\dot{v}_i \equiv -\sum_j [\, (P/\rho^2)_j + (P/\rho^2)_i \,]\nabla_i w_{ij} = -\sum_j \nabla_i w_{ij} \ .$$

Evidently the *continuum* dynamics for this inviscid nonconducting model fluid is *identical* to the *molecular* dynamics for unit-mass particles interacting with the pairwise-additive Lucy potential with range h :

$$\phi(r_{ij} < h) = \phi_{ij} \propto [\ 1 + 3(r/h)\ ][\ 1 - (r/h)\ ]^3 \ ; \ \Phi = \sum_{i<j} \phi_{ij} \ .$$

The next level in complexity for constitutive relations assumes Newtonian bulk and shear viscosity and Fourier conductivity :

$$P = [\ P_{\rm eq} - \lambda \nabla \cdot u\ ]I - \eta[\ (\nabla u) + (\nabla u)^t\ ] \ ; \ Q = -\kappa \nabla T \ .$$

Here I is the unit tensor, with 1 for the diagonal elements and 0 for the off-diagonal ones, so that its multiplier makes a hydrostatic (diagonal) contribution to the pressure tensor. The contributions of the two viscosity coefficients $\{ \eta, \lambda \}$ provide a symmetric pressure tensor. The "shear" viscosity is η . The "bulk" viscosity, which gives an irreversible viscous resistance to isotropic compression or expansion, is $\lambda + \eta$ in two dimensions and $\lambda + (2/3)\eta$ in three. See Problem 3 at the end of this Chapter.

In order to apply the heat-flux formula a "thermal" equation of state, giving the temperature T as a function of the energy per unit mass e can be based on a density and energy-dependent heat capacity. The equilibrium equation of state can be as simple as van der Waals' or as complex as the polynomial equations of state involving dozens of constants.

Any reasonable constitutive equation obeys the Second Law of Thermodynamics. Consider, for instance, the steady shear $[\ \dot{\epsilon} = (du_x/dy)\ ]$ of a periodic simple fluid with shear viscosity η . The pressure tensor $P_{xy} = -\eta(du_x/dy)$ results in a rate of energy density increase proportional to the square of the strain rate :

$$\rho\dot{e} = \eta(du_x/dy)^2 \ .$$

This conversion of work done to internal heating shows that the viscosity coefficient must be positive. An oscillating volume shows that the "bulk viscosity" and the heat conductivity, κ , must likewise be positive in order to obey the Second Law of Thermodynamics.

The latter case is conceptually interesting. Imagine a bar of length L with heat conductivity κ , constant cross-section A and with the ends of the bar thermostated at temperatures $T + (\Delta T/2)$ on the left and $T - (\Delta T/2)$ on the right. Assuming Fourier's Law holds the heat flux in the bar is

$$Q \equiv -\kappa(dT/dx) = \kappa(\Delta T/L) \ .$$

The hot reservoir transfers heat QA per unit time to the bar, giving rise to a rate of entropy *increase* of the bar $QA/[\ T + (\Delta T/2)\ ]$ at the left. There is a larger rate of entropy *decrease* $QA/[\ T - (\Delta T/2)\ ]$ at the right end of the bar. Keeping just the terms linear and quadratic in ΔT , including the dependence of Q on ∇T , gives a (perhaps surprising) overall *decrease* of bar entropy with time, $\dot{S}_{\rm bar} = -(QA/T)(\Delta T/T)$. The overall entropy loss rate due to heat transfer is quadratic in the temperature gradient and proportional to the volume :

$$\dot{S}_{\rm bar} = -\kappa V(d\ln T/dx)^2 \ .$$

Although the Second Law at first appears to be preserved, with the heat reservoir entropy changes just balancing those of the bar, the bar entropy itself is *paradoxical, decreasing* without limit even though the system is in a *stationary* state. The conventional treatment of this problem according to irreversible thermodynamics is to state that there is a phenomenological "entropy production" *within* the bar, just large enough to offset the net entropy loss rate associated with the ends of the bar.

A similar (and equally mistaken) point of view could be offered in the case of a steady shear flow, (du_x/dy) . Consider this flow for a fluid volume element $V = dxdy$. If we view the element as doing work W on its surroundings, rather than the other way around, the element's energy should be expended at a rate $\dot{W} = \eta V(du_x/dy)^2$. As a consequence the energy of the element would necessarily drop *unless* compensating heat flowed steadily *into it* from the surrounding fluid. But such an overall conversion of heat to work is forbidden by the Second Law. In fact the heat is actually generated *within* the fluid element and flows *out* into the surrounding fluid. In this second (correct) view the internally generated heat is equivalent to an entropy production proportional to the square of the velocity gradient and to the volume. In both viscous and conducting flows the entropy production vanishes at equilibrium and is proportional to the square of the driving gradient away from equilibrium.

Let us turn next to another idealized, but inhomogenous, continuum problem, free expansion. In this case entropy isn't produced by the dissipation of gradients or by the flow of nonequilibrium currents. Instead entropy is produced by a constant-energy expansion, increasing the number (or phase volume) of the available microstates ($\propto V^N$) as the expanding fluid comes to occupy a larger volume. This problem can be investigated in detail so as to illustrate the application of smooth-particle ideas to continuum problems.

4.5 Free Expansion Within a Square 128 × 128 Container

The free-expansion problem, in which a high-pressure fluid expands to fill a larger fixed-wall container illustrates a constant-energy irreversible process. No *work* is done by the container walls and no *heat* is transferred through those walls, which are assumed to be motionless and nonconducting. Evidently the bulk and shear viscosities eventually cause the expanding fluid to reach a quiescent state of (*mechanical*) equilibrium with pressure constant. Heat conductivity will likewise eventually produce *thermal* equilibrium as all temperature gradients dissipate.

Along with Harald Posch, we studied the number-dependence of a series of free expansion problems using a smooth-particle algorithm. In such computer simulations it is convenient to contain the expansion within *pe-*

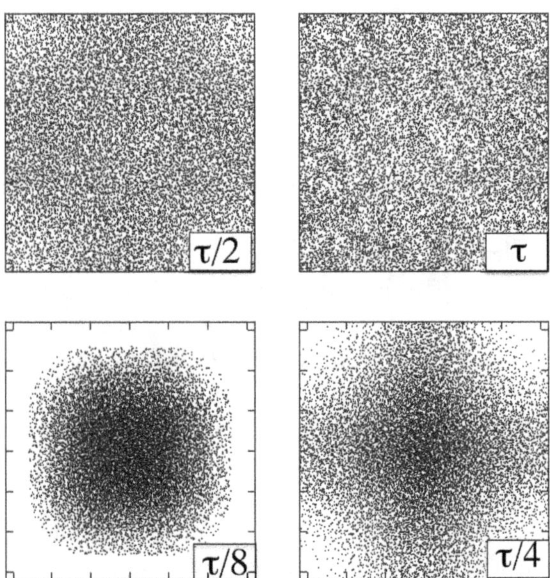

Fig. 4.2 Snapshots of $128 \times 128 = 16\,384$ Lucy particles with $h = 6$ undergoing free expansion, from a density $\rho = 4$ to a density of unity. The adiabatic equation of state $P = (\rho^2/2)$ corresponds to a sound velocity $c = \sqrt{\rho}$ which is 2 in the initial dense state. τ in these snapshots corresponds to the sound traversal time, $64 = (L_x/c)$ for $\rho = 4$.

riodic rather than hard-wall boundaries. We simulated the expansion of a series of N-body compressed gases. Each gas was originally confined to a quarter of its container, at a compressed density $\rho = 4$, four times its final post-expansion equilibrium value. The gas expands, at roughly the speed of sound, and soon fills its container uniformly, at the final density $\rho = 1$.

In all these smooth-particle simulations we used the equilibrium ideal-gas equation of state appropriate to a two-dimensional monatomic gas :

$$P = \rho e = \rho T = (\rho^2/2) \ .$$

In this case the smooth-particle equations of motion generate trajectories "isomorphic" to those from Newton's equations of motion for constant-energy molecular dynamics. The particle trajectories in these two cases, one macroscopic and the other microscopic, are *identical.* In the pairwise-additive molecular dynamics analog, the pair interaction is proportional to the smooth-particle continuum's weight function w :

$$\ddot{r}_i = \dot{v}_i \equiv -\sum_j [\ (P/\rho^2)_j + (P/\rho^2)_i\ ]\nabla_i w_{ij} \longrightarrow \ddot{r}_i = \dot{v}_i \equiv -\sum_j \nabla_i w_{ij} \ .$$

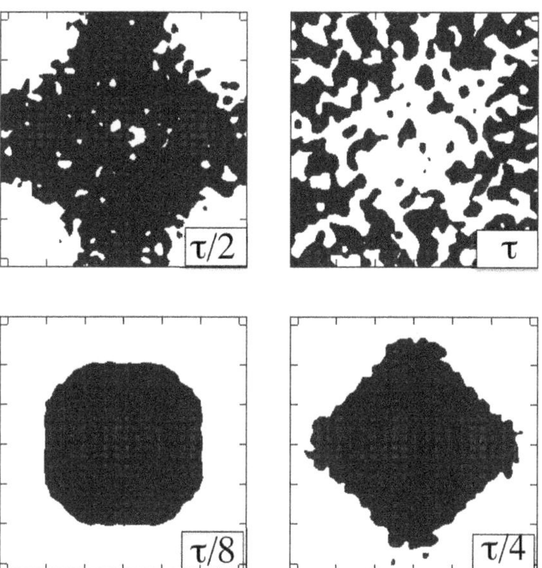

Fig. 4.3 Snapshots of the density evolution for the 16 384 Lucy particles shown in **Figure 4.2** . Grid points with above-average density ($\rho > 1$) are shown here. The time τ in the snapshots corresponds to the sound traversal time, $64 = (L_x/c)$ for $\rho = 4$.

Figure 4.2 shows the particle positions evolved from a cold 128×128 square-lattice particle system with an initial density of 4 (nearest-neighbor spacing 0.5). The potential energy of the system is the sum of the pair contributions to the particle densities, $e_i = (1/2)\sum_j \phi_{ij}$. Snapshots of the density on a 256×256-point grid were computed by summing the contributions from nearby particles, $\rho_g = \sum_j w_{gj}$ using Lucy's weight function with a range $h = 6$. Grid points with densities exceeding the average of unity are shown in black in **Figure 4.3** .

For the most part the motion following the initial violent expansion has died out after one or two sound traversal times. The expansion and equilibration process is completely time-reversible so far as the motion equations are concerned, although the underlying macroscopic expansion should result in increased entropy. Apart from an arbitrary additive constant the two-dimensional monatomic ideal-gas entropy depends on the temperature and density, $(S/Nk) = \ln[\, T/\rho\,]$. Because at constant energy the ideal-gas temperature is unchanged, the entropy increase from expansion should be $Nk \ln 4$, quite close to the measured increase shown in **Figure 4.4** .

We estimated the macroscopic entropy by defining individual smooth-

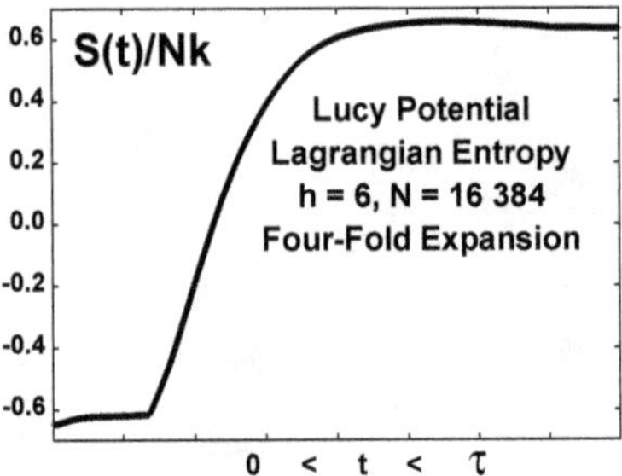

Fig. 4.4 Time dependence of the Boltzmann entropy, $(S/Nk) = \ln(e/\rho)$ where the energy per particle includes the Lagrangian kinetic energy : $(1/2)[\ \langle\ v^2\ \rangle - \langle\ v\ \rangle^2\ ]$. The interparticle forces and the density are both calculated from Lucy's weight function $w(r < h = 6)$. The energy per particle e includes small random values of $\{\ p_x, p_y\ \}$. The initial velocities are chosen so that the initial energy per particle is 2.00 . The initial Boltzmann entropy $(S/Nk) = -\ln(2)$, increases to its final value $+\ln(2)$ corresponding to a four-fold isoenergetic expansion.

particle entropies and summing them up, getting the time history of the entropy evolution, $S(t) = \sum s_j$. This entropy increase, for a system whose motion is perfectly reversible, illustrates Loschmidt's paradox, the incompatibility of microscopic reversibility with macroscopic irreversibility as described by the Second Law of Thermodynamics. We will come back to the details of this apparent paradox in Chapter 8 .

4.6 Oscillation of a Two-Dimensional Drop

At sufficiently low energies and pressures the release of a confined liquid results in a single oscillating drop rather than a rapidly-expanding gas filling its container. A smooth-particle simulation of the motion of an initial square of fluid at density $\rho = 4$ was illustrated in **Figures 4.2 and 4.3.** Here we use a condensed-phase equation of state (applied according to each particle's density) stabilizing a liquid-gas interface, with the liquid at a relatively-high density, near 4 :

$$\{\ e_i = (1/2)(\rho_i - 4)^2 \longleftrightarrow P_i = \rho_i^3 - 4\rho_i^2\ ;\ \rho_i = \sum_j w_{ij}\ \} .$$

$$\{\ \ddot{r}_i = \dot{v}_i \equiv -\sum_j [\ \rho_i + \rho_j - 8\ ]\nabla_i w_{ij}\ \} .$$

These motion equations result in adiabatic roughly-elliptical oscillations. Due to the smooth-particle attractive potential there is no tendency for the

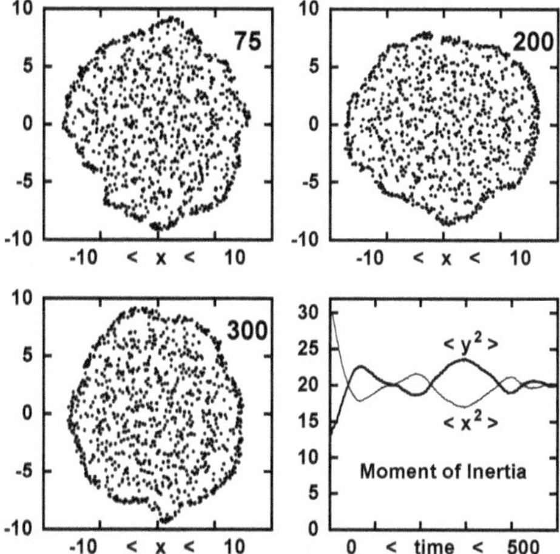

Fig. 4.5 Snapshots for a 1024-particle oscillating drop. The potential energy is $\sum(1/2)(\rho - 4)^2$ where each particle's density is computed from Lucy's weight function with $h = 3$. The drop is centered at the origin so that the sums $\sum(x^2, y^2)$ correspond to components of the moment of inertia, with their difference indicating the ellipticity of the drop. The snapshots span a very rough estimate of one drop vibration time.

liquid drop to fill the entire square container. See **Figure 4.5** for a series of snapshots illustrating the vibration of the drop. Initially the circular drop was stretched 20% in the x direction and similarly compressed in the y direction. The time-dependence of the drop's oscillations can be related to surface tension, which favors a circular drop.

4.7 Shockwaves and Fluxes, Irreversibility, Time Delay

Steady shockwaves present a time series of nonequilibrium states, all with the same mass, momentum, and energy fluxes, confined by *equilibrium* boundary conditions. Just as in the case of Joule-Thomson flow, where hot compressed fluid expands irreversibly, shockwaves focus our attention on *fluxes* because *all* the local fluxes are constant in such steady flows. The simplicity of the shockwave problem, even though it is very far from equilibrium, commends its study. To summarize, a shockwave is the nonequilibrium region joining a rapidly flowing cold fluid (or solid) and a slower hot one. The width or "thickness" of the shock is on the order of the mean free path, for gases, or the interparticle spacing, for liquids and solids. **Figure 4.6** shows the shockwave from both the continuum and the atomistic

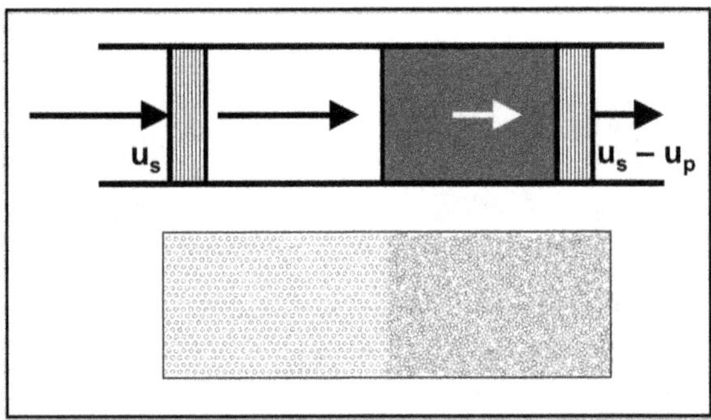

Fig. 4.6 Stationary shockwave geometry (above) : cold fluid or solid enters at the left, undergoes compression and heating in the shock process, and exits at the right. The mass, momentum, and energy fluxes are constant throughout the shock. Particle view of a strong shockwave (below) : a cold crystal is compressed two-fold and slowed to half its initial speed by the shock process. The thickness of the shock front is only a few atomic spacings.

viewpoints.

The particle snapshot in the lower half of the Figure shows a fast-moving cold zero-pressure triangular lattice entering at the left, with a mass flux $\rho_{\rm cold}u_s$. This fast-moving material is slowed to half its initial speed, melts, and doubles in density as it contacts the hot slower-moving fluid on the right side of the shock region. The hot dense fluid then exits at the right, with mass flux $\rho_{\rm hot}(u_s - u_p)$. Not only are the cold and hot values of the mass flux equal, the mass flux *within the shockfront* necessarily has this same constant value, as follows from the Eulerian form of the continuity equation, applied to a stationary state :

$$(\partial\rho/\partial t) = -\nabla \cdot (\rho u) \equiv 0 \rightarrow (\rho u) = \text{constant} \longrightarrow$$

$$\rho_{\rm cold}u_s = \rho(x)u(x) = \rho_{\rm hot}(u_s - u_p) \ .$$

The mass flux is easy to visualize and could be determined locally by watching a movie of the flow. Any plane normal to the x axis is crossed by particles carrying their mass. A time-averaging process can determine the mass flux, the flow of mass per unit area and time. To make this conceptual picture into a definite description of the flow requires the definition of a "weight function". When the time comes we will adopt Lucy's weight function, not just for the mass flux, but also for the momentum and energy fluxes, which are likewise constant in a steady flow.

The local density, velocity, and mass flux ρu depend (weakly) on the form and (significantly) on the range of the weight function. Despite this dependence we are guaranteed that the resulting definitions $\{ \rho, u, \rho u \}$ satisfy both the Eulerian and the Lagrangian versions of the continuity equation. The equation of motion and the energy equation also require definitions of the velocity and temperature gradients. These gradients depend more significantly on the range h of $w(r < h)$. This is an advantage of the smooth-particle approach. The range of the weight function can be chosen so as to simplify the resulting macroscopic constitutive relations. We will consider this optimization in more detail in Chapter 7. Here we wish only that the reader keep in mind the ambiguities intrinsic to expressing macroscopic quantities in terms of microscopic coordinates, velocities, and accelerations.

Now consider the momentum flux. It is composed of two parts, the "convective" part, ρu^2, associated with the moving particles, and the "comoving part" P_{xx} due to the action-at-a-distance linking pairs of particles. The convective contributions are distributed over the particle size according to the weight function $w(x < h)$. Although the conservation relations specifically involve the longitudinal coordinate x we wish to emphasize — on physical grounds — that such local quantities as density, velocity, energy, temperature, and pressure *all* require a circular average in two dimensions or a spherical average in three, and that the choice of the range of w is an *opportunity* to optimize the connections linking microscopic dynamics with continuum mechanics. Both the convective and comoving parts of the pressure tensor and the flux of energy share this arbitrary nature.

In the shockwave problem it is mandatory to distinguish the longitudinal pressure-tensor component P_{xx} from the transverse component(s). These two can be quite different. For a Newtonian fluid the difference is twice the shear viscosity times the strain rate :

$$P_{xx} = P_{yy} - 2\eta(du/dx) \ .$$

The constant *longitudinal* momentum flux, where the velocity and pressure tensor vary in the x direction, follows from the Eulerian form of the longitudinal equation of motion for a stationary flow :

$$(\partial/\partial t)(\rho u_x) = -(\partial/\partial x)(\rho u_x^2 + P_{xx}) \equiv 0 \longrightarrow$$

$$[\ P_{xx} + \rho u_x^2\ ] = \text{ constant } \longrightarrow$$

$$[\ P_{xx} + \rho u_x^2\ ]_{\text{cold}} = [\ P + \rho u_x^2\ ]_{\text{cold}} = [\ P + \rho u_x^2\ ]_{\text{hot}} = [\ P_{xx} + \rho u_x^2\ ]_{\text{hot}} \equiv$$

$$P_{xx}(x) + \rho(x) u_x^2(x)\ [\text{ For Any } x\ ] \ .$$

Finally, the energy flux follows from the Eulerian *energy* equation. In the stationary case the spatial variation is entirely in the x direction :

$$(\partial/\partial t)(\rho e + \tfrac{1}{2}\rho u_x^2) = -(\partial/\partial x)[\ \rho u_x e + \tfrac{1}{2}\rho u_x^3 + u_x P_{xx} + Q_x\ ] \equiv 0 \longrightarrow$$

$$(\rho u_x)[\ e + (P_{xx}/\rho) + \tfrac{1}{2}u_x^2\ ] + Q_x = \text{constant} \longrightarrow$$

$$(\rho u)[\ e + (P/\rho) + \tfrac{1}{2}u_x^2\ ]_{\text{cold}} \equiv (\rho u)[\ e + (P/\rho) + \tfrac{1}{2}u_x^2\ ]_{\text{hot}} \equiv$$

$$(\rho u)[\ e + (P_{xx}/\rho) + \tfrac{1}{2}u_x^2\ ] + Q_x\ [\text{ For Any } x\ ]\ .$$

Notice that the comoving heat flux $Q_x(x)$ and the viscous parts of the pressure tensor vanish in the cold and hot regions, but *not* in the interior of the shockwave. We turn next to simple illustrative solutions of the three conservation laws subject to shockwave boundary conditions.

4.7.1 *A Simple Model for Shockwave Structure*

To illustrate the steps necessary for the quantitative treatment of shockwaves from the standpoint of continuum mechanics consider a model with the equilibrium equation of state :

$$P = \rho e\ ;\ e = (\rho/2) + 2T\ .$$

Boundary conditions which satisfy conservation of mass, momentum, and energy and correspond to two-fold compression from an initial density of unity and temperature of zero are the following :

$$\rho : 1 \rightarrow 2\ ;\ T : 0 \rightarrow (1/8)\ ;\ e : (1/2) \rightarrow (5/4)\ ;\ P : (1/2) \rightarrow (5/2)\ ;\ u : 2 \rightarrow 1\ .$$

The mass, momentum, and energy fluxes for these boundary conditions have the following values :

$$\rho u = 2\ ;$$

$$P_{xx} + \rho u^2 = (9/2)\ ;$$

$$\rho u[\ e + (P_{xx}/\rho) + (u^2/2)\ ] + Q_x = 6\ .$$

It is noteworthy that this formulation of the conservation laws doesn't allow for *nonequilibrium* contributions to the energy per unit mass e .

The detailed internal structure of the shockwave requires constitutive relations for the pressure tensor and the heat-flux vector. For simplicity we

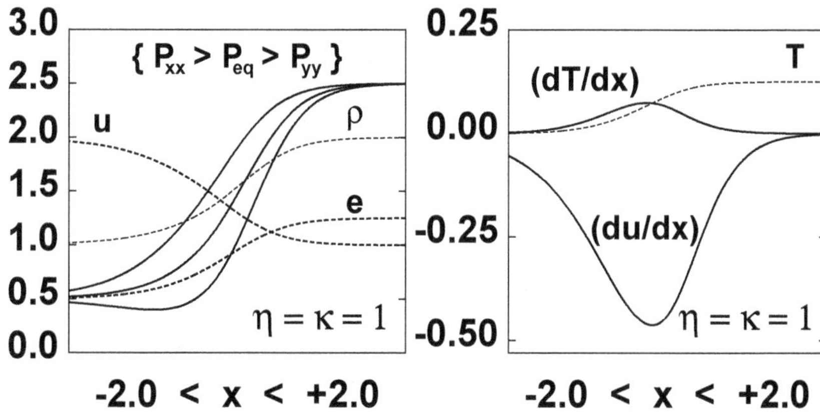

Fig. 4.7 Mechanical and thermal variables according to the Navier-Stokes equations with zero bulk viscosity: $\lambda = -\eta$. The shear viscosity and heat conductivity are both unity: $P_{xx} = P_{\rm eq} - \eta(du/dx)$ and $P_{yy} = P_{\rm eq} + \eta(du/dx)$; $Q_x = -\kappa(dT/dx)$.

use Newtonian viscosity and Fourier heat conductivity in this illustration :

$$P_{xx} = (9/2) - (4/\rho) = (\rho^2/2) + 2\rho T - (\lambda + 2\eta)(du/dx) \; ;$$

$$e + (P_{xx}/\rho) + (u^2/2) - (\kappa/2)(dT/dx) = 3 \; .$$

The *structure* of the shockwave requires the simultaneous solution of these two ordinary differential equations, one for (du/dx), the other for (dT/dx) . Solutions are shown in the **Figures 4.7 and 4.8** for equal values of the shear viscosity and heat conductivity, with $\lambda = -\eta$, corresponding to vanishing bulk viscosity, and for $\lambda = 0$.

4.7.2 *Irreversibility, Time Delay, Locality*

Before leaving this brief discussion of idealized one-dimensional shockwaves it is worthwhile to review the differences between the continuum approach illustrated in **Figures 4.7 and 4.8** and ordinary molecular dynamics as interpreted with Lucy's weight function. In molecular dynamics the time reversibility of the motion equations can *never* match an irreversible constitutive model like Newtonian viscosity or Fourier heat conduction. Newton's microscopic motion equations guarantee that the pressure tensor is invariant to time reversal while the heat flux changes sign.

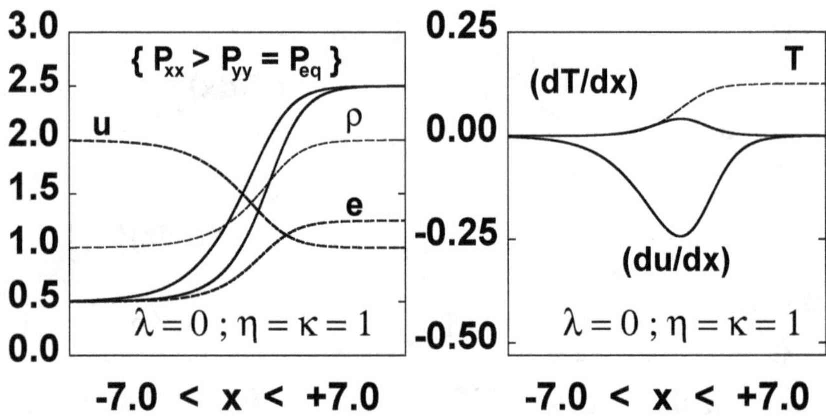

Fig. 4.8 Mechanical and thermal variables according to the Navier-Stokes equations with "Stokesian" ($\lambda \equiv 0$) viscosity: $P_{xx} = P_{\rm eq} - 2\eta(du/dx)$ and $P_{yy} = P_{\rm eq}$. The shear viscosity and heat conductivity are both unity. Notice the increased shock width relative to that in **Figure 4.7** due to the presence of both bulk and shear viscosities.

Newton's viscosity shows that P_{xy} changes sign on reversing a shear flow. Fourier's law indicates that time reversal leaves the heat flux unchanged. Contrarily, Newton's microscopic motion equations show that the heat flux changes sign when the velocity is reversed. Additionally the usual continuum constitutive equations are "local" and "instantaneous", with the stress and heat flux at a particular point and time proportional to the velocity and temperature gradients *at* that same point and same time. Even in the simplest case, with pair forces, the stress tensor and heat flux vector from molecular dynamics are nonlocal. The precise manner in which the nonlocal flux contributions are *defined* (half at each particle's location or smoothly distributed *between* the members of each pair) affects the resulting constitutive relation derived from molecular dynamics.

These contradictions between the microscopic and macroscopic formulations of transport theory are true and cannot be changed. But they can be understood from the standpoint of linear reponse theory, a perturbation theory giving the first-order deviations from Gibbs' equilibrium results. Linear response theory shows (quite correctly) that the response of the pressure tensor to the strain rate is delayed by an effective "collision time" so that the paradoxical results of continuum mechanics involve its instantaneous nature. There is no logical end to models intended to bridge the gap between time-reversible simulations and irreversible descriptions of the results. Nonlinearity is also a concern. Periodic simulations of simple shear show that viscosity decreases with increasing strain rate. Simulations of shockwaves show the opposite rate dependence. Much remains to be done.

At the 2014 Melbourne Memorial Conference honoring the late Ian Snook, his friend and colleague Peter Daivis pointed out that Daivis' PhD student, Andrew Nicholas Charles, has recently published his dissertation : "Smoothed Particle Modelling of Liquid-Vapour Phase Transitions", available as a .pdf file. That work points out the usefulness of Smooth Particle Applied Mechanics in treating complex phase-equilibrium problems. Simulations aimed at determining the *nonequilibrium* contributions to internal energy in strong shocks are another very promising research direction.

4.8 Summary

The conservation of mass, momentum, and energy lead to the continuum continuity, motion, and energy equations. These partial differential equations relate the time dependence of field quantities such as density, velocity, energy, stress, and heat flux to spatial gradients of field quantities. The equations can be solved on stationary "Eulerian" grids, comoving "Lagrangian" grids, or on particulate grids composed of representative interpenetrating "smooth particles". A variety of far-from-equilibrium problems can be solved and described with these techniques and related to more detailed atomistic simulations. The linking together of these different points of view makes possible the multiscale simulations used to describe biomedical reactions described in Chapter 9 .

4.9 References

The free-expansion results, as well as a comparison of them with continuum simulations, can be found in Wm. G. Hoover, H. A. Posch, V. M. Castillo, and C. G. Hoover, "Computer Simulation of Irreversible Expansions *via* Molecular Dynamics, Smooth Particle Applied Mechanics, Eulerian, and Lagrangian Continuum Mechanics", Journal of Statistical Physics **100**, 313-326 (2000).

The simple shockwave model is described and analyzed in Wm. G. Hoover, C. G. Hoover, and F. J. Uribe, "Flexible Macroscopic Models for Dense-Fluid Shockwaves: Partitioning Heat and Work; Delaying Stress and Heat Flux; Two-Temperature Thermal Relaxation", Proceedings of XXXVIII International Summer School-Conference APM 261-273 (2010) = arχiv 1005.1525 .

The molecular dynamics simulations are detailed in Wm. G. Hoover and C. G. Hoover, "Well-Posed Two-Temperature Constitutive Equations for Stable Dense Fluid Shockwaves using Molecular Dynamics and Generalizations of Navier-Stokes-Fourier Continuum Mechanics", Physical Review E **81**, 046302 (2010) = arχiv 1001.1015 .

Smooth-particle simulations go back to Lucy's "A Numerical Approach to the Testing of the Fission Hypothesis" and the Gingold-Monaghan work, "Smoothed Particle Hydrodynamics: Theory and Application to Nonspherical Stars" in 1977, when all three men were together at Cambridge. "Dissipative Particle Dynamics" was developed in 1992 by Hoogerbrugge and Koelman. These methods are similar, with the latter one emphasizing stochastic and rheological, as opposed to astrophysical, applications. See, for instance, Español and Warren's 1995 Europhysics Letter, "Statistical Mechanics of Dissipative Particle Dynamics" as well as Bill's World Scientific 2006 Book, *Smooth Particle Applied Mechanics: The State of the Art.*

4.10 Problems

1. Consider a two-dimensional fluid, initially of unit density, with a steady purely-radial velocity, $u(r) = \dot{\epsilon}r$. Compare the Eulerian and Lagrangian time derivatives of the velocity, $(\partial u/\partial t)$ and $\dot{u}$, and as well as the two time derivatives of the density ρ .

2. Compute the angular acceleration of a square element of area $dxdy$ due to an asymmetric pressure tensor: $P_{xy} \neq P_{yx}$. Show that the acceleration diverges in the limiting case that $dx = dy \rightarrow 0$.

3. Beginning with Newton's viscous pressure-tensor definition

$$P = [\ P_{\rm eq} - \lambda\nabla\cdot u\ ]I - \eta[\ \nabla u + (\nabla u)^t\ ]\ ,$$

where " t " indicates transpose, show that the bulk viscosity has the values given in the text, $\lambda + \eta$ in two dimensions and $\lambda + (2/3)\eta$ in three, by considering isotropic compression/expansion cycles of both two- and three-dimensional "Newtonian fluids" :

$$(\partial\dot{x}/\partial x) = (\partial\dot{y}/\partial y) = \dot{\epsilon}/2$$

in two dimensions and

$$(\partial\dot{x}/\partial x) = (\partial\dot{y}/\partial y) = (\partial\dot{z}/\partial z) = \dot{\epsilon}/3$$

in three dimensions. Here $\dot{\epsilon}$ is the "volume strain rate", $(\dot{V}/V)$, for these deformations and is negative in compression.

4. Show, by following the derivation of the virial theorem, that a many-body Hamiltonian with potential energy $\Phi = \sum \phi_{i<j}$ implies that $P_{xy} = P_{yx}$.

5. Compute a fourth-order Runge-Kutta solution of the shockwave problem with $P_{\rm eq} = (1/2)\rho^2 + 2\rho T = \rho e$ for $\kappa = \eta = 1$ and for $\kappa = 0$. Compare your results to those of **Figures 4.7 and 4.8** .

6. Show that the sound velocity c of a gas with the equation of state $P = (\rho^2/2)$ is $\sqrt{\rho}$.

7. Investigate the entropy production in the free expansion problem by solving the smooth-particle equations with shear viscosity added to the pressure tensor and with heat conductivity added to the heat flux.

Chapter 5

Numerical Molecular Dynamics and Chaos

Topics

Formulations of Particle Mechanics / Leapfrog Method and Hamilton's Least-Action Principle / Runge-Kutta Integration Method / Finite-Precision Solutions and Periodic Orbits / Levesque and Verlet's Bit-Reversible Algorithm / Random Symplectic Algorithms and Time Reversibility / Ergodicity and Lyapunov Instability with $\{\dot{q}, \dot{p}, \dot{\zeta}, \dot{\xi}\}$ / Dimensionality Loss in Oscillator Problems / Many-Body Lyapunov Instability / Coordinate Dependence of Lyapunov Spectra /

5.1 Formulations of Particle Mechanics

In its simplest, most useful, and most usual forms, particle mechanics can be formulated in Newtonian, Lagrangian, or Hamiltonian terms :

$$\{ \ m\ddot{r} = F(r) = -\nabla\Phi(r) \ \} \ ; \ \text{Newtonian} \ ;$$

$$\{ \ (d/dt)(\partial\mathcal{L}(q,\dot{q})/\partial\dot{q}) = (\partial\mathcal{L}(q,\dot{q})/\partial q) \ \} \ ; \ \text{Lagrangian} \ ;$$

$$\mathcal{L} \equiv K(q,\dot{q}) - \Phi(q) \ ;$$

$$\{ \ \dot{q} = +(\partial\mathcal{H}(q,p)/\partial p) \ ; \ \dot{p} = -(\partial\mathcal{H}(q,p)/\partial q) \ \} \ ; \ \text{Hamiltonian} \ ;$$

$$\mathcal{H} \equiv K(p) + \Phi(q) \ .$$

In this Chapter we lay out two useful approaches to solving these ordinary differential equations of atomistic mechanics. Because continuum problems can be formulated in terms of smooth particles these same microscopic methods can be suitably modified to describe large-scale macroscopic problems.

Special methods, of which the "Leapfrog" algorithm of Section 2.4 is prototypical, can be developed for second-order differential equations such as Newton's. In the next Section we discuss two ways in which the Leapfrog algorithm can be improved. Because second-order equations can typically be written as two first-order equations ,

$$\{ \ddot{r} \equiv \dot{v} \ ; \ m\ddot{r} = F(r(t)) \ \} \longrightarrow \{ \ m\dot{v} = F(r(t)) \ ; \ \dot{r} = v \ \} \ ,$$

it is enough (and is typically more useful) to be able to solve *first*-order equations. Because describing constrained or driven systems typically involves first-order time derivatives, methods for first-order equations are both necessary and sufficient to reach our goal of solving a wide range of problems in particle mechanics. Because the Leapfrog approximation for velocity, $v(t) = [\ r(t+dt) - r(t-dt)\ ]/(2dt)$, is inaccurate, other approaches have been discovered. Of these the "Velocity-Verlet" Algorithm has been promoted as an improvement over the Leapfrog approach.

This Chapter begins with two relatively simple integration algorithms – one of them Velocity-Verlet – and illustrates their use for both few-body and many-body problems. Precision, accuracy, and stability of the solutions is discussed in sufficient detail for the reader to be able to apply these concepts to practical problems in particle and continuum mechanics.

5.2 Leapfrog Method and the "Velocity-Verlet Algorithm"

The "Leapfrog Method" for solving Newton's equations of motion is based on a Taylor's series expansion of the second time derivative of the Cartesian coordinate r . The second derivative $\ddot{r}$ is also the acceleration a and the first time derivative of the velocity v :

$$a = \dot{v} = \ddot{r} \ ;$$

$$v_{+1/2} \simeq (r_{+1} - r_0)/dt \ ; \ v_{-1/2} \simeq (r_0 - r_{-1})/dt \ ;$$

$$a_0 \simeq (v_{+1/2} - v_{-1/2})/dt = [\ r(t+dt) - 2r(t) + r(t-dt)\ ]/dt^2 \ \longrightarrow$$

$$r(t \pm dt) \equiv 2r(t) \mp r(t-dt) + a(t)dt^2 \ .$$

[Coordinate Leapfrog]

The Leapfrog name imagines successive leaps in time alternating between coordinate updates and velocity updates. Alternatively it can refer to leaping from $r(t \mp dt)$ to $r(t \pm dt)$ over the current coordinate $r(t)$.

The final form above contains no velocities. It reproduces $\ddot{r}$ with an error of order $(\ddddot{r}\ /12)dt^2$. Because the underlying Newtonian equation is second-order *two* initial values need to be specified to start the coordinate version of the algorithm. Then as many more as desired can be computed — in the past or in the future — by solving iteratively for $r(t \pm dt)$ and $r(t \pm 2dt)$ and $r(t \pm 3dt)$ ... :

$$r(t \pm dt) = -r(t \mp dt) + 2r(t) + (dt)^2 F(r)/m \ .$$

Figure 2.6 on page 38 shows that the algorithm is "second-order" in the sense that the coordinate error at a fixed time is proportional to the square of the timestep dt .

In applying the Leapfrog algorithm to Hamiltonian problems it is desirable to have a formulation for the momenta $\{ \ p(t) \ \}$ *at the same set of times* $\{ \ t \ \}$ as the coordinates $\{ \ q(t) \ \}$. One approach to this problem is the Cartesian "Velocity-Verlet" algorithm :

$$x(t + dt) \equiv x(t) + v(t)dt + (1/2m)F(t)dt^2$$

$$\longleftrightarrow$$

$$x(t - dt) \equiv x(t) - v(t)dt + (1/2m)F(t)dt^2 \ .$$

Subtracting the second version from the first reveals how this velocity is defined :

$$x(t + dt) - x(t - dt) \equiv 2v(t)dt \ .$$

The Velocity-Verlet velocity is just the naïve centered-difference approximation. *Adding*, rather than *subtracting*, the two Velocity-Verlet equations shows that they are precisely equivalent to Leapfrog :

$$x(t + dt) + x(t - dt) \equiv 2x(t) + (1/m)F(t)dt^2 \ .$$

[Coordinate Leapfrog]

The first-order approximation $p(0) \equiv m[\ q(+dt) - q(-dt)\ ]/(2dt)$ can be improved by adopting a higher-order difference approximation to the momentum. A useful algorithm for generating the momenta, again based on summing Taylor's series, can be developed by combining the coordinate data for four times centered on the time of interest, here $t = 0$:

$$p(t = 0)/m \equiv (4/3)\frac{[\ q(+dt) - q(-dt)\ ]}{2dt} - (1/3)\frac{[\ q(+2dt) - q(-2dt)\ ]}{4dt} \simeq$$

$$(4/3)[\ \dot{q} + (dt^2/6)\ \dddot{q}\ + (dt^4/120)\ \overset{.....}{q}\ ] -$$

$$(1/3)[\ \dot{q} + (4dt^2/6)\ \dddot{q}\ + (16dt^4/120)\ \overset{.....}{q}\ ] \simeq$$

$$\dot{q} - (dt^4/30)\ \overset{.....}{q}\ .$$

A check of the computer programming can be based on the harmonic oscillator problem, which has an analytic solution for the Leapfrog implementation. For a specially symmetric solution we choose the initial oscillator coordinate at a turning point, say $q = 1$:

$$q(ndt) = e^{in\alpha} \rightarrow e^{+i\alpha} - 2 + e^{-i\alpha} = -(dt)^2 \longrightarrow$$

$$\alpha = \arccos[\ 1 - (dt^2/2)\ ]\ .$$

Figure 5.1 compares the two formulations of the leapfrog momenta for the initial condition $(q,p) = (1,0)$ to the analytic orbit, $q^2 + p^2 = 1$.

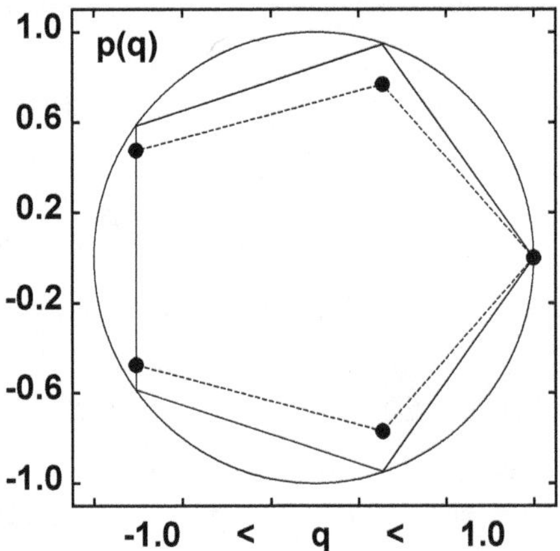

Fig. 5.1 Leapfrog oscillator (q,p) points using the first-order and third-order approximations to the momentum with $dt = (2\pi/5)$. The circle represents the analytic orbit, a circle of radius unity. The filled circles and dashed lines indicate the first-order solution.

The central-derivative formulæ on page 99 of Milne's *Numerical Calculus* provide derivatives accurate to fifth and seventh orders in dt , but applying these to the leapfrog algorithm provides no noticeable improvement over our third-order approach because the underlying leapfrog algorithm has coordinate errors of order dt^4 .

We pointed out the interesting connection between the Leapfrog algorithm and Hamilton's "Least-Action" Principle in Section 2.5.3 . Gillilan and Wilson, in their 1992 Journal of Chemical Physics paper "Shadowing, Rare Events, and Rubber Bands. A Variational Verlet Algorithm for Molecular Dynamics" showed that first-order approximations to the kinetic and potential energy integrals in Least-Action integration give the Leapfrog Algorithm.

Our higher-order approach to velocities is apparently a definite improvement over the $v \equiv [\ dq/dt\ ]$ definition of Velocity-Verlet, and gives an order-of-magnitude better energy conservation for the oscillator problem of **Figure 5.1** . Nevertheless the improved velocity is evidently of the same *order* as the $[\ dq/dt\ ]$ approximation. For smaller timesteps the oscillator problem shows that the improvement is close to a factor of three for small dt rather than an order of magnitude. Let us turn to a (much) more accurate approach.

5.3 Runge-Kutta Integration Method

Though the Leapfrog method just described is adequate for the *second-order* (in the time) motion equations of Newtonian molecular dynamics, it has disadvantages. Many implementations of molecular dynamics, including most nonequilibrium simulations, are governed by *first-order* motion equations. This is the typical situation when Lagrange multipliers or friction coefficients are used to drive systems away from equilibrium or to constrain systems in nonequilibrium states.

More general integration methods, of higher accuracy, are needed in many applications. The local single-timestep phase error of the leapfrog algorithm is of order dt^3 . Reducing this error to the roundoff level of 10^{-16} would require a timestep smaller than 0.00001 . A good fourth-order method, of which there are many, incurs a single-step error of order $dt^5/5!$ which reaches double-precision roundoff error with $dt = 0.001$.

Among fourth-order methods the Runge-Kutta algorithms are a good choice. Consider the first-order differential equation :

$$\dot{f} \equiv (d/dt)f = \texttt{rhs(f)} \ ,$$

where we use `f(t)` for the time-dependent solution and `rhs` for the time derivative of `f` . We assume that the initial value `f0` and the form of the righthand side function `rhs(f)` are known. f can be thought of as a vector, representing the solution of many coupled first-order differential equations.

Our goal is to find a solution which is accurate to order dt^4 at the end of one timestep `dt` . A combination of four time derivatives,

{ fp0, fp1, fp2, fp3 } , chosen to reproduce the fourth-order Taylor's series for $f(t)$ is as follows :

```
fp0 = rhs(f0)
f1 = f0 + (dt/2)*fp0
fp1 = rhs(f1)
f2 = f0 + (dt/2)*fp1
fp2 = rhs(f2)
f3 = f0 + dt*fp2
fp3 = rhs(f3)
f4 = f0 + (dt/6)*(fp0 + 2*fp1 + 2*fp2 + fp3)
```

The underlying idea here is to combine four different estimates of the derivative so as to reproduce the fourth-order calculation of f4 with an error $(\overset{.....}{f}/120)dt^5$. Compare the analytic results for an oscillator started out motionless at $Q = 1$:

$$\dot{Q} = +P \; ; \; \dot{P} = -Q \longrightarrow Q = +\cos(t) \; ; \; P = -\sin(t) \; ,$$

with a one-step numerical solution (from $t = 0$ to $t = dt$) . We indicate the fourth-order Runge-Kutta algorithm result by using lower case variables, q and p with a timestep dt = 0.10 :
Q = +0.9950041652780257 ; P = -0.0998334166468282 ;
q = +0.9950041666666667 ; p = -0.0998333333333333 .
The numerical errors, 10^{-9} and 10^{-7} are consistent with the expected series truncation error $dt^5/5! \simeq 10^{-7}$ and agree with the analytic solution of the Runge-Kutta algorithm given for this problem on page 15 of *Computational Statistical Mechanics.*

It is seldom the case that a simulation problem will require an integrator more sophisticated than the fourth-order Runge-Kutta algorithm. Such a situation *does* arise for a *triply*-thermostated oscillator problem treated in Chapter 6, where the coupled set of five differential equations :

$$\{ \dot{q} = p \; ; \; \dot{p} = -q - \zeta p - \xi p^3 - \varsigma p^5 \; ; \; \dot{\zeta} = p^2 - 1 \; ; \; \dot{\xi} = p^4 - 3p^2 \; ; \; \dot{\varsigma} = p^6 - 5p^4 \} \; ,$$

tends toward instability using a fixed Runge-Kutta timestep.

The instability can be evaded by estimating the integration error and adjusting dt up or down to bring the error into the desired range (perhaps from 10^{-16} to 10^{-12}) . The error can be estimated by comparing the fourth-order Runge-Kutta integration with a *fifth*-order Runge-Kutta integrator.

The fifth-order value of $f(dt) \simeq$ `f6` has an error of order $(\stackrel{......}{f}/720)dt^6$. One (of many) fifth-order integrators is :

```
fp0 = rhs(f0)
f1 = f0 + (dt/2)*fp0
fp1 = rhs(f1)
f2 = f0 + (dt/16)*(3*fp0 + fp1)
fp2 = rhs(f2)
f3 = f0 + (dt/2)*fp2
fp3 = rhs(f3)
f4 = f0 + (dt/16)*( - 3*fp1 + 6*fp2 + 9*fp3)
fp4 = rhs(f4)
f5 = f0 + (dt/7)*(fp0 + 4*fp1 + 6*fp2 - 12*fp3 + 8*fp4)
fp5 = rhs(f5)
f6 = f0 + (dt/90)*(7*fp0 +32*fp2 +12*fp3 +32*fp4 +7*fp5)
```

If the difference between the fourth-order and fifth-order solutions is greater than 10^{-12} the timestep is cut in half and the step repeated. If the difference is less than 10^{-16} the step is accepted and the timestep is doubled. This adaptive Runge-Kutta algorithm performs very well with relatively-stiff nonlinear dynamics problems like the triply-thermostated oscillator described above.

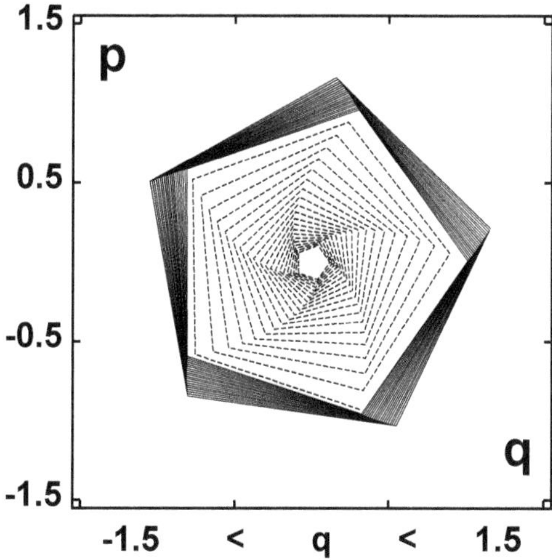

Fig. 5.2 Runge-Kutta trajectories for the harmonic oscillator showing fourth-order energy loss and fifth-order energy gain for a (very large) timestep $dt = (2\pi/5)$.

A comparison of the fourth- and fifth-order Runge-Kutta integrators for the harmonic oscillator problem $\{ \dot{q} = +p \; ; \; \dot{p} = -q \}$ is informative. **Figure 5.2** shows 100 timesteps, starting with $\{ q,p \} = \{ 1,0 \}$ using a timestep of $(2\pi/5)$, chosen large enough to illustrate the integration errors. For the purely-linear oscillator problem the fourth-order integrator approaches $(q,p) \rightarrow (0,0)$ while the fifth-order integrator diverges.

A word of caution applies with adaptive integrators. The proper weighting of data from an adaptive integerator is *nonuniform*. States generated by the integrator must be assigned weights proportional to the timestep dt associated with those states. Alternatively, the need for nonuniform weights can be avoided by interpolating the set of dependent variables to equally-spaced time intervals.

5.4 Finite-Precision Solutions and Periodic Orbits

There is a fascinating qualitative difference between *numerical* solutions of the motion equations and those of continuous analysis. Numerical solutions have a finite precision (the number of digits kept in "solving" the equations) . The usual "double-precision" computation, with 48 binary bits of accuracy, incurs roundoff and truncation errors of the order of $(1/2)^{48} \simeq 3.55 \times 10^{-15}$. Evidently an N-dimensional double-precision phase space consists of approximately $\mathcal{N} = 10^{15N}$ discrete points. In this case the probability that one of the $n(n+1)/2$ pairs formed from the first n trajectory points is composed of two identical points (so that the trajectory linking them is periodic) should be of order unity when

$$n(n+1)/2 \simeq 10^{15N} \rightarrow n \simeq 10^{7.5N} .$$

Assuming that the trajectory jumps *randomly* from point to successive point both the average transient interval until duplication occurs and the average length of the subsequent periodic orbits turn out to be the same :

$$\langle n_{\rm transient} \rangle = \langle n_{\rm period} \rangle = \sqrt{\pi \mathcal{N}/8} .$$

For a variety of numerical examples consistent with this analysis (carried out by Celso Grebogi, Edward Ott, and James Yorke in the 1988 Physical Review A) see Christoph Dellago and Bill's 2000 Physical Review E article, "Finite-Precision Stationary States At and Away from Equilibrium" . There is an earlier work along these lines for maps, rather than flows: "Numerical Study of Discrete Plane Area-Preserving Mappings" by F. Rannou, in Astronomy and Astrophysics **31**, 289-301 (1974).

In principle any stable numerical solution of atomistic mechanics using a digital computer must eventually reach a periodic orbit. As noted above, there can also be a transient trajectory before the final periodic orbit is reached. It is both interesting and paradoxical, to note that the "transient" (as opposed to periodic) part of a "many"-body trajectory requires more than the age of the Universe (4×10^{17} seconds) to compute once the number of (three-dimensional) particles exceeds more than a few. The moral of this observation is that one ought not view trajectories as actually sampling *all* accessible states. Instead, the reason why averaging over such collections of typical states ("ensembles") works so well is that the usual fluctuations are small enough to be neglected.

Evidently Zermélo's Paradox (that irreversible-looking evolutions must ultimately return near their initial conditions) has no real-world applicability. On the other hand Loschmidt's Paradox (that irreversible-looking evolutions can equally well be run backward) deserves more thought as the "bit-reversible" algorithm described next amply demonstrates.

5.5 Levesque and Verlet's Bit-Reversible Algorithm

Newton's motion equations are precisely reversible. If one's arithmetic operations were performed exactly, without any truncation or roundoff errors, the Leapfrog algorithm could likewise be reversed exactly :

$$r(t + dt) = 2r(t) - r(t - dt) + F(t)(dt^2/m)$$

$$\longleftrightarrow$$

$$r(t - dt) = 2r(t) - r(t + dt) + F(t)(dt^2/m) \ .$$

Notice that choosing *integer* values for the coordinates can actually lead to a perfectly-reversible algorithm provided that the rounding of $F(t)(dt^2/m)$ to an integer is treated identically in the "forward" and "backward" time directions. The simplest procedure, just truncating the combination $[\ F(t)(dt^2/m)\ ]$ to an integer (so that $[\ 2.5\ ]$ becomes 2 and $[\ -2.5\ ]$ becomes -2 , for instance) works perfectly well. If instead the combination $[\ 2r(t) - r(t+dt) + F(t)(dt^2/m)\ ]$ were converted to an integer the reversibility of the algorithm would be destroyed. See **Figure 5.3** for the harmonic oscillator example detailed below.

For a Newtonian problem, just as for a bit-reversible problem, there is no distinction between the trajectory going forward and its reverse.

Loschmidt's reversibility paradox applies here, where any movie of the motion, played backward, satisfies exactly the same dynamical law as did the movie when created forward in time. Let us detail an oscillator example.

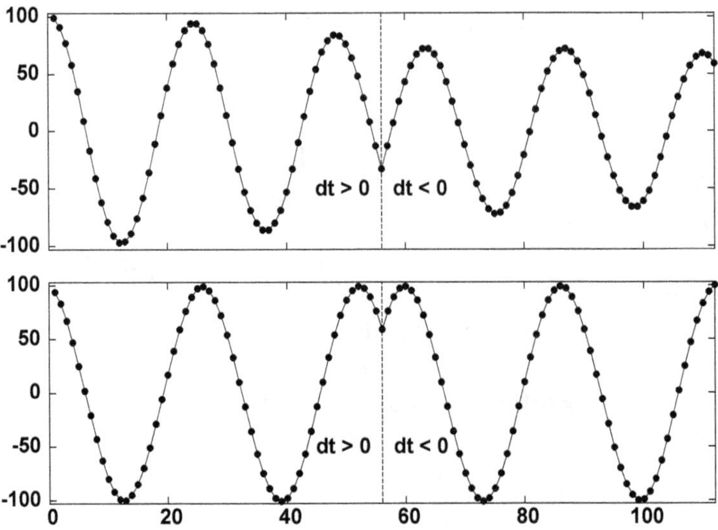

Fig. 5.3 A truly bit-reversible oscillator trajectory (below) and a faulty implementation (above) run for 56 timesteps with $dt = (1/4)$ and then reversed.

5.5.1 *Bit-Reversible Harmonic Oscillator*

A short program for a harmonic oscillator described by six-digit coordinates and with a timestep $dt = 0.15$ follows. We choose initial conditions adjacent to a "turning point", a maximum in the coordinate, at the time $(dt/2)$. Just before and just after this turning point, the coordinates `jnow` are chosen equal to 1 000 000 . The program then iterates forward for thirty steps and exchanges the last two coordinates (this exchange is equivalent to a time reversal). The exchange is followed by another 30 steps (now going "backward") and recovers the initial conditions exactly.

Figure 5.3 illustrates the resulting forward and backward bit-reversible harmonic oscillator solution. Over a million iterations all the integer values of the coordinate q and the increment from acceleration, $-qdt^2$ lie in the integer ranges :

$$\{ -1\ 003\ 798 \leq q(t) \leq +1\ 003\ 789\ ;\ -22\ 585 \leq -q(t)dt^2 \leq +22\ 585 \}\ .$$

Here are the steps implementing the initialization and forward iteration of a 30-step bit-reversible program. The initial data define a "turning point", with two successive equal coordinates `iold` and `jnow` :

```
iold = 1000000 !  first coordinate
jnow = 1000000 !  second coordinate
dt = 0.15d00
do 10 it = 1,30 !  carry out 30 Leapfrog steps forward
iforce = -dt*dt*jnow !  getting integer accelerations
knew = jnow + jnow - iold + iforce
time = it*dt
write(12,12) it,time,jnow
12 format(i10,f10.2,i10)
iold = jnow !  shift now --> old
jnow = knew !  shift new --> now
10 continue
```

After 30 forward steps the present and past coordinates are permuted :

```
ii = jnow
jj = iold
iold = ii
jnow = jj
```

Finally, 30 steps backward lead to the reversed initial conditions :

```
do it = 31,60
iforce = -dt*dt*jnow
knew = jnow + jnow - iold + iforce !  Leapfrog
time = it*dt
write(12,12) it,time,jnow
iold = jnow
jnow = knew
```

The data file begins with

```
1 0.15 1000000
2 0.30  977500
3 0.45  933007
```

and ends with

```
58 8.70  933007
59 8.85  977500
60 9.00 1000000
```

A longer variation on this theme, designed to discover the Poincaré recurrence time, is given as Problem 2 at the end of this Chapter.

5.6 Optimized Symplectic Algorithms - Time Reversibility

It is sometimes suggested that "symplectic" (meaning "intertwined") algorithms, algorithms that agree with Liouville's Theorem, are superior to those which don't. Liouville's Theorem for Hamiltonian systems states that phase volume is a constant of the motion, $\dot{\otimes} \equiv 0$. Typical symplectic algorithms alternate series of coordinate and momentum updates so that both $(\partial \dot{q}/\partial q)$ and $(\partial \dot{p}/\partial p)$ vanish. A third-order integrator of this type, from Stephen Gray, Donald Noid, and Bobby Sumpter's 1994 Journal of Chemical Physics article, "Symplectic Integrators for Large Scale Molecular Dynamics Simulations: A Comparison of Several Explicit Methods" iterates the following sequence of steps when applied to the oscillator equations $\{ \ \dot{q} = +p \ ; \ \dot{p} = -q \ \}$:

$$dp = -0.2683301 q dt \ ; \ dq = +0.9196615 p dt \ ;$$

$$dp = +0.1879916 q dt \ ; \ dq = -0.1879916 p dt \ ;$$

$$dp = -0.9196615 q dt \ ; \ dq = +0.2683301 p dt \ .$$

The maximum first-period energy errors using $dt = (2\pi/32)$ and $(2\pi/64)$ are 0.000045 and 0.0000056 . Rounding the integrator coefficients to four-figure accuracy leads to slight increases, 0.000049 and 0.0000076 . Evidently precise coefficients are not necessary.

Motivated by this work, Bill, Oyeon Kum, and Nancy Owens showed that integrators of this type can be developed using Monte Carlo choices for the coefficients, chosen to minimize the energy error in a single-period harmonic oscillator simulation. See their 1995 Journal of Chemical Physics article, "Accurate Symplectic Integrators *via* Random Sampling". This

Monte Carlo method dramatically improved the accuracy with a slightly more complicated time-reversible integrator :

$$dq = +0.005904pdt \ ; \ dp = -0.171669qdt \ ;$$

$$dq = +0.515669pdt \ ; \ dp = +0.516595qdt \ ;$$

$$dq = -0.021573pdt \ ; \ dp = -1.689852qdt \ .$$

$$dq = -0.021573pdt \ ; \ dp = +0.516595qdt \ ;$$

$$dq = +0.515669pdt \ ; \ dp = -0.171669qdt \ ;$$

$$dq = +0.005904pdt$$

With $dt = (2\pi/100)$ this method has a first-period oscillator energy error of about 10^{-9}.

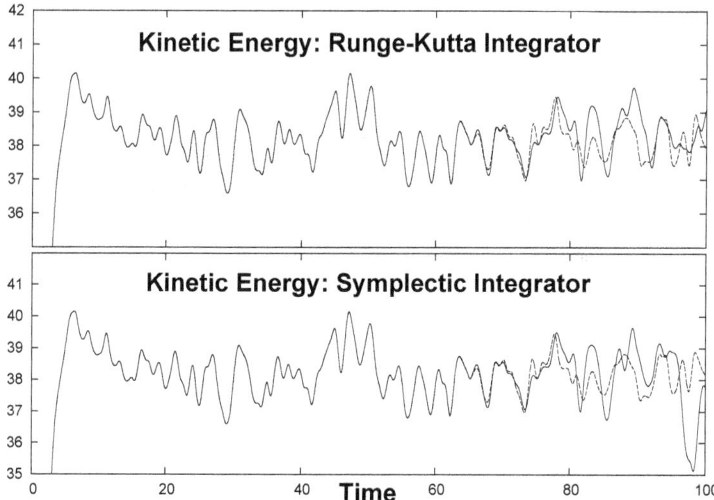

Fig. 5.4 Comparison of energies using fourth-order symplectic and Runge-Kutta integrators. Energies for both $dt = 0.005$ (dashed) and $dt = 0.010$ (solid) are plotted. All four Lucy-potential simulations conserve total energy to eight-figure accuracy for runs fifty times longer than those shown here. The container is a periodic 16×16 box.

In order to check the usefulness of the symplectic integrator for a many-body system we considered a periodic 64-atom system of Lucy-potential particles, with

$$\phi_{\rm Lucy}(r < 3) = (5/9\pi)[\ 1 + r\ ][\ 1 - (r/3)\ ]^3 \ \longrightarrow \int_0^3 2\pi r \phi_{\rm Lucy}(r) dr \equiv 1 \ .$$

The initial arrangement was a square lattice with nearest-neighbor spacing 2 and with kinetic and potential energies of 24 and 21.40 . The velocities were first chosen randomly and then shifted and scaled :

$$\texttt{px(i)} \rightarrow \texttt{px(i)} - (\texttt{1/N})\sum_{\texttt{j=1}}^{\texttt{N}} \texttt{px(j)} \ ; \ \texttt{px(i)} \rightarrow \sqrt{\texttt{24/}\sum \texttt{px(j)} * \texttt{px(j)}} \ .$$

The { `py(i)` } were adjusted in the same way. The result gives x and y kinetic energies of 12 . Exactly the same initial conditions were used to compare results for two different timesteps using the dynamical Monte Carlo integrator. In **Figure 5.4** we compare the time-dependence of the kinetic energy using $dt = 0.010$ and 0.005 . Visually apparent minor deviations between the two curves appear almost immediately, after about 7000 of the larger timesteps. On the other hand the total energy stays constant to eight-figure accuracy throughout both runs. This disparity is a consequence of the "Lyapunov instability" of the dynamics. This type of instability is illustrated and discussed further in the next Section.

For the runs shown here, with a million timesteps with $dt = 0.005$ and a half million with a timestep of 0.01 , there is no essential advantage to either integrator. The symplectic integrator takes a little more programming and is a little slower (with five force evaluations per step, compared to Runge-Kutta's four) but there is no essential difference in the results obtained by these approaches. On the other hand, the symplectic integrator could be implemented with integer coordinates { `q` } and impulses { `pdt` } , so that a *strictly-reversible* algorithm, like Dominique Levesque and Loup Verlet's but more nearly accurate, could be applied to situations where an exact reversibility is desirable. Because the symplectic property requires a Hamiltonian system, such algorithms are not useful for dealing with nonequilibrium steady-state problems (which typically involve non-Hamiltonian feedback forces or constraints) .

5.7 Ergodicity and Lyapunov Instability with $\{\dot{q}, \dot{p}, \dot{\zeta}, \dot{\xi}\}$

The many-body example just considered is "Lyapunov unstable". This means that two nearby solutions separate exponentially fast, $\propto e^{+\lambda t}$, from one another. The simplest system of time-reversible differential equations showing this behavior is *quadratic*, rather than linear :

$$\dot{q} = p \ ; \ \dot{p} = -q - \zeta p \ ; \ \dot{\zeta} = [\ (p^2/T) - 1\ ]/\tau^2 \ .$$

The equations represent a harmonic oscillator with kinetic-energy integral control using the control variable ζ . It is easy to see the time reversibility. Any solution $\{ q(t), p(t), \zeta(t) \}$ can be reversed and run backward by changing the signs of the time t, the momentum p and the control variable ζ . These simple, time-reversible motion equations describe the "Nosé-Hoover oscillator", and were motivated by Shuichi Nosé's development of a somewhat more complicated approach to a canonical-ensemble dynamics.

5.7.1 *The* $\{\dot{q}, \dot{p}, \dot{\zeta}\}$ *Model* $\longrightarrow$ *Mixed Chaos and Regularity*

The Nosé-Hoover oscillator motion equations are precisely consistent with the canonical Gaussian distribution for all three variables :

$$f(q,p,\zeta) \propto e^{-q^2/2T} e^{-p^2/2T} e^{-\tau^2\zeta^2/2} .$$

To see that this is true it is sufficient to consider the flow of probability density into and out of a fixed phase-volume element $dqdpd\zeta$:

$$(\partial f/\partial t) \equiv -(\partial(f\dot{q})/\partial q) - (\partial(f\dot{p})/\partial p) - (\partial(f\dot{\zeta})/\partial\zeta) =$$

$$-f[\ (\partial\dot{q}/\partial q) + (\partial\dot{p}/\partial p) + (\partial\dot{\zeta}/\partial\zeta)\ ] - \dot{q}(\partial f/\partial q) - \dot{p}(\partial f/\partial p) - \dot{\zeta}(\partial f/\partial\zeta) .$$

Substituting the Gaussian form for f and using the Nosé-Hoover oscillator motion equations, we show that $(\partial f/\partial t)$ vanishes identically, so that the Gaussian distribution is stationary :

$$(\partial f/\partial t) = -f[\ -\zeta\ ] - \dot{q}(-qf/T) - \dot{p}(-pf/T) - \dot{\zeta}(-\zeta f\tau^2) =$$

$$-f\left[\ -\zeta + (pq/T) - (qp/T) + (\zeta p^2/T) - [\ (p^2/T) - 1\ ]\zeta\ \right] \equiv 0 .$$

Notice that the friction coefficient ζ has a Gaussian distribution here. Like q and p it is a time-dependent variable, sometimes positive and sometimes negative. ζ furnishes an "integral feedback" force driving twice the kinetic energy p^2 toward the temperature T with a rate governed by the timescale parameter τ . These motion equations were developed in order to make a dynamical model for the canonical distribution ,

$$f(q,p,\zeta) \propto e^{-E/kT} .$$

Clint Sprott was led to exactly these same facing-page $(\dot{q}, \dot{p}, \dot{\zeta})$ equations in the course of a systematic search of quadratic equations with the twin goals of simplicity and chaos. In the original work on this model the equilibrium temperature T was constant. Sample chaotic and regular solutions of the three motion equations were illustrated in **Figure 1.6** on page 13 .

The details of this model turned out to be highly complex, as is described in Bill's 1986 Physical Review A paper with Harald Posch and Franz Vesely, "Canonical Dynamics of the Nosé Oscillator: Stability, Order, and Chaos". We revisited a heat-conducting generalization of this model, introduced with Harald Posch in 1997, in a new work carried out with Clint Sprott: "Heat Conduction, and the Lack Thereof, in Time-Reversible Dynamical Systems: Generalized Nosé-Hoover Oscillators with a Temperature Gradient", Physical Review E **89**, 042914 (2014). The cross sections of the chaotic dynamics, shown at the upper left of **Figure 1.6** show that the Nosé-Hoover oscillator is far from ergodic. Its dynamics covers only a part of the complete canonical distribution, where *the* part (of which there are an infinite number!) depends on the initial conditions and even on the integration algorithm.

To emphasize this latter point **Figures 5.5 and 5.6** compare *two* fourth-order and *two* fifth-order Runge-Kutta laptop solutions of the motion equations with a relaxation time $\tau = (1/5)$:

$$\{ \dot{q} = p \; ; \; \dot{p} = -q - \zeta p \; ; \; \dot{\zeta} = 25(p^2 - 1) \} \; .$$

All the calculations have the same initial conditions $(q, p, \zeta) = (0, \pm 5, 0)$. In addition to the "classic" fourth-order algorithm of Section 5.3 we used another, likewise "classic" and of about the same vintage :

```
fp0 = rhs(f0)
f1 = f0 + (dt/3)*fp0
fp1 = rhs(f1)
f2 = f0 - (dt/3)*fp0 + dt*fp1
fp2 = rhs(f2)
f3 = f0 + dt*(fp0 - fp1 + fp2)
fp3 = rhs(f3)
f4 = f0 + (dt/8)*(fp0 + 3*fp1 + 3*fp2 + fp3)
```

We also constructed a simple alternative to the fifth-order integrator of Section 5.3 by factoring out the common denominator of the fractions on the righthand sides of the integrator. This changes the details of the multiplying and dividing operations on the righthand sides. Even such a miniscule change is enough to obtain different "solutions" to chaotic problems !

The exact solution to this Nosé-Hoover equilibrium oscillator solution *is* evidently chaotic. For a relatively small timestep, $dt = 0.001$, three of the four algorithms produce similar Poincaré sections, like the one shown

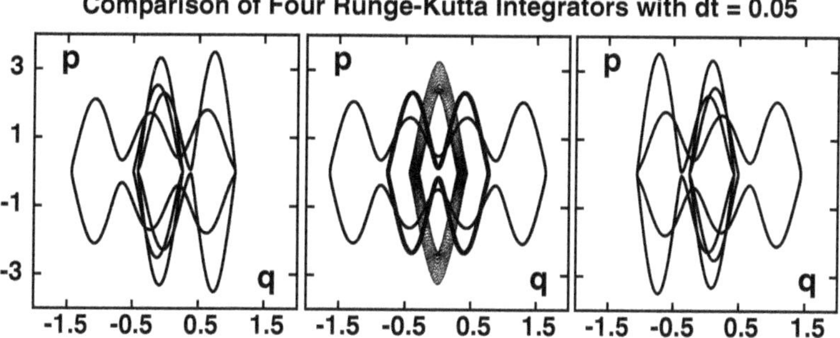

Fig. 5.5 Comparison of equilibrium Nosé-Hoover oscillator (q, p) projections for two fourth-order and two fifth-order Runge-Kutta integrators with a timestep $dt = 0.05$. Differences remain for $dt = 0.01$ and all four integrators agree (as is shown in **Figure 5.6**) when dt is reduced to 0.001 . The two symmetric periodic solutions in the central panel, one with two maxima, the other with four, were generated with fourth-order integrators. The single-maximum solution in the central panel comes from the fifth-order integrator on page 109 . The more-elaborate side-panel solutions with seven maxima are from exactly that same integrator with the fractions treated differently, as explained in the text. The two side panels were generated with initial conditions $(0, \pm 5, 0)$.

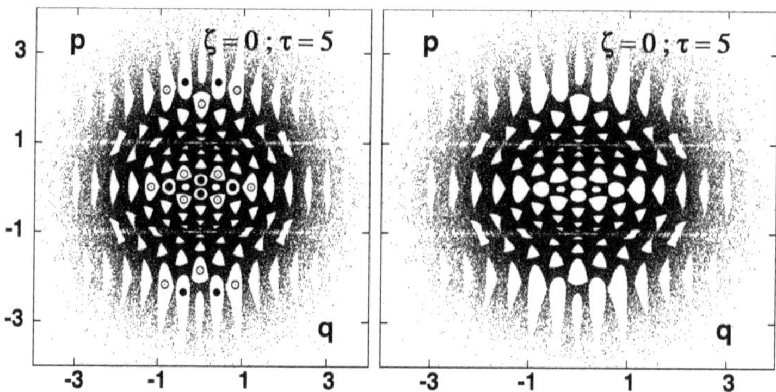

Fig. 5.6 With a timestep $dt = 0.001$ the four integrators agree with the accurate Poincaré section shown to the right. With $dt = 0.01$ the fifth-order integrators still agree but the two fourth-order integrators give the two limit cycles indicated by the six filled circles and the twelve open circles, all with the initial condition $(qp\zeta) = (0, 5, 0)$.

at the right of **Figure 5.6**. We would expect overall agreement with this timestep because the integration errors, of order $dt^5/120$ or $dt^6/720$, are all of them smaller than the double-precision roundoff errors. A timestep ten or fifty times larger, $dt = 0.010$ or 0.050 , results in an assortment of stable periodic orbits, noticeably different to what we believe to be a stable chaotic sea. The even larger timestep, $dt = 0.05$, gives either limit cycles or a torus ! The Poincaré sections in **Figure 5.6** all correspond to about 200,000 cross-section points, for which $|\ \zeta\ | < 0.001$. The limit cycles for the less-accurate algorithms with $dt = 0.05$ are all of them stable.

Though all of this dynamics is interesting, and even æsthetic, it is a distraction far from our goal of realizing the canonical distribution with a time-reversible dynamical method. A way to address the shortcoming of nonergodicity (holes in the distribution along with tori to fill them) is to control *more* than just the second moment of the velocity distribution. We turn to two versions of that idea next, the Hoover-Holian and the Martyna-Klein-Tuckerman oscillators, each of them described by four ordinary differential equations..

5.7.2 Two $\{\dot{q}, \dot{p}, \dot{\zeta}, \dot{\xi}\}$ Models $\longrightarrow$ Four-Dimensional Chaos

It is most interesting to see that adding in a *second* feedback variable makes solutions of the motion equations appear *simpler* rather than more complex. We illustrate this idea by supplementing the control of $\langle\ p^2\ \rangle$ with a new variable ξ chosen to manage the *fourth* moment of the velocity distribution $\langle\ p^4\ \rangle$. Once again the corresponding motion equations, all four of them, are time-reversible :

$$\dot{q} = p\ ;\ \dot{p} = -q - \zeta p - \xi p^3\ ;$$

$$\dot{\zeta} = [\ (p^2/T) - 1\ ]/\tau_\zeta^2\ ;\ \dot{\xi} = [\ (p^4/T^2) - 3(p^2/T)\ ]/\tau_\xi^2\ .$$

Changing the signs of the time as well as (p, ζ, ξ) changes the signs of both sides of the $\dot{q}$ equations just given while leaving both sides of the other three evolution equations [for $(\dot{p}, \dot{\zeta}, \dot{\xi})$] unchanged.

So long as the temperature is constant, these motion equations are consistent with canonical phase-space solutions for the doubly-thermostated oscillator, along with Gaussian distributions for both friction coefficients :

$$f(q, p, \zeta, \xi) \propto e^{-q^2/2T} e^{-p^2/2T} e^{-\zeta^2\tau_\zeta^2/2} e^{-\xi^2\tau_\xi^2/2}\ .$$

In this form, with explicit thermostat relaxation times, the two friction coefficients (ζ, ξ) have temperature-independent Gaussian distributions. A canonical temperature dependence for ζ and ξ can be achieved with a slightly modified set of control equations. Simply set $\tau_\zeta^2 = (1/T)$ and $\tau_\xi^2 = (1/T)$:

$$\dot{q} = p\ ;\ \dot{p} = -q - \zeta p - \xi p^3\ ;\ \dot{\zeta} = [\ p^2 - T\ ]\ ;\ \dot{\xi} = [\ p^4 - 3p^2 T\ ] \longrightarrow$$

$$f(q, p, \zeta, \xi) \propto e^{-q^2/2T} e^{-p^2/2T} e^{-\zeta^2/2T} e^{-\xi^2/2T}\ .$$

Just as in the simpler Nosé-Hoover case it is straightforward to evaluate the four-dimensional flow equations $\{ \ (\partial f(q,p,\zeta,\xi)/\partial t) \ \}$ for both sets of motion equations. $(\partial f/\partial t)$ vanishes for both so that either model is consistent with an augmented canonical distribution.

The dynamics for doubly-thermostated oscillators is fascinating. Fixing *just* the (time-averaged) second moment or *just* the time-averaged fourth moment leads to a complexity of solutions, both regular and chaotic. But the *combination* of the two controls provides uniform chaos as well as a good smooth canonical distribution. Harald Posch and Bill studied the ergodicity of the harmonic oscillator by choosing 2500 initial ($0 < q,p < 2$) conditions with (ζ,ξ) = ($0,0$) , comparing the Lyapunov instability for all these different evolutions using 10^8 timesteps for each. The results were nicely consistent with uniform chaos and ergodicity.

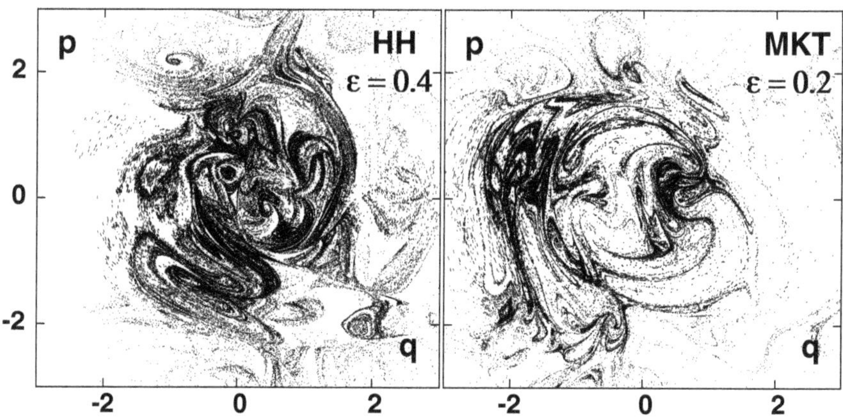

Fig. 5.7 (qp) points near the plane $\zeta = \xi = 0$ for doubly thermostated oscillators in a temperature gradient, $T = 1 + \epsilon\tanh(q)$. The Hoover-Holian and Martyna-Klein-Tuckerman thermostats are described in the text. See the next Section 5.7.3. The simulations use integral feedback based on the differences [$p^2 - T(q)$] rather than [$(p^2/T) - 1$] . Note the absence of nested tori in these two-dimensional cross sections.

The Hoover-Holian equilibrium oscillator model can be converted to two different *nonequilibrium* systems, with integral feedback based upon either [$p^2 - T(q)$] or [$(p^2/T) - 1$] . Away from equilibrium the temperature $T(q)$ is an explicit function of the coordinate q . A form which smoothly interpolates the temperature between $1-\epsilon$ and $1+\epsilon$ is :

$$T(q) = 1 + \epsilon\tanh(q) \ ; \ T(q \to \pm\infty) = 1 \pm \epsilon \ .$$

This provides a mechanism for the transfer of energy from hot to cold. The dynamics takes advantage of this opportunity by shrinking onto a fractal attractor or a limit cycle. **Figure 5.7** shows solution cross sections for both the HH and MKT versions of the $(qp\zeta\xi)$ model with the relaxation times set equal to unity and integral feedback based on $[\ p^2 - T(q)\ ]$. The values of ϵ, the maximum temperature gradient, are 0.4 for the Hoover-Holian thermostat and 0.2 for the Martyna-Klein-Tuckerman thermostat, chosen to give roughly comparable dynamics. Let us consider the source of the strange appearance of the cross-sectional distributions of **Figure 5.7** .

5.7.3 *Dimensionality Loss in Oscillator Problems*

In the equilibrium case, with $T = 1$, the largest Lyapunov exponents for the Hoover-Holian and Martyna-Klein-Tuckerman motion equations are $\lambda_{HH} = 0.068$ and $\lambda_{MKT} = 0.066$. In the *nonequilibrium* cases shown in cross-section in **Figure 5.7** with $T = 1 + \epsilon\tanh(q)$ the motion equations are as follows :

$$\{\ \dot{q} = p\ ;\ \dot{p} = -q - \zeta p - \xi p^3\ ;\ \dot{\zeta} = p^2 - T(q)\ ;\ \dot{\xi} = p^4 - 3p^2T(q)\ \}\ [\ \text{HH}\ ]\ ;$$

$$\{\ \dot{q} = p\ ;\ \dot{p} = -q - \zeta p\ ;\ \dot{\zeta} = p^2 - T(q) - \zeta\xi\ ;\ \dot{\xi} = \zeta^2 - T(q)\ \}\ [\ \text{MKT}\ ]\ .$$

The *nonequilibrium* largest Lyapunov exponents corresponding to the Figure are $\lambda_{HH} = 0.088$ and $\lambda_{MKT} = 0.069$. In the corresponding phase-space cross-sections of **Figure 5.7** we see no trace of the tori or limit cycles which are typical of three-dimensional phase spaces. The two sections shown in the Figure instead reveal the intricate structure of "strange attractors" to be discussed more fully in Chapter 7. The attractors are distributions which, like their equilibrium isothermal relatives with $T = 1$ everywhere, appear to be independent of the initial conditions $\{\ q, p, \zeta, \xi\ \}$. Unlike their equilibrium relatives the distributions of **Figure 5.7** exhibit "dimensionality loss". The correlation dimension of the Hoover-Holian attractor shown at the left is 3.38 . That of the Martyna-Klein-Tuckerman attractor is almost the same, 3.39 . This dimension describes the dependence of the number of pairs of attractor points within a distance r as a power law $\propto r^{3.38}$ or $r^{3.39}$ rather than the equilibrium result r^4 .

This dimensionality loss is a key ingredient of our current understanding of the Second Law of Thermodynamics, and will be discussed in more detail in Chapters 7 and 8 . The complex structures of strange attractors in four or more dimensions, hinted at in the Figure, are well worth investigating in detail. It is a current intellectual challenge to identify and name the nonequilibrium topological features necessary to characterize such four-dimensional flow problems.

In both the four-dimensional cases we have chosen moderately nonequilibrium temperature profiles : $T(q) = 1 + \epsilon \tanh(q)$. The solutions of these equations are "stiff", and require timesteps of 0.005 or 0.0025 for stability. The stiffness can be reduced by increasing the control-variable relaxation times :

$$\dot{\zeta} = p^2 - 1 \longrightarrow (1/4)[\, p^2 - 1 \,] \; ; \; \dot{\xi} = \zeta^2 - 1 \longrightarrow (1/4)[\, \zeta^2 - 1 \,]$$

Figure 5.8 compares short-time plots of $\xi(\zeta)$ for relaxation times of 1 and 2 , with the latter about one third the equilibrium Hamiltonian oscillator period of 2π and making the friction coefficients much less intrusive.

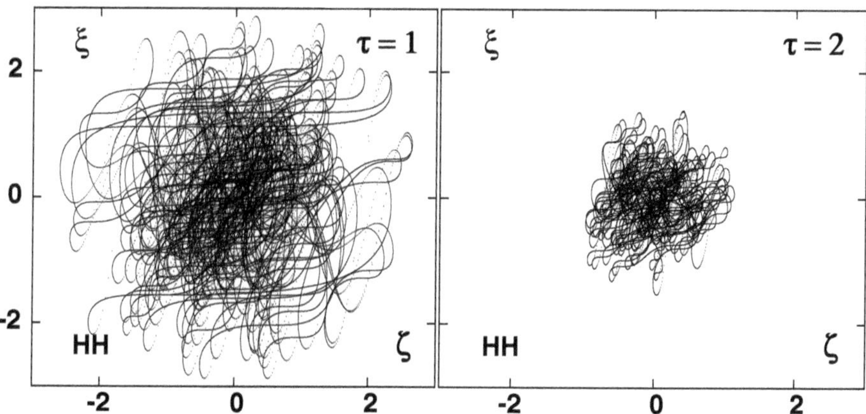

Fig. 5.8 Hoover-Holian thermostat variables for the equilibrium oscillator with two different values of the relaxation time. $\dot{\zeta} = (p^2 - 1)/\tau^2$; $\dot{\xi} = (p^4 - 3p^2)/\tau^2$.

5.7.4 *Characterizing Chaos for the* $\{\dot{q}, \dot{p}, \dot{\zeta}, \dot{\xi}\}$ *Models*

The $\{\dot{q}, \dot{p}, \dot{\zeta}, \dot{\xi}\}$ models typically exhibit chaos (sensitive dependence on initial conditions, with perturbations growing as $e^{+\lambda t}$) both at and away from equilibrium. A Runge-Kutta implementation of the dynamics needs a timestep of 0.005 for stability in typical run lengths of billions of timesteps. [Because the range of the equilibrium Gaussian distributions is *infinite* it can be expected that stability will require a smaller timestep, or an adaptive integrator, as the run length is increased.]

Measuring the (largest) Lyapunov exponent for the model requires tracking the dynamics of two oscillators with the separation between them constrained to a fixed phase-space separation :

$$\delta \equiv \sqrt{dq^2 + dp^2 + d\zeta^2 + d\xi^2} \ .$$

A good choice for δ is 0.000001, as in the FORTRAN fragment below. The constant-separation constraint and the calculation of the instantaneous Lyapunov exponent `yap` is implemented by executing the following steps just after the Runge-Kutta integration of both trajectories. Here the "reference trajectory" is `q1,p1,z1,x1` and the nearby satellite trajectory is `q2,p2,z2,x2` :

```
dq = q2 - q1
dp = p2 - p1
dz = z2 - z1
dx = x2 - x1
dd = dsqrt(dq*dq + dp*dp + dz*dz + dx*dx)
scale = 0.000001d00/dsqrt(dd)
q2 = q1 + scale*dq
p2 = p1 + scale*dp
z2 = z1 + scale*dz
x2 = x1 + scale*dx
yap = -(1.0d00/dt)dlog(scale)
```

Averaging `yap` over a sufficiently long run gives the largest Lyapunov exponent, 0.68 [HH] and 0.066 [MKT] at equilibrium ; 0.089 [HH] and 0.069 [MKT] for the two nonequilibrium cases of **Figure 5.7** , with $T(q) = 1 + \epsilon \tanh(q)$.

5.7.5 *Characterizing Chaos for a Hamiltonian Cell Model*

Another chaotic model with just four equations provides a nice and highly-informative example of Hamiltonian chaos. It is the "cell model" , the dynamics of a single particle occupying a square periodic "cell" with four scattering particles fixed at the four corners of the cell. **Figure 5.9** shows the geometry of the 2×2 cell, along with a short trajectory fragment, $0 < t < 20$. The time reversibility of the trajectory can be checked by changing the sign of dt halfway through the run. With a fourth-order Runge-Kutta timestep of 0.0001 the double-precision reversibility of the trajectory is accurate to about ten decimal places. For the short-ranged smooth pair potential, $\phi(r < 1) = (1 - r^2)^4$, the forces on the moving

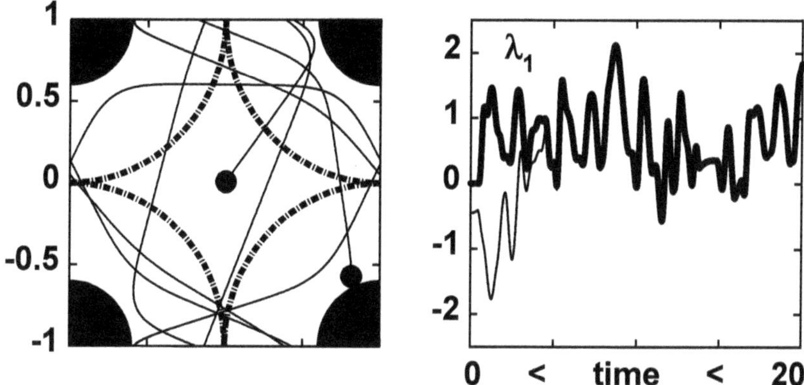

Fig. 5.9 An (x, y) trajectory for $0 < t < 20$ is shown with four scatterers at the corners of a periodic 2×2 cell. The dashed lines show the range of the pair potential and the filled quarter circles show the energetically-excluded portions of the unit cell. The local Lyapunov exponent is shown at the right for the forward (the heavier line width) and reversed versions of the trajectory. The initial velocity is $(\dot{x}, \dot{y}) = (0.6, 0.8)$.

particle have the form :

```
fx1 = fx1 + 8.0d00*dx*(1.0d00 - rr)**3
fy1 = fy1 + 8.0d00*dy*(1.0d00 - rr)**3
```

Just as in the $qp\zeta\xi$-model examples we find the largest Lyapunov exponent by evolving two nearby trajectories and rescaling their separation just after the Runge-Kutta integration :

```
dx = x2 - x1
dy = y2 - y1
dpx = px2 - px1
dpy = py2 - py1
dd = dx*dx + dy*dy + dpx*dpx + dpy*dpy
scale = 0.000001d00/dsqrt(dd)
x2 = x1 + scale*dx
y2 = y1 + scale*dy
px2 = px1 + scale*dpx
py2 = py1 + scale*dpy
yap = -(1.0d00/dt)dlog(scale)
```

Following the rescaling the periodic boundary is imposed :

```
if(x1.lt.-1.0d00) then
x1 = x1 + 2.0d00
x2 = x2 + 2.0d00
endif

if(x1.gt.+1.0d00) then
x1 = x1 - 2.0d00
x2 = x2 - 2.0d00
endif
```

with exactly similar programming for `y1` and `y2` . Following these rescaling and periodicity operations the computation carries on with the next Runge-Kutta timestep.

In this simple Hamiltonian model we see that the largest Lyapunov exponent (plotted on the right side of **Figure 5.9** with a value near 0.7) reverses perfectly as `+dt` $\rightarrow$ `-dt` because both the reference and the satellite trajectories are separately reversible. As a useful and thought-provoking exercise the reader should convince himself that because the motion equations *are* time-reversible, setting `+dt` $\rightarrow$ `-dt` generates exactly the same $\{ x, y \}$ trajectory as does changing the signs of the momenta with `dt` unchanged.

Because reversing any two nearby trajectories reverses their tendency to separate one might well expect that each forward Lyapunov exponent would be "paired" with a backward twin, one a positive Lyapunov exponent and the other negative. What actually happens is a little more complicated. Positive and negative Lyapunov exponents have a tendency toward "pairing" (for Cartesian Hamiltonian systems) but with the pairs $\{ \pm\lambda(t) \}$ forward in time quite different from those with time reversed. The constant-area deformation of a circle in phase space would be described by a pair of exponents, equal in magnitude and opposite in sign. Generalizing this idea to a deforming hyperellipsoid suggests a set of exponent pairs describing the comoving deformation of any infinitesimal extension in phase $\otimes$.

It is certainly true that along a reversed trajectory, any reversed satellite which is separating with increasing time will istead *approach* the reference when time is reversed. But this reversibility for the simple cell model of **Figure 5.9** extends only until a time of about 15. This is because the chosen offset between the reference and satellite trajectory is accurate to only six decimal digits, and so is destroyed in a time of order

$$e^{\lambda_1 t} = 10^6 \longrightarrow t \simeq 20 \ .$$

Once this well-reversed period has passed the instantaneous Lyapunov exponents backward in time are freed to become paired again. But even if the reference trajectory is stored and reused, or calculated in the two time directions bit-reversibly, the reversed Lyapunov exponent pairs bear no relationship to the forward exponent pairs. The observation that the forward and backward pairs of exponents are unrelated reflects the fact that each exponent reflects its past rather than its future.

The fact that Hamiltonian systems like to "pair" their exponents is very useful. Given that $\lambda_1(\pm t) = -\lambda_N(\pm t)$, for the forward and reversed trajectories one only needs to compare $\lambda_1(\text{forward})$ to $\lambda_1(\text{backward})$. The observation that the local values of these exponents are quite different shows that the forward and backward pairs are not simply related to one another. Again, this is because each reflects its past and not its future.

One would guess that a time-dependent external force could destroy exponent pairing. Accordingly, we anticipated that exponent pairing would *not* be observed in systems which are far from equilibrium. Perhaps the separation of the local exponents from pairing could serve as a metric for nonequilibrium character? But perhaps not. Let us explore these ideas for larger systems where the effects of irreversibility can be clarified.

5.8 Lyapunov Instability Algorithm for Many Bodies

Numerical studies of few-body dynamics (even a single thermostated oscillator or three gravitating bodies) can be complex and have fascinated mathematicians and physicists alike for half a century. The incredibly complex thermostated oscillator motions suggested by **Figures 5.6 and 5.7** are far from the gentle Gaussian distributions of Boltzmann and Gibbs. The chaos illustrated in the **Figures** might well suggest that the overly cautious could confine all of their investigations to systems with only two- or three-dimensional state spaces. Instead let us be adventurous !

The study of Lyapunov instability for many-body systems opens up a fascinating window on the nature of the chaos underlying phase-space bifurcations, mixing, and irreversibility. The Second Law of Thermodynamics rightly describes the tendency of mechanical systems to move from special unlikely phase-space states toward the overwhelmingly much more numerous (and therefore more probable) equilibrium states described by thermodynamics. Lyapunov instability furnishes diagnostic tools to discover, quantify, and understand the mechanisms underlying irreversible processes.

The description of Lyapunov instability can be made (much) more detailed than the summary information provided by the largest Lyapunov exponent. A whole "spectrum" of exponents, along with their dependence on time and on the chosen coordinate system, can be defined, computed, interpreted, and promoted. The *number* of these exponents is equal to the number of phase-space dimensions. For a Hamiltonian system the phase-space dimensionality is twice the number of degrees of freedom, $2N$ for N one-dimensional particles, $4N$ for N two-dimensional particles, and $6N$ for three-dimensional particles. For simplicity we focus on a two-dimensional system with N particles and a phase-space dimensionality of $4N$. Our "reference trajectory" is $4N$-dimensional.

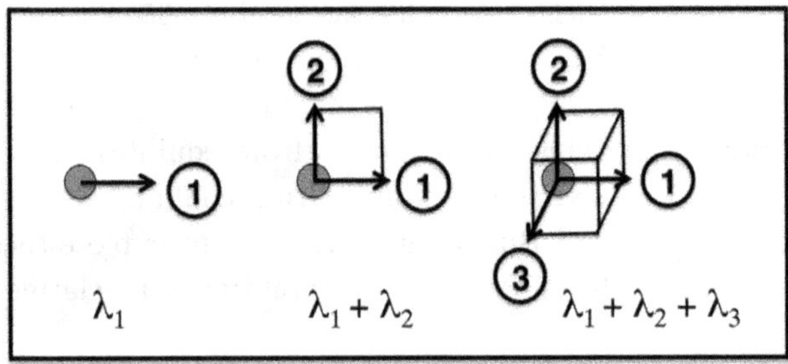

Fig. 5.10 The Lyapunov exponents $\{ \lambda_1, \lambda_2, \lambda_3 \}$ are respectively the growth rates $(\dot{\delta}/\delta)$ of small orthogonal vectors $\{ \delta_1, \delta_2, \delta_3 \}$ in one-, two-, three-dimensional subspaces of n-dimensional phase space. If the vectors are allowed to grow during each timestep then they are *rescaled* in length at the end of each timestep by implementing the Gram-Schmidt algorithm, which also maintains their orthogonality. If the vectors are instead constrained to constant length with Lagrange multipliers $\{ \lambda_1, \lambda_2, \lambda_3 \}$ those multipliers are identical to the "local" (time-dependent) Lyapunov exponents.

Benettin's algorithm for finding the Lyapunov spectrum involves the simulation of $4N$ additional "satellite" trajectories in order to find all $4N$ "local" (meaning time-dependent) Lyapunov exponents. The first exponent describes the rate of separation of two trajectories. The second exponent, when added to the first, describes the growth rate of the area defined by three nearby trajectories. The sum of the first three exponents describes the growth rate of the volume defined by four nearby trajectories. And so on. See **Figure 5.10**. Presently we will describe the programming details of the local exponents' computation. An example calculation of a full spectrum of exponents for an equilibrium fluid is shown in **Figure 5.11**. The spectrum looks much like a solid-state Debye-model spectrum.

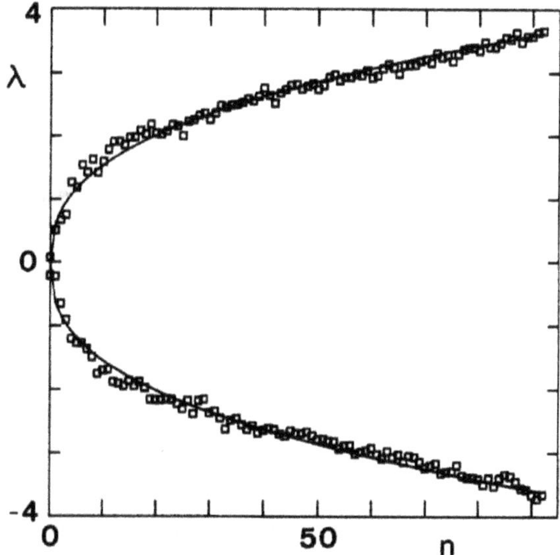

Fig. 5.11 Lyapunov spectrum for an isokinetic 32-body fluid at equilibrium. Simulation results are indicated by squares, whereas the smooth line is a fit to the positive exponents.

5.8.1 *Lyapunov Instability for Thirty-Eight Bodies*

To begin, let us solve a challenging demonstration problem — consider the headon collision of two similar crystals — or two similar drops, or masses of gas — as described by Hamiltonian particle mechanics. We simplify the problem somewhat by imposing an antisymmetry on the $2N$ particles :

$$\{+x_i, +y_i, +\dot{x}_i, +\dot{y}_i\} \equiv \{-x_j, -y_j, -\dot{x}_j, -\dot{y}_j\} \ ,$$

where $j = i + N$. The motion is symmetric about the coordinate origin at $(0, 0)$. To maintain this symmetry it is essential to symmetrize the accelerations ,

$$\{ \ +\ddot{x}_i, +\ddot{y}_i \ \} \equiv \{ \ -\ddot{x}_j, -\ddot{y}_j \ \}$$

at every timestep. Otherwise roundoff errors from different summation orders in the two colliding bodies grow exponentially and soon destroy the intended symmetry. See **Figure 5.12** for snapshots from a 38-body simulation. The initial conditions were chosen such that two 19-body drops coalesced to form a single liquid drop.

As two bodies coalesce to become one (or several bodies, depending on the relative velocity) the leading edges transmit sudden accelerations to their neighbors. A compression wave moves through both bodies at about the speed of sound. Soon the two bodies either equilibrate as one or burst into fragments, depending upon the form of the forces, the energy of the initial coordinates and velocities, and the boundary conditions.

Our illustration problem uses a many-body collision with initial center-of-mass velocities of ± 0.145095 and with sufficient thermal velocities to give each drop a laboratory-frame energy of unity. We choose a density-dependent attractive potential and a repulsive pair potential to govern the dynamics. The *density* at each particle is computed from Lucy's weight function with range $h = 3$ and includes the "self" contribution to density, $(5/\pi h^2)$.

$$\rho_{\text{Lucy}} \equiv (5/\pi h^2) \sum_{i \leq j} [\ 1 + 3z\ ][\ 1 - z\ ]^3 \text{ for } z = (r/h) < 1\ .$$

In addition to the attractive potential $(1/2)(\rho - 1)^2$, *pairs* of particles closer than unit distance repel one another with the pairwise-additive core potential $\phi(r < 1) = (1 - r^2)^4$. The reference trajectory for the simulation is computed with Levesque and Verlet's bit-reversible leapfrog algorithm. The exact time-reversal of such a problem (changing the sign of dt at the midpoint of the run) is a useful test of the bit-reversible integrator. A suitable reference-to-satellite phase-space separation is $\delta = 0.0001$ with a fourth-order or fifth-order Runge-Kutta timestep of 0.001 .

Figure 5.12 shows four snapshots (two forward and two backward, with opposite velocities) from a bit-reversible simulation. Two drops, with $3 + 4 + 5 + 4 + 3 = 19$ particles each, collide with an initial center-of-mass velocity of $\langle\ \dot{x}\ \rangle_{\text{com}} = \pm 0.145095$. The $4 \times (19 + 19) = 152$ satellite trajectories use fourth-order Runge-Kutta integration while the reference trajectory is generated with a bit-reversible leapfrog integrator. The reference values of the particle momenta (given to the Runge-Kutta integrator at the start of each timestep) were computed from the leapfrog integrator using a momentum definition from Section 5.2 correct to third-order in dt :

$$p_t \equiv (4/3)(q_{t+dt} - q_{t-dt})/(2dt) - (1/3)(q_{t+2dt} - q_{t-2dt})/(4dt)\ .$$

At the halfway point ($t = 600\,000dt = 600$) of a 1.2 million timestep forward/backward cycle, the dynamics of the reference trajectory and the 4×38 satellite trajectories were *all* reversed by setting $+dt \to -dt$. Thus the two pairs of forward/backward snapshots in the figure have the *same* coordinates but *reversed* momenta.

Throughout the forward and reversed collision processes the ith nearby satellite trajectory is constrained to maintain a fixed small phase-space

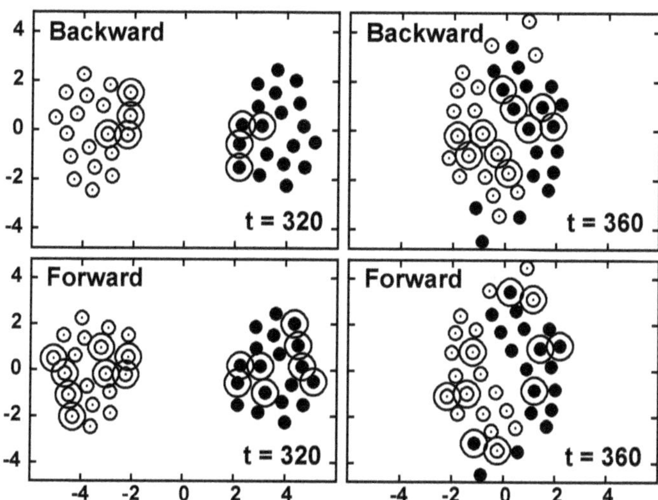

Fig. 5.12 Particles making above-average contributions to $\lambda_1(t)$ are shown for the forward and backward trajectories at times 320 and 360. The smaller filled/open circles indicate particles originally in the left/right-moving drops. In the initial condition of this problem the center-to-center drop separation was 100 . The larger circles indicate all particles making an above-average contribution to the largest Lyapunov exponent. It is noteworthy that the "important" particles are very different in the forward and reversed versions of the reference trajectory.

separation, $\delta = 0.0001$, from the reference system. Its tendency to separate from the reference trajectory—the virtual separation rate in the absence of constraints—is the instantaneous value of the ith Lyapunov exponent $\lambda_i^f(t)$. At any particular time there is also a backward separation rate $\lambda_i^b(t)$ from the reversed trajectory with the *same* coordinates but opposite momenta. Because dt is negative in the reversed trajectory all the reference-trajectory coordinates and velocities are visited in reversed order.

This forward *versus* backward information is enough to identify the motion direction consistent with Second Law Irreversibility. Only such an annotated snapshot, not the entire movie, is required.

Runge-Kutta integration can only provide an accurate reversal for times of order 10, just 10 000 timesteps with $dt = 0.001$. By contrast, the bit-reversible integration is precisely reversible for an arbitrarily-long time period. For the relatively long-time problem of **Figure 5.12** , with two forward and backward cycles of 1.2 million timesteps each, it would have been feasible to *store* the forward "reference trajectory" ($152 \times 600\ 000 = 91\ 200\ 000$ double-precision numbers), playing them backward for the entire reversed analysis. Although reversing a fourth- or fifth-order Runge-Kutta simulation is simpler than the bit-reversible approach it cannot do so accurately even for 100 000 steps.

The "pairing" associated with simple Hamiltonian mechanics, as in the one-body cell model problem, can be, and is, destroyed in far-from-equilibrium situations. The drop or crystal collision problems are good examples of this symmetry breaking. But despite the apparent *irreversible* nature of such an inelastic collision — kinetic energy has been dissipated as heat — the underlying dynamics *is* officially time-reversible. By using Levesque and Verlet's bit-reversible algorithm (with *integer* values of the reference-trajectory coordinates) rather than Runge-Kutta, the computation can be made precisely time-reversible in fact.

But *despite* this reversibility the instantaneous (or "local") Lyapunov exponent(s) can be used to distinguish a forward-moving snapshot from its backward twin. This is because both these "instantaneous" exponents carry with them information about their system's past history but have no clue as to their still unexplored future. Only one of the two shows an evolution in the direction we associate with the Second Law of Thermodynamics.

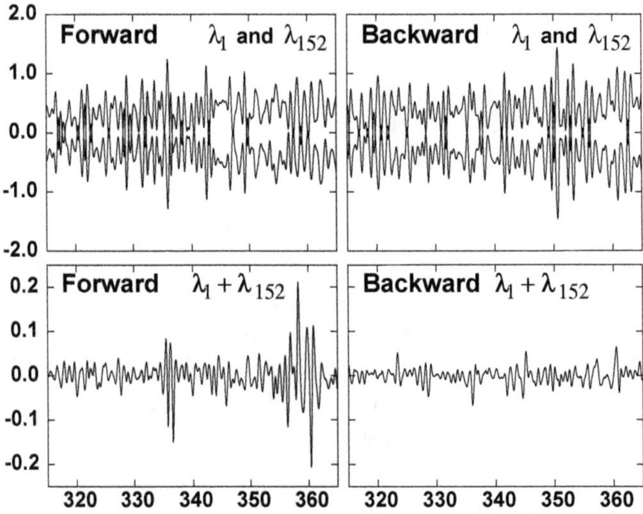

Fig. 5.13 The first and last Lyapunov exponents for the inelastic collision of Figure 5.10 are shown. The pairing is interrupted, forward in time, particularly near a time of 360. The reversed trajectory behaves differently, with uniform fluctuations in the sum $\lambda_1 + \lambda_{152}$.

The local values of the Lyapunov exponents converge in a time of order

50. The results from a run of length 2400 , with reversals at 600, 1200, and 1800 , show that the important particles in **Figure 5.12** at times of 320 and 360 repeat exactly 1.2 million timesteps later at times 1520 and 1560. In the backward version of the inelastic collision the important particles are those which have recently separated. The forward important particles are more homogeneously distributed prior to the collision.

Figure 5.13 shows the pairing (quite consistent with Hamilton's motion equations) of the first and the last Lyapunov exponents There are two brief nonpairing episodes, near times 335 and 360 , associated with the collision forward in time. In the reversed motion these episodes are absent. In the reversed trajectory the summed exponents show relatively small fluctuations. This asymmetric time dependence is real. Forward in time the summed exponents, $\lambda_1 + \lambda_{152}$, reach 0.2 near times 360 and 1560, four or five times larger than the same exponent sum in the reversed direction.

Because the lack of pairing in this 38-body problem is only an order of magnitude above the numerical noise level of the calculation we next study a similar problem, but with only 14 bodies, so as to reduce the computational fluctuations.

5.8.2 *Lyapunov Instability for Fourteen Bodies*

Although the underlying sets of ordinary differential equations *are* time-reversible a *symmetry-breaking*, due to inhomogeneity, can destroy the forward-backward pairing characteristic of time-averaged equilibrium Hamiltonian Lyapunov spectra. Let us investigate the lack of pairing for a relatively-slow collision of two cold hexagonal crystallites. Each crystallite is made up of seven particles, with a short-ranged repulsive pair potential and a density-dependent many-body attractive potential for each particle's density :

$$\Phi = \sum_{i<j} \phi_R(r_{ij}) + \sum_{i=1}^{14} \phi_A(\rho_i) \ ; \ \rho_i = \sum_{j=1}^{14} w_{\rm Lucy}(r_{ij} < h) \ ;$$

$$\phi_R(r < 1) = (1 - r^2)^4 \ ; \ \phi_A(\rho) = (1/2)(\rho - 1)^2 \ ;$$

$$w_{\rm Lucy}(r < h) = (5/\pi h^2)(1 + 3z)(1 - z)^3 \ ; \ z = (r/h) \ ; \ h = 3.5 \ .$$

As before, we symmetrize the coordinates, velocities, and accelerations relative to the origin of the coordinate system.

We divide up the bit-reversible reference-system dynamics into ten batches of 10 million-timestep cycles, with half of each cycle forward in time and half backward. Thus the time dependence of the fourteen reference-particle coordinates are identical in the forward and backward directions. The 56 nearby satellite trajectories become quite reproducible starting with the second cycle, and so can be accurately characterized with just two or three of the ten million-timestep cycles. **Figure 5.14** shows the behavior of the first and last Lyapunov exponents, forward in time, for four of the ten cycles. Notice that there is a short window in which the pairing is noticeably imperfect. The particle plot in **Figure 5.15** shows that this lack of pairing occurs as the central triplet row of each hexagon shears over its two neighboring rows of neighbors. Evidently "pairing" can be lost during such noticeably irreversible events. It is interesting that in the reversed direction of time the same window reveals near-perfect pairing. For more details see our 2014 arχiv 1405.2485 paper, "What is Liquid? Lyapunov Instability Reveals Symmetry-Breaking Irreversibilities Hidden within Hamilton's Many-Body Equations of Motion".

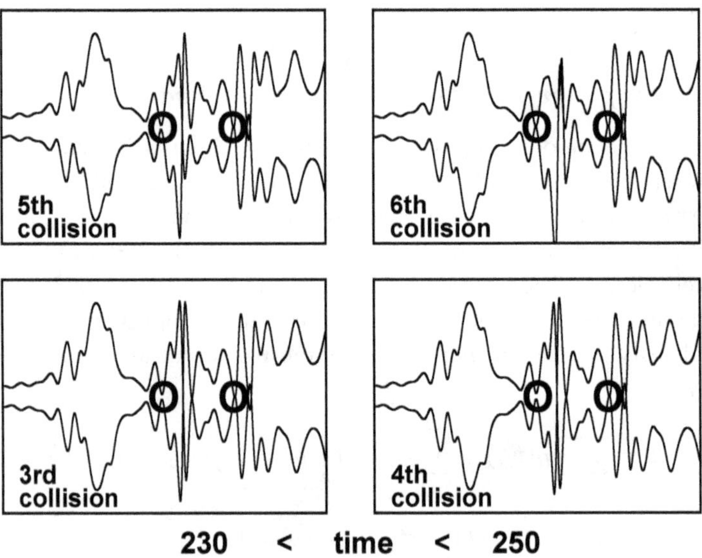

Fig. 5.14 First and last Lyapunov exponents for four repetitions of the collision process for two seven-particle crystallites. Pairing is nearly perfect except for this window where the particles shear over one another, as shown on the next page.

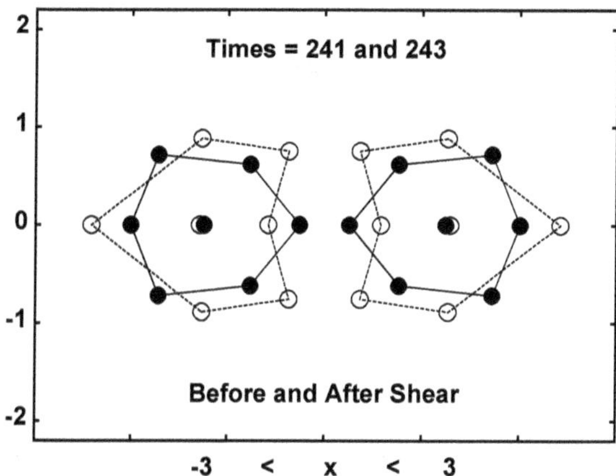

Fig. 5.15 Snapshots at times of 241 and 243, before (filled circles) and after (open circles) the cooperative shearing motion of the central row of particles.

5.8.3 *Benettin's N-Body Lyapunov-Spectrum Algorithm*

The Lyapunov instability just discussed, in which the leading forward and backward exponents and vectors are quite different, can be explored by Runge-Kutta integration if the required time is not too long. Here we outline the steps required to compute the instantaneous Lyapunov spectra for an N-body system. This method is based on Gram-Schmidt orthonormalization. For clarity we again consider a two-dimensional case.

As sketched in Section 5.8 we choose an offset vector length δ and begin by generating an orthogonal set of vectors, one for each phase space dimension, linking satellite vectors to the reference trajectory, which has phase-space coordinates `xr(ip),yr(ip),pxr(ip),pyr(ip)` for each of the N particles. For the $4N$ offset vectors $\delta_1 \ldots \delta_{4N}$ appropriate to a two-dimensional problem we choose initial conditions with a FORTRAN fragment that generates $4N$ orthogonal phase-space vectors, each of length δ :

```
do jv = 1,4*N
do ip = 1,N
dx(ip,jv) = 0.0d00
dy(ip,jv) = 0.0d00
dpx(ip,jv) = 0.0d00
dpy(ip,jv) = 0.0d00
```

```
if(jv.eq.ip+0*N) dx(ip,jv) = delta
if(jv.eq.ip+1*N) dy(ip,jv) = delta
if(jv.eq.ip+2*N) dpx(ip,jv) = delta
if(jv.eq.ip+3*N) dpy(ip,jv) = delta
enddo
enddo
```

Thus the $4N$ components of satellite vector `jv` correspond to the $4N$ components for each Particle `i = 1 ... N` :

```
xr(i)+dx(i,jv),yr(i)+dy(i,jv),pxr(i)+dpx(i,jv),pyr(i)+dpy(i,jv)
```

The initial orientation of the vectors isn't important. In a few "Lyapunov times" (say $10/\lambda_1$) the whole set of orthogonal vectors becomes oriented as dictated by the dynamics of the neighboring reference trajectory.

The reference trajectory and the $4N$ satellite trajectories, each treated as an evolving $4N$-dimensional vector, are then integrated iteratively, using the same Runge-Kutta routine for each of the trajectories, from time t to time $t + dt$. The resulting change in the length of the first vector gives the first local Lyapunov exponent $\lambda_1(t)$:

$$| \ \delta_1(t+dt) \ | = \sqrt{\sum_{i=1}^{N} [\ \delta x_i^2 + \delta y_i^2 + \delta px_i^2 + \delta py_i^2 \ ]} \ .$$

$$\lambda_1(t) = (1/dt) \ln[\ | \ \delta_i(t+dt) \ |/ \ \delta \ ] \ .$$

Following this, the first offset vector is rescaled :

$$\delta_1(t+dt) \longrightarrow \delta_1(t+dt) \times [\ \delta/| \ \delta_1(t+dt) \ | \ ] \ .$$

Next, the projections of the $N-1$ remaining vectors in the direction of δ_1 are removed. Then $\delta_2(t+dt)$ is rescaled with the change in its length giving the second local exponent.

$$\delta_2(t+dt) \longrightarrow \delta_2(t+dt) \times [\ \delta/| \ \delta_2(t+dt) \ | \ ] \ .$$

The projections of the remaining vectors onto δ_2 are next removed, following which δ_3 is rescaled to give the third local exponent. Once this Gram-Schmidt projection and rescaling process has been repeated for the remaining vectors $\delta_4 \ldots \delta_{4N}$, the $4N$ vectors are once again orthogonal and of equal length δ so that the Runge-Kutta integrator can propagate the vectors through another timestep.

For clarity we illustrate the operations required for δ_4 just after the rescaling of δ_3 and the removal of the projections onto it from $\delta_4 \dots \delta_{4N}$:

$$\lambda_4(t) = (1/dt) \ln[\sqrt{\delta_4 \cdot \delta_4}/\delta] ;$$

$$\delta_4 \longrightarrow \delta_4 \times (\delta/\sqrt{\delta_4 \cdot \delta_4}) .$$

As each of the $4N$ vectors is scaled and the projections of the remaining vectors are removed from it the computational effort increases as N^3 . Early work showed that the (time-averaged) spectrum of exponents :

$$\{ \lambda_i \} \equiv \{ \langle \lambda_i(t) \rangle \} ,$$

is relatively featureless, describable by a power-law distribution. See again **Figure 5.11** on page 129 . In practice, at least for Hamiltonian systems, the leading time-dependent Lyapunov exponent, $\lambda_1(t)$, and the vector $\delta_1(t)$ associated with it, appear to describe most of the interesting physics associated with Lyapunov instability. Lyapunov vectors near the middle of the spectrum, with small growth rates, are associated with spatially-extended structures resembling “normal modes”, but different in that these modes are exponentially unstable rather than periodic in the time.

One interesting aspect of the time-dependent Gram-Schmidt vectors is their connection to the change of phase volume. Because the local Lyapunov exponents give the stretching rates in $4N$ orthogonal phase-space directions their sum reproduces the instantaneous rate-of-change of a small comoving volume element ,

$$(d/dt) \ln \otimes = \sum_{i=1}^{4N} \lambda_i(t) .$$

In the Hamiltonian case this sum vanishes, as Liouville's Theorem shows. In the Cartesian case we have seen that the exponents are typically paired. In thermostated nonequilibrium situations, unless the system is homogeneous, pairing can be destroyed. In any case, away from equilibrium the time-averaged negative exponents necessarily overwhelm the positive ones. The summed spectrum gives the rate at which the phase volume contracts onto its strange-attractor phase-space distribution, $(d \ln \otimes/dt) = \sum_i \lambda_i(t)$.

5.8.4 *Vector Rotation, Covariance, and Scale Dependence*

For Hamiltonian systems Liouville's Theorem shows that the sum of the Gram-Schmidt Lyapunov exponents vanishes, a useful check on the computation. For Cartesian coordinates the pairing of the vectors and exponents is another valuable check. The *direction* associated with δ_1 corresponds to the direction of maximum growth for two trajectories constrained to maintain a fixed separation while otherwise following the equations of motion. But such a vector typically *rotates* much more rapidly than it decays or shrinks. As the number of degrees of freedom increases so do the rotation rates, while the growth and shrinkage rates are relatively stable. For this reason the local exponents and their vectors can incur huge fluctuations. Despite all the rapid rotation the vectors themselves soon pick out the instantaneous phase-space directions of comoving instability appropriate to the chosen coordinate system.

An alternative description of the unstable growth and decay of phase-space vectors beyond the leading ones forward and backward in time can be based on "covariant vectors", as discussed in a series of papers by Harald Posch and many others. The covariant vectors are formally expressed as linear combinations of the forward and backward Gram-Schmidt vectors. The numerical algorithms to implement these ideas require either the storage of "forward" trajectories or the use of a bit-reversible algorithm to generate precise "backward" trajectories. The covariant vectors are chosen to follow the equations of motion but without being additionally restricted to be orthogonal. In the Gram-Schmidt case the vectors forward in time bear no simple relation to those backward in time. In the "covariant" case a unique set is chosen, the same in both the forward and backward directions.

Because the maximum growth directions become maximum decay directions in the reversed flow, the first and last covariant vectors correspond to the first Gram-Schmidt vectors forward and backward in time. The sum of the corresponding covariant Lyapunov exponents has no simple connection to the rate of change of phase volume. The covariant vectors have also lost any connection to the notions of "past" and "future".

A prominent fly in the ointment, for both the Gram-Schmidt and the covariant exponents, is the sensitive dependence of the local Lyapunov spectrum on the phase space coordinates. If an arbitrary phase-space direction is chosen at a phase point the spatial variation of the leading Lyapunov exponent is typically fractal. A different integrator, or timestep, or direction changes the details of the local spectrum. Only in the direction parallel to the reference trajectory is the variation regular.

5.8.5 *Coordinate Dependence of Oscillator Spectra*

The exponents and their vectors also depend on the chosen coordinate system. This can be seen for even the simplest problem, a one-dimensional harmonic oscillator. This problem has no exponential instability and no longterm tendency for trajectories to separate. Because the oscillator equations of motion are *linear* we can display faithfully-accurate finite enlargements of the two infinitesimal offset vectors :

$$\{ \ \delta(t) = (q,p)_{\rm Satellite} - (q,p)_{\rm Reference} \ \} \ .$$

In this example problem we will choose a vector length of unity, $\mid \delta \mid = 1$.

Consider, for an oscillator, the phase-space perturbation (δ_q, δ_p) , kept constant in length by a Lagrange multiplier (the multiplier is equal to the "largest" Lyapunov exponent, $\langle \ \lambda_1(t) \ \rangle = \lambda_1 = 0$, when time-averaged) :

$$\{ \ \dot q = +p \ ; \ \dot p = -q \ \} \longrightarrow \{ \ \dot\delta_q = +\delta_p - \lambda_1(t)\delta_q \ ; \ \dot\delta_p = -\delta_q - \lambda_1(t)\delta_p \ \} \ .$$

The condition that the perturbation is of constant length, $\delta_q^2 + \delta_p^2$ constant, is satisfied automatically (by a *vanishing* multiplier) so that the "largest" Lyapunov exponent vanishes, as does also the "smallest", as their sum is necessarily zero. Accordingly, the oscillator has two vanishing Lyapunov exponents corresponding to the two Lyapunov vectors parallel and perpendicular to the oscillator orbit :

$$\langle \ \lambda_1(t) \ \rangle = \lambda_1 = 0 \ ; \ \langle \ \lambda_2(t) \ \rangle = \lambda_2 = 0 \ .$$

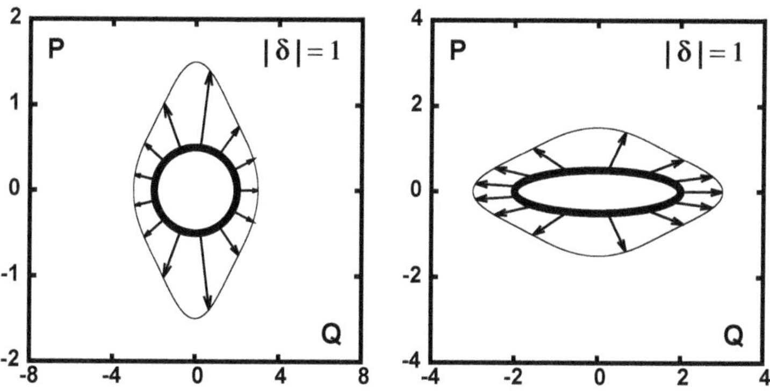

Fig. 5.16 Two views of a scaled oscillator orbit, with scale factor $s = 2$ and $\mid \delta \mid = 1$, are given at the left and right. The arrows connect corresponding points on the reference and satellite trajectories at 13 equally-spaced times and are all of unit length. $\mathcal{H} \equiv 1$.

Now suppose we describe the oscillator with a different system of units, much like switching from meters and kilograms to kilometers and grams.

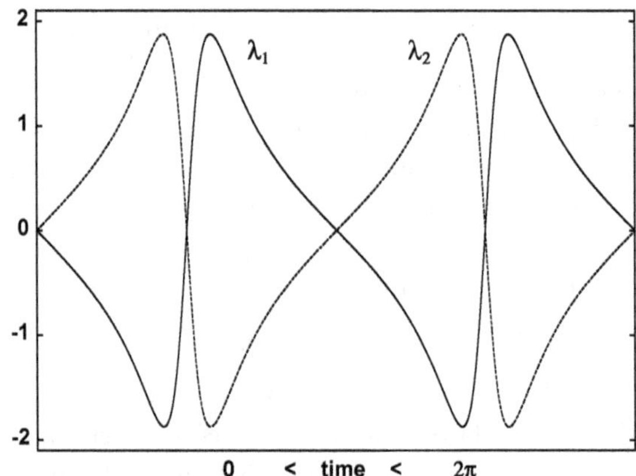

Fig. 5.17 Time variation of the two Lyapunov exponents for one oscillator period. The Lyapunov exponents sum to zero and oscillate at *twice* the oscillator frequency.

For simplicity we scale the coordinates and momenta inversely, so that in terms of the new units, $Q = (qs)$ and $P = (p/s)$, the Hamiltonian and the equations of motion derived from it become :

$$q^2 + p^2 = 2\mathcal{H} = (Q/s)^2 + (Ps)^2 \longrightarrow$$

$$\{ \dot{Q} = +Ps^{+2} \ ; \ \dot{P} = -Qs^{-2} \} \text{ or } \{ \dot{q} = p \ ; \ \dot{p} = -q \} \ .$$

For simplicity we choose the value of the Hamiltonian equal to unity. Two views of the oscillator orbit are illustrated in **Figure 5.16** . Because the (QP) orbit is an ellipse the separation between a reference and satellite trajectory would vary with time unless constrained by a Lagrange multiplier. We illustrate for the particular choice $s = 2$:

$$\dot{Q} = 4P \ ; \ \dot{P} = -(Q/4) \ \longrightarrow \dot{\delta}_Q = 4\delta P - \lambda\delta_Q \ ; \ \dot{\delta}_P = -(\delta P/4) - \lambda\delta P \ .$$

Now use the condition that $\delta \cdot \dot{\delta}$ vanishes (for the right choice of λ) :

$$\delta_Q\dot{\delta}_Q + \delta_P\dot{\delta}_P = (15/4)\delta_Q\delta_P - \lambda \equiv 0 \longrightarrow \lambda \equiv (15/4)\delta_Q\delta_P \ .$$

The time dependence of the Lyapunov exponents (which necessarily sum to zero) is shown in **Figure 5.17** . For $s = 2$ or any value of s other than unity the varying separation between the reference and satellite trajectories generates oscillating local exponents. The local exponents *fluctuate*, as in **Figure 5.17** . An analytic *time average* over a complete oscillator period gives a root-mean-squared value of both the Lyapunov exponents equal to

$$\langle \lambda_1^2 \rangle = \langle \lambda_2^2 \rangle = (s^{+1} - s^{-1})^2/2 \ .$$

This scaled-oscillator problem is instructive. Because simply *changing the units of measurement* changes the numerical values of the local exponents there can be no precise results independent of coordinate-system changes.

A similar coordinate-frame dependence can be seen on comparing the Cartesian and polar descriptions of a Hooke's-Law pendulum. For some of the details see page 31 of our 2012 book *Time Reversibility, Computer Simulation, Algorithms, Chaos.*

Evidently by choosing appropriate units one could emphasize either the coordinates or the momenta as providing the "important" particles for instability. In fact these two limiting choices provide very similar results as shown in Joseph Ford and Spotswood Stoddard's pioneering work, and confirmed later for a nonequilibrium "color conductivity" problem in Bill's 1998 Physical Review E work with Kevin Boercker and Harald Posch, "Large-System Hydrodynamic Limit for Color Conductivity in Two Dimensions".

With the Lyapunov exponents and vectors we have a useful tool for the investigation of dynamic instabilities. In order to apply them to a wide range of problems we need to confront the inability of Hamiltonian mechanics to deal with many problem types, including nonequilibrium steady states as well as some nonsteady problems, like turbulence. The control variables necessary to such problems are the subject of the next Chapter.

5.9 Summary

Numerical methods for solving the ordinary differential equations of molecular dynamics are based on series expansions and can be developed analytically or by Monte Carlo exploration. Although chaos appears in some three-dimensional problems it appears that four-dimensional problems are simpler still and are for that reason specially likely to stimulate new and interesting ideas. Lyapunov spectra, or even just the largest local Lyapunov exponent, provide a rich source of diagnostics for chaotic systems. Because the spectra are only formally time-reversible, but are actually mostly not, comparisons of forward and backward analyses have potential value for the discussion of thermodynamic irreversibility in far-from-equilibrium processes.

5.10 References

Ford and Stoddard's work showed that either the coordinates or the momenta could be used to characterize Lyapunov instability: S. D. Stoddard and J. Ford, "Numerical Experiments on the Stochastic Behavior of a Lennard-Jones Gas System", Physical Review A **8**, 1504-1512 (1973). They used Runge-Kutta integration.

The "bit-reversible" version of the Leapfrog method is described in D. Levesque and L. Verlet, "Molecular Dynamics and Time Reversibility", Journal of Statistical Physics **72**, 519-537 (1993) and was applied to Lyapunov spectrum calculations in M. Romero-Bastida, D. Pazó, J. M. López, and M. A. Rodríguez, "Structure of Characteristic Lyapunov Vectors in Anharmonic Hamiltonian Lattices", Physical Review E **82**, 036205 (2010).

Lyapunov spectrum calculations are detailed in G. Benettin, L. Galgani, A. Giorgilli, and J. M. Strelcyn, "Lyapunov Characteristic Exponents for Smooth Dynamical Systems and for Hamiltonian Systems; a Method for Computing All of Them. Part 1 : Theory", Meccanica **15**, 9-20 (1980). See also I. Shimada and T. Nagashima, "A Numerical Approach to Ergodic Problem of Dissipative Dynamical Systems", Progress of Theoretical Physics **61**, 1605-1616 (1979) and the references cited therein. The alternative Lagrange multiplier approach is described in W. G. Hoover and H. A. Posch, "Direct Measurement of Lyapunov Exponents", Physics Letters A **113**, 82-84 (1985).

The covariant vectors are described in H. Bosetti, H. A. Posch, Ch. Dellago, and Wm. G. Hoover, "Time-Reversed Symmetry and Covariant Lyapunov Vectors for Simple Particle Models In and Out of Thermal Equilibrium", Physical Review E **82**, 046218 (2010) = arχiv:1004.4473 and H. A. Posch, "Symmetry Properties of Orthogonal and Covariant Lyapunov Vectors and Their Exponents", Journal of Physics A: Mathematical and Theoretical **46**, 254006 (2013) = arχiv:1107.4032 (2012). More details can be found in F. Ginelli, P. Poggi, A. Turchi, H. Chaté, R. Livi, and A. Politi, "Characterizing Dynamics with Covariant Lyapunov Vectors", Physical Review Letters **99**, 130601 (2007) and C. L. Wolfe and R. M. Samelson, "An Efficient Method for Recovering Lyapunov Vectors from Singular Vectors", Tellus **59A**, 355-366 (2007).

5.11 Problems

1. Find the dependence of the error on dt in the following approximation for the momentum at time t using coordinates generated by the leapfrog method :

$$p_t \equiv (4/3)[\ q_{t+dt} - q_{t-dt}\ ]/(2dt) - (1/3)[\ q_{t+2dt} - q_{t-2dt}\ ]/(4dt)\ .$$

Check this prediction by carrying out a numerical simulation for 62 harmonic-oscillator timesteps with $dt = 0.1$, beginning with the exact values $q_0 = 1$ and $q_{0.1} = \cos(0.1)$ and using the leapfrog method to generate and store $q_{0.2}, q_{0.3}, \dots q_{6.3}$. The approximate momenta can then be compared to the suggested approximation and to the analytic result: $p(t) = -\sin(t)$.

2. Write a bit-reversible oscillator program using `dt = 0.5d00` and with the first two oscillator coordinates `q(1) = -3` and `q(2) = +3` . Compare two versions of the code :

$$q_{t+dt} = [\ 2q_t - q_{t-dt} + F_t dt^2\ ] \text{ and } q_{t+dt} = 2q_t - q_{t-dt} + [\ F_t dt^2\ ]\ ,$$

where the terms in square brackets are rounded to integers.

Show that always rounding "up" or always rounding "down" gives a periodic orbit with a Poincaré recurrence length of 50 timesteps, corresponding to four oscillations, with an oscillator period of 6.25 , while rounding in the usual way, by truncation, soon leads to a periodic orbit :

$$\{\ 0, +1, +1, 0, -1, -1, 0, +1, +1, 0, -1, -1, 0, \dots\ \}\ .$$

This orbit has period 3 , due to roundoff error, and is also an exact solution for the Leapfrog algorithm for $dt = 1$ corresponding to a period of 6 .

3. Show that the "higher-order" velocity definition :

$$v(t=0) \equiv (4/3)\frac{[\ q(+dt) - q(-dt)\ ]}{2dt} - (1/3)\frac{[\ q(+2dt) - q(-2dt)\ ]}{4dt}\ ,$$

reduces the Velocity-Verlet oscillator's energy error by about a factor of three for small timesteps.

4. Show that the normalized four-dimensional Gaussian distribution

$$f(q,p,\zeta,\xi) = e^{(-1/2)(q^2+p^2+\zeta^2+\xi^2)}/(2\pi)^2\ ,$$

is the stationary solution of the four ordinary differential equations given on page 120 when the temperature and the relaxation times are set equal to unity .

5. Tuckerman's "Chain Thermostats" add additional quadratic differential equations to the Nosé-Hoover thermostat. The simplest version of this idea, for the harmonic oscillator, gives the following set :

$$\{ \dot{q} = p \ ; \ \dot{p} = -q - \zeta p \ ; \ \dot{\zeta} = p^2 - 1 - \xi\zeta \ ; \ \dot{\xi} = \zeta^2 - 1 \} \ .$$

Determine whether or not these equations are ergodic. Use the Gaussian distribution to find the root-mean-squared velocity of the four-dimensional trajectory in phase space :

$$v \equiv \sqrt{\dot{q}^2 + \dot{p}^2 + \dot{\zeta}^2 + \dot{\xi}^2} \ ,$$

using the fact that the stationary probability density for these equations is the same as that in Problem 5.3 above. This ergodicity problem was the Ian Snook Prize Problem for 2014 .

6. Show that changing the ordering in one line of the Runge-Kutta fifth-order integrator described on page 109 from
`f4 = f0 + (dt/16)*( - 3*fp1 + 6*fp2 + 9*fp3)` to
`f4 = f0 + (dt/16)*( + 9*fp3 - 3*fp1 + 6*fp2)`
changes the numerical solution of the Nosé-Hoover equilibrium oscillator. Choose initial conditions $(qp\zeta) = (050)$. Does the solution also change if the problem is a harmonic oscillator?

Chapter 6

Time-Reversible Particle Thermostats

Topics

Temperature, Contact Temperature, and Thermostats / Nosé's Hamiltonian Thermostats / Nonequilibrium Nosé-Hoover Oscillator / Patra-Bhattacharya Oscillator / Campisi's Logarithmic Hamiltonian Thermostat / Hoover-Leete Isokinetic Hamiltonian Thermostat / Landau-Lifshitz Configurational Hamiltonian Thermostat / Feedback / Gaussian and Nosé-Hoover Thermostats / Two-Moment, Three-Moment, and Chain Thermostats / Dimensionality Loss / Time-Reversible Stochastic Thermostats? / Ergostats and Barostats /

6.1 Temperature, Contact Temperature, and Thermostats

In Chapters 1 and 3 we emphasized that *temperature* is the new state variable which distinguishes thermodynamics from mechanics. Temperature makes it possible to study heat flow and to define entropy. Here we seek more, the management and control of temperature, first in atomistic mechanics and then in statistical mechanics. The statistical mechanics of temperature was clarified by Boltzmann, Gibbs, and Maxwell. Because only low-density gases (and harmonic oscillators) were feasible in the days of analytic theory it was natural to stress the importance of the "ideal" gas, a gas with low enough density that the potential energy can be ignored relative to the much larger kinetic energy, $K = \sum(p^2/2m) \;>>\; \Phi$.

In mechanical models, such as kinetic theory and statistical mechanics, temperature appears in the "thought experiment" detailed in Section 3.2, measuring the pressure exerted on the walls of an ideal gas' container. Defined in this way pressure and temperature reflect the kinetic energy of the gas, $K = (DkT/2)$ in D dimensions :

$$P \equiv (NkT/V) = (2K/DV) \longleftrightarrow T \equiv (PV/Nk) = (2K/NkD) \ .$$

Considering gases of hard parallel (constrained to be rotationless) squares or cubes, with distinct (necessarily conserved) pressure-tensor components, P_{xx} , P_{yy} , and P_{zz} , makes it possible to formulate separate mechanical definitions of the diagonal temperature-tensor components $\{ T_{ii} \}$. A mechanical definition of temperature through an ideal-gas pressure measurement requires no new concepts. Force per unit area is enough.

Alternatively, rather than starting out with temperature defined in terms of ideal-gas pressure, one could begin with Gibbs' relatively-abstract notion of *entropy*. Gibbs' entropy is the product of Boltzmann's constant k and the logarithm of the number of available phase-space "states" ,

$$S \equiv k \ln \int \ldots \int \prod (dqdp) \ \ [\text{ available states }] \ .$$

Maximizing the entropy of two thermally-interacting systems, eventually produces "thermal equilibrium". At the entropy maximum the two interacting systems necessarily share a common value of $(\partial S/\partial E)_V$ and likewise $(\partial E/\partial S)_V$.

Then the trick is to choose one of the two systems to be a monatomic D-dimensional ideal gas, for which the entropy is easy to calculate :

$$S = Nk \ln[\ (V/N)(E/N)^{D/2} \ ] + \text{constant} \ .$$

Maximizing the system-plus-ideal-gas entropy gives Gibbs' definition of temperature :

$$(\partial E/\partial S)_V = 1/(\partial S/\partial E)_V = (2E/NDk) \equiv T \ .$$

From this statistical-mechanical viewpoint temperature emerges as the derivative of energy with respect to entropy, $T = (\partial E/\partial S)_V$.

This entropic view of temperature seems to us artificial. We prefer the mechanical approach, carrying out a "thought experiment" of a pressure measurement. It seems to us simpler to start out with kinetic theory. That approach shows that when a mass M comes to thermal equilibrium with an ideal-gas thermometer, with no long-term tendency to heat up or cool down, the mass' mean squared velocity is (DkT/M) , where T is the temperature of the thermometer. For the one-dimensional case see Problem 6.1 . For a more general case see Wm. G. Hoover, B. L. Holian, and H. A. Posch, "Comment I on 'Possible Experiment to Check the Reality of a Nonequilibrium Temperature' ", Physical Review E **48**, 3196-3198 (1993) .

In one, two, or three dimensions the kinetic temperature of the mass M comes to match that of the ideal-gas thermometer particles as the result of collisions. This kinetic temperature has an additional advantage. It is readily adapted to *nonequilibrium* systems where entropy isn't clearly defined. Away from equilibrium there is no Zeroth Law of Thermodynamics, but the ideal-gas thermometer still provides a reproducible readily-measurable "temperature", even in the case where T_{xx} and T_{yy} differ so that temperature is a tensor quantity.

At equilibrium Boltzmann's proportionality constant k relates the kinetic temperature scale to the mechanical pressure scale by choosing one of two interacting systems to be an ideal gas. Boltzmann's kinetic-theory equation, *The Boltzmann Equation*, for the comoving rate of change of the one-body phase-space density, in (qp) space, is :

$$(df/dt) = \dot{f}(q,p,t) = (\partial f/\partial t)_{\text{collisions}} \ .$$

When the collisions have done their work, so that f no longer changes, its logarithm can only depend on the three quantities conserved by collisions: mass, momentum, and energy. This argument shows that the comoving maximum-entropy velocity distribution is :

$$f_{\text{eq}}(p_x, p_y, p_z) \propto e^{-p_x^2/2mkT} e^{-p_y^2/2mkT} e^{-p_z^2/2mkT} \ .$$

In order to go beyond these equilibrium considerations, so as to treat simple shear and heat flows, or more complex situations with nonlinear transport, like shockwaves, it is necessary to make a definite choice in order to assign *nonequilibrium* temperatures. The second moment definitions ,

$$\langle\ T_{xx},\ T_{yy},\ T_{zz}\ \rangle \equiv \langle\ (p_x^2/mk)\ ,\ (p_y^2/mk)\ ,\ (p_z^2/mk)\ \rangle\ ,$$

are specially appropriate because their sum is conserved by (two-body) particle collisions, while the other even moments are not. In **Figure 6.1** we show the trajectories of two particles interacting with the pair potential $\phi(r<1) = (1-r^2)^4$. The initial velocities are $\dot{x} = \pm 0.5$ and the impact parameter (which would be the closest-approach distance if there were no forces) is 0.3624 . After collision the kinetic energy is accurately in the y direction. The collisional transfer from T_{xx} to T_{yy} is rapid and local.

The various "configurational temperatures" which come from the canonical ensemble, don't share such a conservation property and don't have a mechanical definition. The Landau-Lifshitz' configurational temperature :

$$kT_{xx} = \langle\ F_x^2\ \rangle / \langle\ \nabla_x^2 \mathcal{H}\ \rangle\ ;\ kT_{yy} = \langle\ F_y^2\ \rangle / \langle\ \nabla_y^2 \mathcal{H}\ \rangle\ ,$$

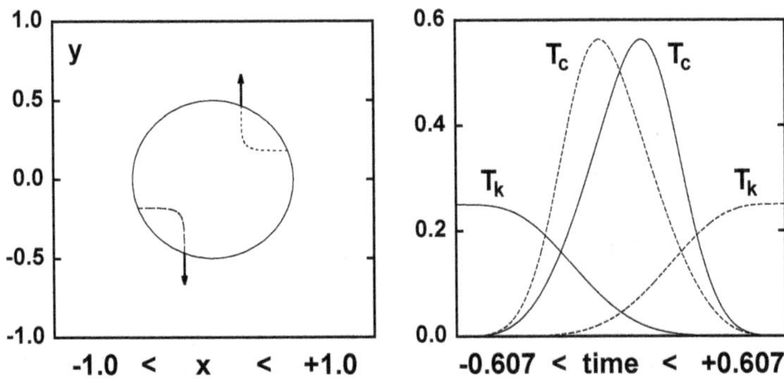

Fig. 6.1 Time-dependent temperatures in a two-body collision. On the left : a collision with initial values of $\sum(p_x^2, p_y^2) = (0.50, 0.00)$ and final values $(0.00, 0.50)$. Initially $y = \pm 0.181176$. This choice converts horizontal to vertical motion using the repulsive pair potential $\phi = (1 - r^2)^4$. Although the individual moments are not conserved the summed-up second moments *are* preserved by two-body collisions. On the right : the time dependence of the configurational and kinetic temperatures. The solid lines show horizontal T_{xx} components and the dashed lines show vertical T_{yy} components.

is the simplest choice. The second derivative, in the denominator, waves a caution flag. This definition can in fact give unbounded contributions, both positive and negative, to temperature in situations near the zeros of $\nabla_x^2 \mathcal{H}$ and $\nabla_y^2 \mathcal{H}$. The time histories of the two configurational temperatures corresponding to the same soft-disk collision above are also shown in **Figure 6.1** . They bear no obvious relationship to their kinetic relatives.

At equilibrium *any* useful thermometer has to satisfy the Zeroth Law of Thermodynamics, "If one body is in thermal equilibrium with two others then those two are likewise in thermal equilibrium with each other". So far as our (classical) models are concerned the separation of the Hamiltonian into kinetic and potential parts, $\mathcal{H} = K + \Phi$, guarantees that time-averaged equilibrium temperatures satisfy the Zeroth Law.

In "Extended Irreversible Thermodynamics" it is assumed, without any compelling argument, that one can quantify a *nonequilibrium* entropy. A recent objection (there are plenty of older ones) to nonequilibrium entropies is based on the existence of fractal nonequilibrium phase-space distributions. Because Gibbs' entropy measures the mean logarithm of the phase-space distribution, which diverges when the distribution is fractal, there are strong reasons to doubt the utility of nonequilibrium entropy.

Wolfgang Muschik and David Jou, among many others, have stressed the importance of the mechanical link between the thermometer and the system being measured. Their "contact temperature" conveys this picture.

One could imagine an anharmonic heat-conducting chain (the "ϕ^4" model would be a good choice) linked to one or more system degrees of freedom by a spring, as *defining* a local system temperature.

Is the idea of contact temperature a good one? From the physical standpoint the concept is plausible. It is hard to imagine measuring temperature without some kind of "contact". At a minimum such contact would require a definite value for a spring constant. Black body radiation is the most plausible noncontact alternative but any numerical implementation would require an ambitious recipe for the interaction of photons with matter.

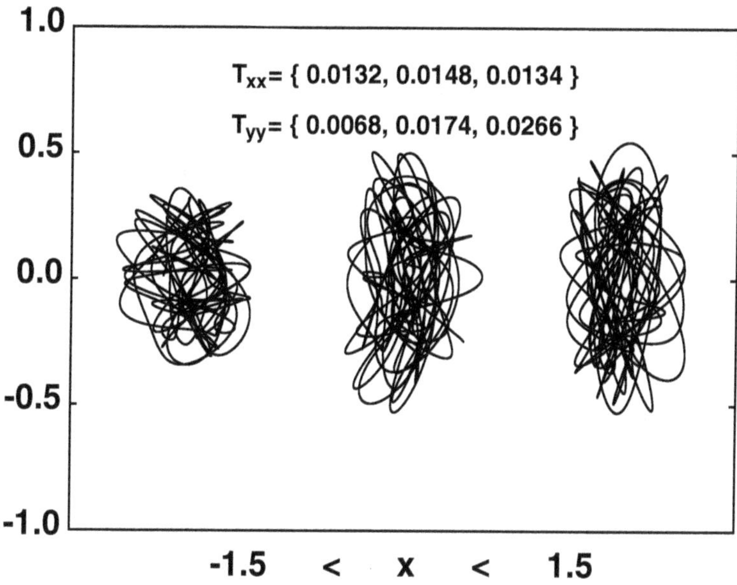

Fig. 6.2 Nonequilibrium ϕ^4 system with Nosé-Hoover control of the cold and hot particles. T_{xx} varies little. The temperature gradient is concentrated in the transverse temperature T_{yy}. The leftmost "cold" particle, with temperature 0.01, has a mainly positive friction, +0.0492 while the rightmost "hot" particle, with temperature 0.02, has a negative friction -0.0246 . The middle Newtonian particle has temperature 0.0161 .

Is contact temperature a useful idea in simulation? Let us consider a simple example: a Newtonian ϕ^4 particle sandwiched between two thermostated ϕ^4 particles, one cold and one hot. Each of the three particles is tethered to its lattice site with a quartic potential, δ_r^4 . Additionally, Hooke's-Law springs link the Newtonian particle to its two thermostated neighbors, $\phi_{\rm Hooke} = (\mid r \mid - 1)^2/2$. Typical trajectories appear in **Figure 6.2** along with the horizontal and vertical time-averaged temperatures.

The kinetic temperatures T_{xx} and T_{yy} of the Newtonian particle and the two thermostated particles can readily be measured. It is convenient to add the six kinetic temperature components to the fourteen differential equations for $\{ x, y, p_x, p_y, \zeta \}$, giving a total of twenty ordinary differential equations.

One could approach the contact idea by installing a contacting thermometer parallel, or perpendicular, or both, comparing the results to those of the kinetic-temperature thought experiment carrried out with ideal-gas thermometers. The interaction would have to be very weak, so as to avoid perturbing the dynamics of the three-particle system. With a sufficiently weak interaction the ideal-gas thermometer in equilibrium with each degree of freedom would simply measure that degree of freedom's kinetic temperature $kT = m\langle v^2 \rangle$. Accomplishing such measurements seems unrewarding because the result is known in advance.

One could also imagine, and even implement, a microcrystalline thermometric particle, complete with hundreds of degrees of freedom and with attractive and repulsive forces giving rise to rotation, vibration, and translation. András Baranyai did exactly this, as is discussed in Chapter 1 .

We think such detailed microscopic models suffer from unnecessary complexity, fluctuations, and inhomogeneities. It is far simpler to imagine a mind's-eye "bag of gas", an ideal-gas thermometer, making sufficiently rapid Maxwell-Boltzmann collisions with the investigated system, to measure its instantaneous kinetic temperature. That will be our point of view. We leave the imperfect inhomogenous thermometric pictures behind now, and adopt instead the kinetic-theory definition of temperature. This choice is, we believe, the simplest, and we are quite willing to interpret nonequilibrium constitutive relations in terms of this choice. Because temperature needs defining in a comoving frame, moving with the local stream velocity, any implementation requires a definition of stream velocity. For that purpose the smooth-particle weight functions are a useful approach. The corresponding velocity definition becomes :

$$u(r,t) \equiv \sum v_i(t)w(r - r_i) / \sum w(r - r_i) \ ,$$

where choosing the range and functional form of the weight function suggest a variety of useful and interesting research projects. See the problems.

With temperature defined we turn next to the task of imposing it on mechanical systems, managing their temperatures. This will help us to generate both equilibrium and nonequilibrium states, homogeneous or not, and time-dependent or not. We undertake a relatively comprehensive review of the thermostat types developed in the past four decades. Many of them are useful while some are not. We make an effort to distinguish the two types. The wide variety, and the stimulating flow of new ideas, make computational thermostats an absorbing field of study. Without question Shuichi Nosé's pioneering work, published in 1984, well deserved the IBM Japan Science Prize it won for him in 1989. Let us follow in his steps.

6.2 Nosé's Hamiltonian Thermostats

Nosé was 32 , working as a post-doctoral Fellow in Canada after his Kyoto PhD, and loosely supervised by Mike Klein. Shuichi developed a dynamical approach to Gibbs' canonical ensemble. If he followed a logical route to this goal its traces are conspicuously absent in his two very clear and very surprising 1984 papers. The useful results can be stated in a couple of lines. Certainly we would enjoy to know canonical-ensemble properties for a system described by a simple Hamiltonian, such as

$$\mathcal{H}(q,p) = K(p) + \Phi(q) \longrightarrow f(q,p) \propto e^{-K(p)/kT}e^{-\Phi(q)/kT} \ .$$

We would additionally be pleased to link those canonical properties as closely as possible to molecular dynamics simulations, macroscopic thermodynamics, and to computational fluid dynamics. Nosé began a path toward achieving all of these goals.

Nosé first of all modified a simple-fluid Hamiltonian by including in it a single new pair of canonical variables, a dimensionless "scale factor" s and its conjugate momentum p_s . Because s is dimensionless this "momentum" has units of action, [mass][length]^2/[time] . We follow Nosé by introducing kT into the Hamiltonian along with an "effective mass" M . This "mass" actually has units [mass][length]^2 :

$$\mathcal{H}_{\text{Nosé}} \equiv K(p/s) + \Phi(q) + \#kT\ln s + (p_s^2/2M) \ ;$$

$$K(p/s) \equiv \sum^{\#-1}(p^2/2ms^2) \ .$$

The introduction of the scale factor s in the kinetic energy provides a control mechanism for the kinetic temperature. Nosé's Hamiltonian includes an unusual coupling of the coordinate s with the whole set of Cartesian momenta $\{ p \}$. Perhaps it conveys a picture: scaling the momentum distribution so as properly to match the temperature T ? Here # is the total number of degrees of freedom, including the new coordinate-momentum pair (s, p_s) as the last of them. Then, Nosé computes the equations of motion (we consider the Cartesian case for simplicity) and multiplies each of them by s , a trick he calls "scaling the time" :

$$\{ \dot{q} = (p/ms^2) \times s \ ; \ \dot{p} = F \times s \} \ ;$$

$$\dot{s} = (p_s/M) \times s \ ; \ \dot{p}_s = \sum^{\#-1} [\ (p^2/ms^3) - (kT/s) \ ] \times s \ .$$

The multiplication step does not change the (q, p, s, p_s) phase-space trajectory at all, only the speed at which it is followed.

Next, he replaces the combination (p/s) by p [we use the same symbol for the new momentum, making the substitution $(p/s) \longrightarrow p$] :

$$\{ \dot{q} = (p/m) \ ; \ \dot{p} = F - (p_s/M)p \} \ ;$$

$$(\dot{s}/s) = (p_s/M) \ ; \ \dot{p}_s = \sum^{\#-1} [\ (p^2/m) - kT \ ] \ .$$

Finally, Nosé shows that the probability density for these new equations of motion *is* Gibbs' canonical distribution. The lengthy derivation that he gives in his 1984 papers includes some tedious and unnecessary algebra.

We chose our own much simpler and avowedly nonHamiltonian route to the canonical distribution, ending up with similar motion equations, different only in the omission of s and the definition of # . We *begin* with, rather than *derive*, the Nosé-Hoover equations of motion for # Cartesian degrees of freedom ,

$$\{ \dot{q} = (p/m) \ ; \ \dot{p} = F - \zeta p \} \ ; \ \dot{\zeta} = \sum^{\#} [\ (p^2/mkT) - 1 \ ]/\tau^2 \ .$$

For convenience we have replaced Nosé's (p_s/M) by ζ and M by $kT\tau^2$, where ζ is a friction coefficient, with units 1/[time], and τ is the characteristic response time associated with the control variable :

$$\zeta = (p_s/M) \ ; \ \tau^2 = (M/kT) \longrightarrow (p_s/kT) = \zeta\tau^2 \ .$$

We can then confirm that a stationary probability density $f(q, p, \zeta)$ with the scale factor s omitted, and unchanged by the remaining motion equations $(\dot{q}, \dot{p}, \dot{\zeta})$, is

$$f(q, p, s, \zeta) \propto (1/s) e^{-K/kT} e^{-\Phi/kT} e^{-\tau^2\zeta^2/2} \longrightarrow$$

$$f(q, p, \zeta) \propto e^{-K/kT} e^{-\Phi/kT} e^{-\tau^2\zeta^2/2} \ ; \ K \equiv \sum^{\#} (p^2/2m) \ .$$

In the last line the superfluous scale factor s has disappeared. To show that this *is* a stationary solution we consider (as an exercise for the reader) the phase-space flow equation (the many-dimensional analog of the continuum continuity equation) :

$$(\partial f/\partial t) = -\sum(\partial f\dot{q}/\partial q) - \sum(\partial f\dot{p}/\partial p) - \sum(\partial f\dot{\zeta}/\partial\zeta) \equiv 0 \ .$$

Note the absence of $-(\partial f\dot{s}/\partial s)$.

6.2.1 *More General Thermostat Forms*

Wolfgang Bauer, Aurel Bulgac, and Dimitri Kusnezov carried out a wide-ranging exploration of thermostat types using the continuity-equation approach. For small systems the equations of motion $\{ \ \dot{p} = F - \zeta^3 p \ \}$ and $\{ \ \dot{p} = F - \zeta p^3 \ \}$ are the simplest and most useful examples of their generalizations of Nosé's idea. The distribution for the simplest of such harmonic oscillator problems, with $\dot{p} = -q - \zeta^3 p$ at unit temperature, includes a quartic potential $(1/4)\zeta^4$ in the exponent of the stationary phase-space distribution :

$$\{ \ \dot{q} = p \ ; \ \dot{p} = -q - \zeta^3 p \ ; \ \dot{\zeta} = p^2 - 1 \ \} \longrightarrow$$

$$f(q,p,\zeta) \propto e^{-q^2/2} e^{-p^2/2} e^{-\zeta^4/4} \ .$$

Their papers are full of examples and even include a recipe for generating Brownian Motion with time-reversible deterministic thermostat variables.

An interesting and significant improvement on Nosé's Hamiltonian was made by Carl Dettmann and published with Gary Morriss. See their 1996 Physical Review E paper, "Hamiltonian Formulation of the Gaussian Isokinetic Thermostat" for more details. In July 1996 , at a Lyons meeting of CECAM [Centre Européen de Calcul Atomique et Moléculaire], Bill asked Carl about the possibility of obtaining the Nosé-Hoover equations directly from a Hamiltonian, without any time-scaling. About 12 hours later, Dettmann had an answer :

$$\mathcal{H}_{\rm Dettmann} = s \times \mathcal{H}_{\rm Nosé} =$$

$$\sum^{\#}(p^2/2ms) + s\Phi(q) + skT\#\ln s + s(p_s^2/2M) \ \equiv 0 \ !$$

The many people that had scrutinized and pondered Nosé's Hamiltonian over the course of a dozen years never thought of multiplying the Hamiltonian itself by s , rather than following Nosé and scaling the equations of motion. The special trick flourish at the end, setting the Hamiltonian equal to zero, was Carl Dettmann's own contribution to the magical quality of Nosé's work.

6.2.2 *Oscillators–Nosé and Nosé-Hoover Thermostating*

In applications the original Hamiltonian form of Nosé's ideas has some disabling disadvantages. Although it isn't at all evident from the equations of motion, *the phase-space distribution for Nosé's original Hamiltonian turns out to be singular, diverging as* an inverse power of s . Furthermore, for small simple systems such as a single harmonic oscillator, his new thermostat idea is not sufficiently chaotic to generate the entire canonical distribution.

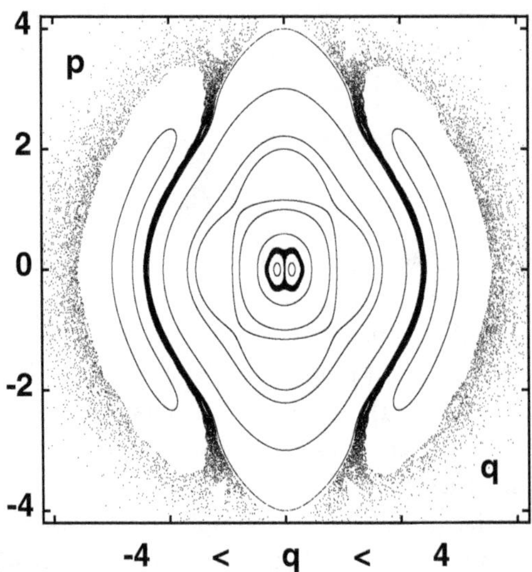

Fig. 6.3 Nosé-Hoover equilibrium oscillator $\zeta = 0$ cross section [equivalent to Nosé's $p_s = 0$ section] for initial (qp) values $(01), (02), (03), (04), (+30), (-30)$, all with $\zeta = 0$. The predominant features are tori, which show up in the figure as closed curves. Points from the conservative fractal, with initial conditions $(q, p, \zeta) = (0, 5, 0)$ are also shown.

Figure 6.3 , a cross section of the harmonic oscillator phase space for $\zeta = p_s = 0$, shows this nonergodicity very well. Just outside a radius of 2, we see a relatively thin chaotic sea of (q,p) or $(q,p/s)$ canonically-distributed points along with two large banana-shaped cavities containing tori. Most of the phase space consists of these tori. They are nonmixing and not chaotic. They prevent Nosé's thermostat from providing the Gaussian canonical equilibrium distribution of the oscillator coordinate and velocity. We will see that even when given an opportunity to dissipate energy, some of the equilibrium tori shown here remain conservative and resist dissipation.

For the harmonic oscillator problem Nosé's original Hamiltonian formulation is :

$$2\mathcal{H} = q^2 + (p/s)^2 + 2\ln s^2 + p_s^2 \longrightarrow f(q,p,s,p_s) \propto e^{-\mathcal{H}} \ ;$$

$$\dot{q} = (p/s^2) \ ; \ \dot{p} = -q \ ; \ \dot{s} = p_s \ ; \ \dot{p}_s = (p^2/s^3) - (2/s) \ .$$

The final "2", coming from the logarithmic term in the Hamiltonian, indicates that the number of degrees of freedom is two, the (q,p) oscillator degree of freedom and the additional (s,p_s) thermostat pair introduced by Nosé.

Consider now the phase-space distribution consistent with the motion equations :

$$f(q,p,s,p_s) \propto (1/s)^2 e^{-q^2/2} e^{-(p/s)^2/2} e^{-p_s^2/2} = e^{-\mathcal{H}} \ .$$

The phase-space flow equation shows that this distribution is a stationary solution of the motion equations. This is nothing more than a consequence of Liouville's Theorem. *Any* function of the Hamiltonian (such as the probability density $f(q,p,s,p_s)$ above) is left *unchanged* by the Hamiltonian equations of motion. Consider the two contributions to the flow equation for $(\partial f/\partial t)$:

$$-f[\ (\partial\dot{q}/\partial q) + (\partial\dot{p}/\partial p) + (\partial\dot{s}/\partial s) + (\partial\dot{p}_s/\partial p_s)\ ] = 0 + 0 + 0 + 0 \equiv 0 \ ;$$

$$-\dot{q}(\partial f/\partial q) - \dot{p}(\partial f/\partial p) - \dot{s}(\partial f/\partial s) - \dot{p}_s(\partial f/\partial p_s) =$$

$$f[\ q(p/s^2) + (p/s^2)(-q) + [\ (2/s) - (p^2/s^3)\ ]p_s + p_s[\ (p^2/s^3) - (2/s)\ ]] \equiv 0 \, .$$

Because both contributions vanish the phase-space distribution *is* stationary. That doesn't mean that the distribution is necessarily very good. The divergence for small s^2 is troubling. There is also no guarantee that the motion equations will fill out the whole distribution. In fact it appears that most of the four-dimensional phase-space flow consists of two-dimensional tori.

Figure 6.3 provides a guide to the initial (q,p) phase points in the $[\ \zeta \propto p_s = 0\ ]$ plane which generate two-dimensional tori rather than a three-dimensional or four-dimensional portion of the canonical distribution. The predominant regions of phase space follow tori rather than chaos. The singular nature of the equations of motion where s vanishes poses no particular difficulty for accurate simulations. The formal solution only indicates that the *original* Nosé Hamiltonian is an inconvenient representation. Numerical work shows that the underlying motion is not ergodic.

But there is a way out. To follow it, first replace (p/s) by p . Then write the momentum p_s as the friction coefficient ζ . As a result, the poorly-behaved equations of motion become transformed to the following set :

$$\{\ \dot{q} = (p/s)\ ;\ \dot{p} = (-q - \zeta p)/s\ ;\ \dot{s} = \zeta\ ;\ \dot{\zeta} = (p^2 - 2)/s\ \}\ .$$

Multiplying *these* equations by a factor of s gives the Nosé-Hoover equations of motion, provided that the last "2" in the equation is replaced by "1".

This replacement can perhaps be rationalized as accounting for a reduction in the number of degrees of freedom from 2 to 1 , ending up with $\dot{\zeta} = (p^2 - 1)$ rather than $\dot{\zeta} = (p^2 - 2)$. It could equally-well be considered to show that Nosé's rescaled motion equations are actually slightly different to Bill's, in the interpretation of the number of degrees of freedom # . It is certainly curious that beginning with a Hamiltonian giving poorly-behaved singular dynamics Nosé was able to derive equations of motion which are not only quite stable but are also particularly valuable for simulating a host of both few- and many-body problems.

Numerical solutions of the *equilibrium* harmonic oscillator problem with a thermostat relaxation time τ included :

$$\{\ \dot{q} = p\ ;\ \dot{p} = -q - \zeta p\ ;\ \dot{\zeta} = [\ p^2 - T\ ]/\tau^2\ \}\ ,$$

result in some incredibly intricate toroidal and fractal structures. The chaotic solution is full of holes, down to infinitesimal in size. An infinite number of nonchaotic torus solutions inhabit these cavities in the chaotic sea. See **Figure 5.6** on page 119 for the additional complexity that can occur when the action of the thermostat is faster, $\dot{\zeta} = [\ p^2 - 1\ ]/\tau^2$. That example, for $\tau = (1/5)$, exhibits on the order of one hundred well-defined holes.

Nested tori (with three vanishing Lyapunov exponents) fill in the holes in the chaotic sea. That sea has a simple nonzero Lyapunov spectrum, $\{\ +\lambda, 0, -\lambda\ \}$. The infinite number of tori could be expected on simple physical grounds—because a chaotic sea looks just like an infinitely-long periodic orbit there can be no doubt that infinitely-many periodic orbits, some stable and some not, lie nearby. The boundary "separating" the sea from the tori is actually incredibly complicated and fractal despite the eight-fold symmetry ($\pm q, \pm p, \pm\zeta$) of the solutions of the equations of motion. These symmetries of the equilibrium oscillator and its sections disappear once a temperature gradient is added, enabling dissipation, to which we turn next.

6.3 Nonequilibrium Nosé-Hoover Oscillator

Even more-intricate complexity results when the oscillator's target temperature is made a function of q . Evidently a coordinate-dependent temperature $T(q)$ provides an opportunity for the dynamics to transport heat in the direction opposite to the temperature gradient. Choosing a hyperbolic tangent form provides a smooth temperature interpolation :

$$T(-\infty) = 1 - \epsilon \ < \ 1 + \epsilon \tanh(q/\lambda) \ < \ T(+\infty) = 1 + \epsilon \ ,$$

with most of the variation occurring over a spatial range of order λ (where *this* λ is a length scale rather than a Lyapunov exponent λ !) :

$$\{ \ \dot{q} = p \ ; \ \dot{p} = -q - \zeta p \ ; \ \dot{\zeta} = [\ p^2 - T(q) \ ]/\tau^2 \ ; \ T \equiv 1 + \epsilon \tanh(q/\lambda) \ \} \ .$$

The maximum temperature gradient (at $q = 0$) is (ϵ/λ) .

Figures 6.4 and 6.5 hint at the complexity of this simple oscillator problem for two of the *simplest* cases, where both τ and λ are equal to unity. In **Figure 6.4** the temperature gradient is moderate, $\epsilon = 0.10$. There is a shift and a dimensionality loss in the chaotic sea as well as an increase in complexity. The invariant tori change shape. Numerical solutions show that these tori are not dissipative. That is, the comoving phase volume change, from Liouville's theorem :

$$(\dot{\otimes}/\otimes) = (\partial\dot{q}/\partial q) + (\partial\dot{p}/\partial p) + (\partial\dot{\zeta}/\partial\zeta) =$$

$$0 - \zeta + 0 = -\zeta = \lambda_1 + \lambda_2 + \lambda_3 \leq 0 \ ,$$

is, on average, zero on the tori and negative in the chaotic sea.

The tori, being conservative equilibrium structures, are evidently reversible in p , stable, and symmetric. The strange-attractor sea of **Figure 6.4** , unlike the conservative chaotic sea of **Figure 6.3** , is necessarily dissipative and intrinsically irreversible. How in the world can these qualitatively different features coexist in one and the same cross section?

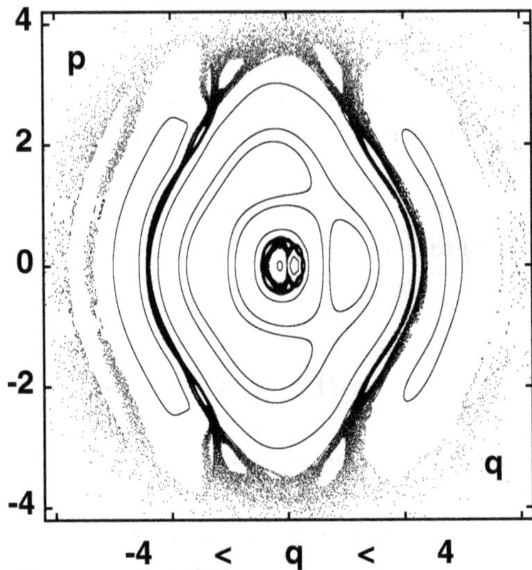

Fig. 6.4 Nosé-Hoover *nonequilibrium* zero-ζ cross section showing conservative tori generated by initial values of $\{ qp\zeta \} = \{ 010,\ 020,\ 030,\ -300,\ +300 \}$ and with a maximum temperature gradient $\epsilon = 0.10$. The distributions of chaotic points, from initial values 040 and 050 , both of which generate the strange attractor, have been included too. All these simulations used the "classic" fourth-order Runge-Kutta integrator described in Section 5.3 with $dt = 0.001$. Note the asymmetry in q and compare this cross section to the equilibrium section displayed in **Figure 6.3** .

See the upper righthand corner of **Figure 8.7** on page 242 for a similar coexistence in the nonequilibrium Galton Board problem. There is an earlier example in the phase plots for a set of three nonlinear laser-physics equations in A. Politi, G. L. Oppo, and R. Badii's "Coexistence of Conservative and Dissipative Behavior in Reversible Dynamical Systems", Physical Review A **33** , 4055-4060 (1986) .

From the physical standpoint it is clear that a shrinking volume, with a positive $\langle \zeta \rangle$, would cause the tori to form limit cycles, and in fact that does happen for stronger temperature gradients. **Figure 6.6** is the limit cycle for $\epsilon = 0.50$. The borderline between the chaos and the limit cycle (where the limit cycle disappears as ϵ is slowly reduced) is in the neighborhood of $\epsilon = 0.40535$. For sufficiently large ϵ the sea and the tori all vanish leaving the limit cycle as the sole solution. The fourth-order Runge-Kutta limit cycle for $\epsilon = 0.50$ is shown in **Figure 6.6** .

The Nosé-Hoover oscillator illustrates the qualitative difference between equilibria and nonequilibrium steady states. In the absence of dissipation the oscillator has a simple Lyapunov spectrum with three long-time-

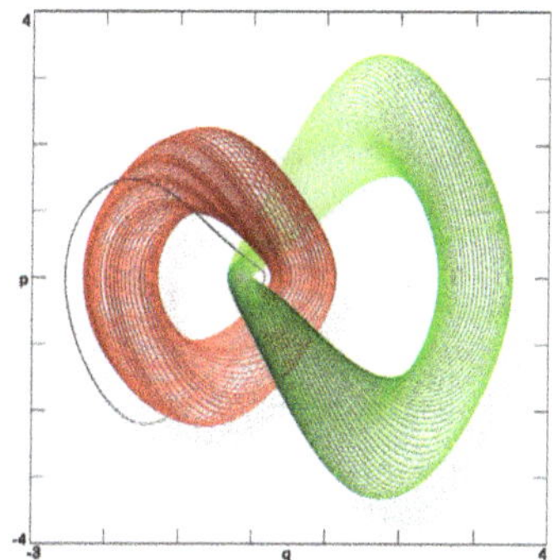

Fig. 6.5 Coexistence of a dissipative limit cycle with two conservative tori. For smaller temperature gradients the phase space can also display a fractal strange attractor. Two interlocking tori are linked with a limit cycle for the nonequilibrium Nosé-Hoover oscillator with a maximum temperature gradient of 0.42 . The tori are "conservative", with vanishing Lyapunov exponents. The limit cycle is "dissipative", acting as a sink for nearby trajectories. For further details see our 2014 Physical Review E paper with Clint Sprott, "Heat Conduction, and the Lack Thereof, in Time-Reversible Dynamical Systems: Generalized Nosé-Hoover Oscillators with a Temperature Gradient".

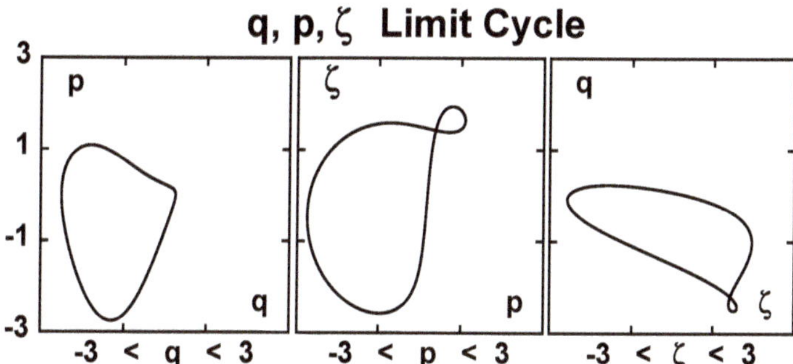

Fig. 6.6 Limit cycle for the Nosé-Hoover nonequilibrium oscillator for $\epsilon = 0.50$.

averaged exponents, $\{ +\lambda, 0, -\lambda \}$ in the chaotic sea and $\{ 0, 0, 0 \}$ on the tori. The symmetry of the chaotic spectrum follows from the time-reversibility of the flow equations through the continuity equation :

$$(\dot{\otimes}/\otimes) = (\partial\dot{q}/\partial q) + (\partial\dot{p}/\partial p) + (\partial\dot{\zeta}/\partial\zeta) = 0 - \zeta + 0 = -\zeta = \lambda_1 + \lambda_2 + \lambda_3 = 0 ;$$

coupled with the observation that the long-time-averaged value of the friction coefficient vanishes for a toroidal (conservative) solution. Gradients larger than 0.50 make dissipation inevitable with a time-averaged positive friction, $\langle \zeta \rangle > 0$. Small gradients give solutions resembling the equilibrium isothermal case. Intermediate gradients with $\epsilon \simeq 0.40$ provide long transients and long periodic orbits, both of which are sensitive to the details of the integration algorithm. Thorough investigations of this model would make excellent thesis projects. See also Section 6.9 .

6.4 Patra-Bhattacharya Thermostat

Puneet Patra and Baidurya Bhattacharya invented a new computational thermostat designed to impose simultaneously both the configurational *and* the kinetic temperature on selected degrees of freedom. Ever since the Nosé-Hoover work of the middle 1980s it has been conventional to modify Hamilton's equation of motion for momentum by adding a friction coefficient ζ designed to reproduce a target kinetic temperature T by "integral feedback" :

$$\{ \dot{p} = F(q) - \zeta p \} \ ; \ \dot{\zeta} \propto [\, p^2 - T \,] \text{ or } [\, (p^2/T) - 1 \,] \longrightarrow$$

$$\zeta(t) \propto \int_0^t [\, p^2 - T \,] dt' \text{ or } \int_0^t [\, (p^2/T) - 1 \,] dt' \ .$$

More recently the desirability of thermostating the configurational temperature T_C by modifying the motion equation $\dot{q} = p$ has been suggested.

$$\{ \dot{q} = p - \xi q = p + \xi F \} \ ; \ \dot{\xi} \propto [\, q^2 - T \,] \text{ or } [\, (q^2/T) - 1 \,] \longrightarrow$$

$$\xi(t) \propto \int_0^t [\, q^2 - T \,] dt' \text{ or } \propto \int_0^t [\, (q^2/T) - 1 \,] dt'$$

See Karl Travis and Carlos Braga's "Configurational Temperature Control for Atomic and Molecular Systems" in the Journal of Chemical Physics **128** (2008) . In both cases, kinetic and configurational, there is a proportionality constant that can be used to control the response time of either thermostat. These two temperatures are defined as in Landau and Lifshitz' text :

$$kT_K \equiv \frac{\langle\, (\nabla_p \mathcal{H})^2 \,\rangle}{\langle\, (\nabla_p^2 \mathcal{H}) \,\rangle} \ ; \ kT_C \equiv \frac{\langle\, (\nabla_q \mathcal{H})^2 \,\rangle}{\langle\, (\nabla_q^2 \mathcal{H}) \,\rangle} \ .$$

Notice that the new thermostated equations of motion, and all those that follow, are "time-reversible" by design. To establish this reversibility simply note that changing the signs of the momentum and the friction coefficients ζ and ξ yields *another* mirror-image solution of the same equations of motion in which the coordinates are reversed in their time order.

At first blush it certainly seems that brute-force "control" of coordinates is unæsthetic. "Scaling" coordinates ($\dot{q} \propto q$) makes no physical sense. Patra and Bhattacharya suggested instead that the coordinates respond to local *forces*. This is an intuitively attractive idea reminiscent of, and included in, "Brownian dynamics" (which ignores inertia) :

$$\dot{q}_{PB} \propto F \ .$$

Apart from a random force this type of motion equation is used to formulate Brownian dynamics. The proportionality constant here would describe the inertia of the force chosen to thermostat the configurational temperature. To examine the usefulness of Patra and Bhattacharya's idea we apply it to the harmonic oscillator.

6.4.1 *Patra-Bhattacharya Oscillator*

The kinetic and configurational temperatures for a harmonic oscillator with unit force constant, mass, and Boltzmann's constant are

$$T_K = p^2 \ ; \ T_C = q^2 \ .$$

We solve the set of Patra-Bhattacharya motion equations in the simplest case, where the temperature T and both relaxation times $\{ \ \tau \ \}$, are set equal to unity :

$$\{ \ \dot{q} = p - \xi q \ ; \ \dot{p} = -q - \zeta p \ ; \ \dot{\zeta} = [\ p^2 - T \ ]/\tau_K^2 \ ; \ \dot{\xi} = [\ q^2 - T \ ]/\tau_C^2 \ \} \ .$$

We can show that the extended Gibbs' canonical probability density ,

$$f(q,p,\zeta,\xi) = e^{-q^2/2}e^{-p^2/2}e^{-\zeta^2/2}e^{-\xi^2/2}/(2\pi)^2 \ ,$$

is stationary for these motion equations. The corresponding phase-space continuity equation is the following :

$$(\partial f/\partial t) = -f[\ (\partial\dot{q}/\partial q) + (\partial\dot{p}/\partial p) + (\partial\dot{\zeta}/\partial\zeta) + (\partial\dot{\xi}/\partial\xi) \ ] +$$

$$f[\ q(p-\xi q) + p(-q-\zeta p) + \zeta(p^2-1) + \xi(q^2-1) \ ]$$

$$= -f[\ -\xi - \zeta + 0 + 0 \ ] + f[\ -\zeta - \xi \ ] \equiv 0 \ .$$

This stationarity of the flow proves that the kinetic and configurational thermostats are "consistent" with Gibbs' canonical distribution. But a particular system may not fill out the full distribution. In order to understand whether or not the motion equations are truly space-filling or "ergodic" we need to have a look at special cases.

When $T = 1$ one can easily find tori like that shown at the left in **Figure 6.7** , with the four variables' initial values ($1.5, 1.0, 0.0, 0.0$) . The periodic orbit in the center begins near ($1.5, 0.0, 0.0, 0.0$) . The chaotic solution at the right has the initial condition ($q, p, \zeta . \xi$) $=$ ($0.0, 5.0, 0.0, 0.0$) .

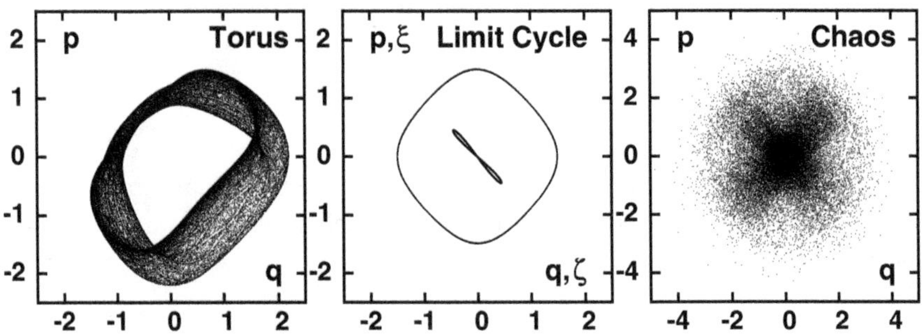

Fig. 6.7 Three types of oscillator solution for the Patra-Bhattacharya thermostat. As there is no dissipation the "limit cycle" in the center is simply a torus with a small minor radius. The nonspherical correlation of q and p seen in the right panel is surprising in view of the four-dimensional spherically symmetric solution of the continuity equation.

The corresponding chaotic orbit is shown at the right in the same **Figure 6.7** . The coexistence of a chaotic sea with infinitely-many periodic orbits and their tori, is reminiscent of the Nosé-Hoover oscillator, and suggests that for small systems the Patra-Bhattacharya thermostat is *not* sufficiently robust to generate Gibbs' canonical distribution. The combined thermostat is not so well behaved as two other doubly-controlled thermostats we have considered for the oscillator, the Hoover-Holian two-moment thermostats and the Martyna-Klein-Tuckerman Chain thermostats.

In fact, the simple harmonic-oscillator example shows that the original Patra-Bhattacharya thermostat induces a strong correlation between the oscillator coordinate and momentum. With the usual choices of unity for the various constants the thermostat produces the correct second moments, $\langle\ q^2, p^2\ \rangle = (1,1)$, but has the flaw that $\langle\ q^2p^2\ \rangle \simeq 1.4$ rather than the canonical value of unity.

Patra himself carried out a comparison of several *different* moment pairs, not just $\langle\ q^2, p^2\ \rangle$. In all cases two different moments were constrained by two thermostat variables for the oscillator. In most cases two thermostats were enough to provide a good approximation to the complete canonical distribution. The simplest such example problem controls one quartic and one quadratic moment. Consider $\langle\ q^4, p^2\ \rangle$:

$$\{\ \dot{q} = p - \zeta q^3\ ;\ \dot{\zeta} = q^4 - 3q^2T\ ;\ \dot{p} = -q - \xi p\ ;\ \dot{\xi} = p^2 - T\ \}\ .$$

Using 10^9 Runge-Kutta timesteps with $dt = 0.001$ gives:

$$\langle\ q^2, q^4, q^6, p^2, p^4, p^6\ \rangle = 1.000,\ 3.000,\ 14.997,\ 1.000,\ 2.999,\ 15.001\ ,$$

strong evidence for the ergodic character of the equations of motion with two constraints.

Because the harmonic oscillator is a relatively difficult system to thermostat, this success indicates the promise of the Patra-Bhattacharya ideas. In the equilibrium case, where T is unity the phase-space distribution is the four-dimensional Gaussian :

$$(2\pi)^2 f = e^{-q^2/2}e^{-p^2/2}e^{-\zeta^2/2}e^{-\xi^2/2} \ .$$

It is amazing that fixing just two moments is enough (apparently, and in most cases) to generate the entire distribution.

To check the space-filling ergodic nature of a particular set of motion equations the simplest approach, and arguably the best method known presently, is to compute finite-time Lyapunov exponents for a large number of different initial conditions, either in a grid or chosen randomly within a phase-space hypervolume. Knowing that the stationary distribution for the oscillator problems is a Gaussian makes it possible also to choose a grid of M points centered in contiguous and space-filling volume elements of equal probability :

$$\int_{q_{\min}}^{q_{\max}} \int_{p_{\min}}^{p_{\max}} \int_{\zeta_{\min}}^{\zeta_{\max}} \int_{\xi_{\min}}^{\xi_{\max}} f(q,p,\zeta,\xi) dq dp d\zeta d\xi \equiv (1/M) \ .$$

Finding the element boundaries is a simple numerical exercise using Runge-Kutta integration.

6.5 Campisi's Logarithmic Hamiltonian Thermostat

In 2012 Michele Campisi suggested a simple "Hamiltonian thermostat" with a *logarithmic* potential, a streamlined version of Nosé's idea :

$$2\mathcal{H}_{\text{Campisi}} = p^2 + T\ln(s^2) \longrightarrow \dot{s} = p \ ; \ \dot{p} = -(T/s) \ .$$

The distribution function implied by these motion equations provides what is apparently the simplest example of Gibbs' canonical distribution :

$$f(s,p) \propto (1/s)e^{-p^2/2T} \longrightarrow$$

$$(\partial f/\partial t) = -(\partial f\dot{s}/\partial s) - (\partial f\dot{p}/\partial p) = f[\ (1/s)p + (p/T)(-T/s)\ ] \equiv 0 \ .$$

Evidently the mean-squared value of p *is* T , as it should be. For Campisi's "thermostat" the configurational temperature introduced in Section 3.6 and detailed below is different to its kinetic brother :

$$T_C \equiv \langle\ [\ \nabla_s T\ln(s)\ ]^2/\nabla_s^2 T\ln(s)\ \rangle = (T/s)^2/(-T/s^2) = -T\ [\ !\ ] \ .$$

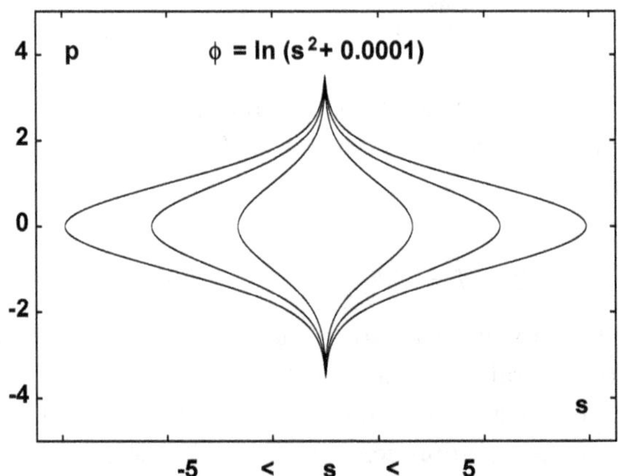

Fig. 6.8 Typical Campisi logarithmic Thermostat orbits. Starting from the inner curve and proceeding to the outer curve the initial values of (s,p) are $(1,1), (2,1)$, and $(3,1)$.

The two-dimensional configurational temperature, left as a problem for the reader, actually diverges ! This difference between the kinetic and configurational temperatures contradicts the universality of temperature as a thermodynamic concept. Any thermostat useful at equilibrium cannot act to destroy the equivalence of T_K and T_Φ . For this reason the "log thermostat" *is* pedagogically useful but thoroughly useless for simulations.

Additionally the divergence of the probability density at $s = 0$ suggests problems similar to those we found for Nosé's original logarithmic-Hamiltonian thermostat. In order to avoid divergence at the origin in numerical work we arbitrarily add a small perturbation so that

$$2\mathcal{H}_{\rm Campisi} \longrightarrow p^2 + \ln(s^2 + 0.0001) \ .$$

For three sample thermostated oscillator orbits see **Figure 6.8** above.

Unlike Nosé, Campisi failed to specify the coupling linking his thermostat variable s to the system being thermostated. His work soon stimulated critical comments from those, ourselves included, who made unsuccessful attempts to apply his thermostat idea to simple heat-flow problems, for which this simplest of Hamiltonian "thermostats" is a dismal failure.

In Marc Meléndez Schofield's 2014 dissertation, "The Theory of Coarse-Graining Without Projection Operators" (Universidad Nacional de Educación a Distancia, Spain) and Sponseller and Blaisten-Barojas' "Failure of Logarithmic Oscillators to Serve as a Thermostat for Small Atomic Clusters", Physical Review E **89**, 021301R (2014) the properties of Campisi's "thermostat" are reviewed. In his thesis Marc suggests a plausible roadmap for Nosé's more-nearly-successful route to the canonical distribution.

Despite all the shortcomings of Campisi's thermostat, the model teaches a useful lesson. Evidently the failure of thermostats based on standard classical mechanics in its Lagrangian and Hamiltonian forms is very general. We will see that using Hamiltonian thermostats to generate "hot" and "cold" regions is also an ineffective failure in that *no heat current flows* in such systems. The failure of Hamiltonian thermostats to simulate steady nonequilibrium flows is quite general. Campisi's Hamiltonian thermostat is certainly the simplest example.

We will next take the time and space to consider two more "conventional" examples, with explicit coupling between thermostats and Newtonian particles, which illustrate this failure and motivate the *non*-Hamiltonian approaches to the management of nonequilibrium simulations. Of course this latter approach has been followed since the 1970s. There are now several examples which demonstrate that the nonHamiltonian path is not only sufficient, but also necessary. We will begin with an interesting example taken from straightforward Lagrangian and Hamiltonian mechanics, the Hoover-Leete isokinetic thermostat.

6.6 Hoover-Leete Isokinetic Hamiltonian Thermostat

In Hamiltonian mechanics the coordinates and momenta are "conjugate" variables. Their equations of motion are "symplectic" (which implies conservation of phase volume through Liouville's Theorem, $\dot{\otimes} = 0$) ,

$$\{ \dot{q} = +(\partial\mathcal{H}/\partial p) \; ; \; \dot{p} = -(\partial\mathcal{H}/\partial q) \} \; .$$

$$(\dot{\otimes}/\otimes) = \sum [\, (\partial\dot{q}/\partial q) + (\partial\dot{p}/\partial p) \,] = \sum [\, (\partial^2\mathcal{H}/\partial q\partial p) - (\partial^2\mathcal{H}/\partial p\partial q) \,] \equiv 0 \; .$$

The *diagonal* second derivatives of $\mathcal{H}$, $(\partial^2\mathcal{H}/\partial q^2)$ and $(\partial^2\mathcal{H}/\partial p^2)$, provide two possible definitions for temperature, configurational and kinetic. They both follow from Gibbs' canonical distribution by an integration by parts :

$$kT\int_{-\infty}^{+\infty}(\partial^2\mathcal{H}/\partial q^2)e^{-\mathcal{H}/kT}dq \equiv \int_{-\infty}^{+\infty}(\partial\mathcal{H}/\partial q)^2 e^{-\mathcal{H}/kT}dq \; ;$$

$$kT\int_{-\infty}^{+\infty}(\partial^2\mathcal{H}/\partial p^2)e^{-\mathcal{H}/kT}dp \equiv \int_{-\infty}^{+\infty}(\partial\mathcal{H}/\partial p)^2 e^{-\mathcal{H}/kT}dp \; .$$

Dividing each of these equations by the appropriate normalization integral, $\int e^{-\mathcal{H}/kT}(dp \text{ or } dq)$, gives the Landau-Lifshitz' formulæ for the kinetic and the configurational temperatures :

$$kT_K \equiv \langle\ (p/m)^2\ \rangle/\langle\ (1/m)\ \rangle = \langle\ (p^2/m)\ \rangle\ ;\ kT_\Phi \equiv \langle\ F^2\ \rangle/\langle\ \nabla^2\Phi\ \rangle\ .$$

The configurational temperature formula is a bit disquieting. Although $(\partial^2\mathcal{H}/\partial q^2)$ is often positive (as in the harmonic oscillator example) it can also vanish or take on *divergent* and *negative* values (as in a sinewave potential or the headon version of the scattering problem of **Figure 6.1**) .

It is amusing that Campisi's "thermostat" of the previous Section 6.5 *always* has a negative configurational temperature in one dimension and a divergent one in two dimensions. In addition, the lack of a *simple* configurational thermometer, analogous to our ideal-gas kinetic-temperature thermometer confirms our caution flag.

By constraining the *kinetic* energy to a fixed value we can simulate an *isothermal* system. In the work described at the Boulder Conference, "Nonlinear Fluid Phenomena", in Physica **118A** , Hoover and Leete showed that the Lagrangian and Hamiltonian formulations of the isokinetic constraint give equivalent dynamical formulations. The Hamiltonian approach is the simpler one :

$$\mathcal{H}_{\rm HL} = 2\sqrt{K_0 K(p)} + \Phi(q) - K_0 \longrightarrow$$

$$\{\ \dot q = (p/m)\sqrt{K_0/K(p)}\ ;\ \dot p = F(q)\ \}\ .$$

$$K_0 = \sum(m\dot x^2 + m\dot y^2)/2\ ;\ K(p) = \sum(p_x^2 + p_y^2)/2m\ .$$

Here K_0 characterizes a fixed temperature $T_K = (2K_0/NDk)$ in D dimensions, where k is Boltzmann's constant. Not just the kinetic energy is fixed. The Hamiltonian is also a constant of the motion. The unusual and unphysical act of fixing volume, energy, *and* temperature violates the thermodynamic principle that two state variables, (E,V) *or* (T,V) uniquely define the state of a fixed amount of material. So long as (E,T,V) are close to consistent with the equation of state this apparent contradiction presents no real difficulties for equilibrium simulations.

But *nonequilibrium* simulations based on the isokinetic Hamiltonian give strange and paradoxical results, well-documented for the ϕ^4 model described in Chapter 7. There is no problem thermostating two sets of particles at different kinetic temperatures, with a Newtonian region linking the thermostated regions. See **Figure 6.9** . With the *kinetic* temperatures fixed in two of the three regions it turns out that the *configurational* temperature equilibrates throughout the system but at the Newtonian value, not necessarily near either of the kinetic reservoir temperatures. There is no tendency whatever for heat to flow from the hot to the cold region! This odd unphysical behavior suggests that there is very little coupling between the two Hamiltonian-based versions of kinetic temperature. Why is this?

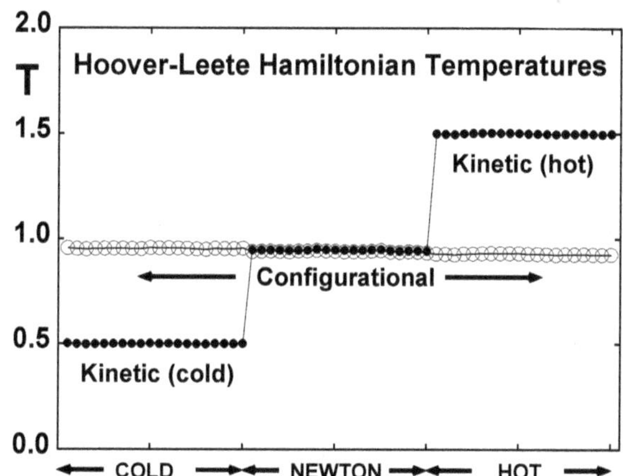

Fig. 6.9 Two kinetic-temperature reservoir regions fail to promote heat flow through a 60-particle ϕ^4 chain. 20 Newtonian particles are sandwiched between 20-particle "cold" and "hot" reservoirs. The temperature profile is an average over 10^9 Runge-Kutta timesteps with $dt = 0.001$. The configurational temperature and kinetic temperature agree in the Newtonian region but do not within either of the "reservoirs", showing yet again the inadequacy of Hamiltonian heat reservoirs.

An observation that offers at least a partial explanation can be found in the phase-space continuity equation as applied to *successful* simulations of nonequilibrium heat flow. Provided that time-reversible *nonHamiltonian* thermostats are used the phase-space probability density invariably concentrates on a fractional-dimensional attractor, with the phase volume continually decreasing in the steady state ,

$$[\langle \dot{\otimes} \rangle < 0] \longrightarrow [\otimes \rightarrow 0] \, .$$

By contrast Hamiltonian mechanics is unable to change phase volume , $[\dot{\otimes}_{\mathcal{H}} \equiv 0]$, and so is intrinsically unable to transfer energy from a "hot" region to a "cold" one. That Hamiltonian mechanics cannot successfully simulate nonequilibrium steady states is an interesting, initially surprising, but fundamentally logical, finding. To gain perspective let us next consider the parallel situation with two configurational temperatures, rather than two kinetic, one constrained to be "cold" and the other "hot".

6.7 Landau-Lifshitz' Configurational Thermostat

The configurational temperature from Landau and Lifshitz' 1952 textbook *Statistical Physics* ,

$$kT_{\Phi} = \langle\ F^2\ \rangle / \langle\ \nabla^2 \mathcal{H}\ \rangle\ ,$$

can be constrained by imposing a holonomic (coordinate-dependent) constraint with straightforward Lagrangian mechanics. The simulation's initial conditions need to chosen to correspond to the correct value of T_{Φ} with the additional requirement that the time derivative $\dot{T}_{\Phi}$ vanishes. The simplest choice imposing $\dot{T}_{\Phi}(t = 0) = 0$ is to set all the system's velocities equal to zero. Then, with T_{Φ} at its desired value and $\dot{T}_{\Phi} = 0$ *isoconfigurational-temperature* dynamics proceeds by constraining $\ddot{T}_{\Phi}$ to zero with a Lagrange multiplier. The detailed programming has a nightmare quality which by itself would be sufficient to change most researchers' goals. Once the Lagrange multiplier has been found the simulation is easy to check, in that both $\mathcal{H}$ and T_{Φ} are constants of the motion.

Karl Travis and Carlos Braga suggested a much simpler "weak-form" implementation of isoconfigurational dynamics. They used integral feedback to impose a *long-time-averaged* configurational temperature rather than an instantaneous one. In their 2008 Journal of Chemical Physics paper, "Configurational Temperature Control for Atomic and Molecular Systems" they used this idea in a successful simulation of an n-decane fluid ($C_{10}H_{22}$) . Their algorithm can be expressed as follows :

$$\dot{q} = p - \zeta F\ ;\ \dot{p} = F\ ;\ \dot{\zeta} \propto [\ F^2 - kT\nabla^2\mathcal{H}\ ]\ ,$$

where the proportionality constant in the definition of the frictional force determines the stiffness of the thermostat variable.

Kinetic temperature is defined and measured in a comoving and corotating frame. An apparent advantage of configurational temperature is that it avoids the need to determine a local flow velocity. But a corollary *disadvantage* can be seen in problems involving rotation. The centrifugal forces would need to be estimated and removed in order to avoid rotation's generating a spurious temperature gradient. The application of hot and cold Hamiltonian constraints on $\ddot{T}_{\Phi}$ to a four-chamber heat flow problem resembles the Hoover-Leete constraints on hot and cold kinetic temperatures T_K. If two of four chambers (periodic in both directions with two thermostated and two Newtonian chambers) have their configurational temperatures fixed then the entire periodic four-chamber system comes to a kinetic-temperature equilibrium, with all four chambers' $\{ T_K \}$ having the same (time-averaged) value. As before, the lack of coupling between the kinetic and configurational temperatures is striking, and argues for restricting configurational temperature to a diagnostic rôle rather than promoting it to service as a control variable.

6.8 Feedback — Gaussian and Nosé-Hoover Thermostats

The Nosé, Campisi, Hoover-Leete, and Landau-Lifshitz Hamiltonian thermostats are all abject failures in modelling heat flow. All of these failures can be rationalized through Liouville's Theorem, which expresses the inability of Hamiltonian systems to show the phase-volume changes associated with heat flow. It is therefore curious that the two most successful thermostats have (somewhat unconventional) Hamiltonian relatives.

Les Woodcock pointed out that a simple rescaling of the particle velocities can maintain a constant kinetic energy. This rescaling, once every timestep :

$$\{ \ v_i(t+dt) \to v_i(t+dt) \times \sqrt{K_0/K(t+dt)} \ \} \ ,$$

can be implemented by adding a continuous feedback variable better suited to analysis. This continuous rescaling is different to the Hoover-Leete rescaling in that it is most simply expressed in terms of a friction coefficient ζ which introduces or extracts energy to keep the kinetic energy of a selected set of degrees of freedom constant :

$$\{ \ \dot{q} = (p/m) \ ; \ \dot{p} = F - \zeta p \ \} \ ; \ \zeta = \sum (F \cdot p) / \sum p^2 \ .$$

The resulting change in kinetic energy vanishes :

$$\dot{K} = \sum (p \cdot \dot{p}/m) = \sum (p/m) \cdot (F - \zeta p) = \sum (p/m) \cdot F - \sum F \cdot (p/m) \equiv 0 \ ,$$

and the equations of motion remain time-reversible, with the control variable ζ and all the momenta $\{ \ p \ \}$ changing sign in a time-reversed trajectory.

Rather than insisting on instantaneous control of the kinetic energy, "isokinetic mechanics", Nosé's mechanics or the Nosé-Hoover modification of it allow for fluctuations in kinetic temperature. In fact the time-dependent kinetic energy $K_t = K_p$ for any desired set of degrees of freedom can be thermostated with a relaxation time τ selected for that set :

$$\{ \dot{q} = (p/m) \; ; \; \dot{p} = F - \zeta p \} \; ; \; \dot{\zeta} = [\, (K_t/K_0) - 1 \,]/\tau^2 \; .$$

Dettmann and Morriss showed that both the isokinetic and the Nosé-Hoover forms of temperature control can be derived from special zero-valued Hamiltonians. Evidently this trick cannot be applied to systems in which two or more temperatures are specified, as in the three- and four-chamber heat flow problems used to demonstrate the failure of conventional Hamiltonian thermostats. But from the pedagogical standpoint it is interesting to see that there is indeed a tenuous connection between Hamiltonian mechanics, with $\mathcal{H} \equiv 0$, and useful thermostats.

6.9 Two-Moment, Three-Moment, and Chain Thermostats

The lack of ergodicity for the Nosé and Nosé-Hoover harmonic oscillator problems shows the need for *judgment* in simulation. The better-behaved Nosé-Hoover equations of motion for the oscillator coordinate q, momentum p, and friction coefficient ζ are :

$$\{ \dot{q} = p \; ; \; \dot{p} = -q - \zeta p \; ; \; \dot{\zeta} = p^2 - 1 \} \; .$$

Here we arbitrarily choose a thermostat-variable temperature of unity. The harmonic-oscillator solutions turn out to have infinitely many periodic orbits, the simplest of which has initial conditions $(q, p, \zeta) = (0, 1.55, 0)$. Although *all* the periodic orbits necessarily reproduce the second moment, $\langle \, p^2 \, \rangle = 1$, the other time-averaged even moments, $\langle \, p^4, \; p^6, \; \ldots \, \rangle$, are in error for the canonical ensembles (and are far too small for the example orbit shown in **Figure 6.10**) .

Because distributions *are* characterized by their moments it is natural to consider controlling more of them. The fourth and sixth moments can be controlled by adding on two new friction coefficients ξ and ς :

$$\dot{q} = p \; ; \; \dot{p} = -q - \zeta p - \xi p^3 - \varsigma p^5 \; ;$$

$$\dot{\zeta} = p^2 - 1 \; ; \; \dot{\xi} = p^4 - 3p^2 \; ; \; \dot{\varsigma} = p^6 - 5p^4 \; .$$

The stationary canonical distribution for this three-moment problem is a five-dimensional Gaussian :

$$f(q, p, \zeta, \xi) \propto e^{-q^2/2} e^{-p^2/2} e^{-\zeta^2/2} e^{-\xi^2/2} e^{-\varsigma^2/2} \; .$$

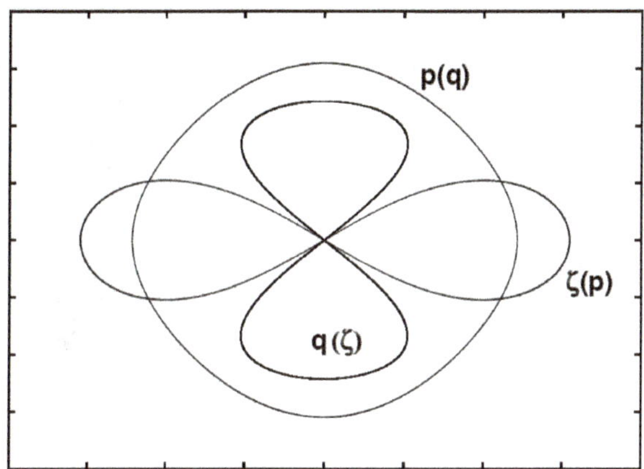

Fig. 6.10 The three projections of the simplest of the equilibrium Nosé-Hoover-oscillator periodic orbits, with period 5.5789 and $\langle\ (p^2/1),\ (p^4/3),\ (p^6/15)\ \rangle = 1.0000,\ 0.5696,$ and 0.2227 . Initially $(q,p,\zeta) = (0.00, 1.55, 0.00)$. In 1986 Posch, Vesely, and Hoover reported an oscillator period of 5.58 for this orbit. See their "Canonical Dynamics of the Nosé Oscillator: Stability, Order, and Chaos", Physical Review A **33** , 4253-4265 . The ordinate and abscissa scales chosen here both cover the range from -2 to $+2$.

As usual we can verify the solution by applying the phase-space flow equation for a stationary state :

$$(\partial f/\partial t) = -f[\ (\partial\dot{q}/\partial q) + (\partial\dot{p}/\partial p) + (\partial\dot{\zeta}/\partial\zeta) + (\partial\dot{\xi}/\partial\xi) + (\partial\dot{\varsigma}/\partial\varsigma)\ ]$$

$$-\dot{q}(\partial f/\partial q) - \dot{p}(\partial f/\partial p) - \dot{\zeta}(\partial f/\partial\zeta) - \dot{\xi}(\partial f/\partial\xi) - \dot{\varsigma}(\partial f/\partial\varsigma) \equiv 0 =$$

$$= f[\ \zeta + 3p^2\xi + 5p^4\varsigma\ ] +$$

$$f[\ q(p) + p(-q - \zeta p - \xi p^3 - \varsigma p^5) + \zeta(p^2 - 1) + \xi(p^4 - 3p^2) + \varsigma(p^6 - 5p^4)\ ]\ .$$

With two or three control variables, and in a four-dimensional or five-dimensional phase space, the solutions of the multi-moment oscillator problems are qualitatively different to the simpler Nosé-Hoover oscillator equations. Consider again the two-moment Hoover-Holian nonequilibrium oscillator problem of **Figure 5.7** on page 121 . **Figure 6.11** shows the variation with the magnitude of the temperature gradient of the (q,p) plane section with $(\zeta,\xi) = (0,0)$. The structures are amazingly intricate, so much so that so far we have no names fitting the various features displayed in the **Figure** . Piotr Pieranski kindly pointed out the similarity of these figures to Stanislaw Witkiewicz' portraits ! Here both the second and the fourth velocity moments have been constrained in the presence of two different temperature profiles, with maximum gradients $\epsilon = 0.20$ and 0.30 .

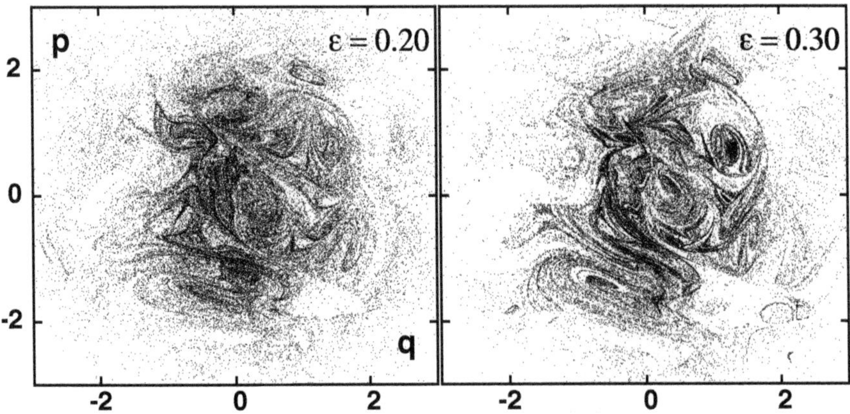

Fig. 6.11 Hoover-Holian nonequilibrium (q,p) sections in the four-dimensional phase space where both $\{ p^2, p^4 \}$ are controlled. In this plot $|\ \zeta\ | < 0.001, |\ \xi\ | < 0.001, \epsilon = 0.20$ and 0.30 . For more details see arXiv:1401.1762 or Physical Review E **89**, 042914 (2014).

With three thermostat variables the time averages of the $\dot{\zeta}$, $\dot{\xi}$, and $\dot{\varsigma}$ equations establish that the second, fourth, and sixth moments can all be reproduced *provided that the equations of motion can be solved.* The even moments of the Gaussian distribution are products of the odd integers :

$$\langle\ p^2\ \rangle = 1\ ;\ \langle\ p^4\ \rangle = 1 \times 3\ ;\ \langle\ p^6\ \rangle = 1 \times 3 \times 5\ ;\ \langle\ p^8\ \rangle = 1 \times 3 \times 5 \times 7\ \ldots\ .$$

Working with Harald Posch and Brad Holian in 1995 we concluded that simultaneous control of all three moments was either impossible or impractical, even with tiny Runge-Kutta timesteps. More recently we reinvestigated these problems and found that the three-moment equations *can* be solved by using the adapative hybrid fourth-and-fifth-order Runge-Kutta program described in Chapter 5. In fact there is no real need for such a three-moment algorithm. All three of the two-moment approaches provide good five-figure values of $\langle\ p^2,\ p^4,\ p^6\ \rangle$. Controlling just the second and fourth moments is considerably less demanding than using any of the algorithms controlling $\langle\ q^6\ \rangle$ or $\langle\ p^6\ \rangle$. Control of just a single moment is not enough, and produces complexity of the Nosé-Hoover type, with both chaotic and toroidal solutions. Here are the chaotic-solutions' moments :

$$\langle\ p^2\ \rangle = 1 \text{ controlled} \longrightarrow 1.00,\ 4.78,\ 39.34\ ;$$

$$\langle\ p^4\ \rangle = \langle\ 3p^2\ \rangle \text{ controlled} \longrightarrow 0.96,\ 2.89,\ 14.81\ ;$$

$$\langle\ p^6\ \rangle = \langle\ 5p^4\ \rangle \text{ controlled} \longrightarrow 1.05,\ 3.22,\ 16.09\ .$$

An alternative approach to achieving harmonic-oscillator ergodicity uses a "chain" thermostat, introduced by Martyna, Klein, and Tuckerman. Let us illustrate this idea for the same harmonic oscillator problem :

$$\dot{q} = p \ ; \ \dot{p} = -q - \zeta p \ ; \ \dot{\zeta} = p^2 - 1 - \xi\zeta \ ; \ \dot{\xi} = \zeta^2 - 1 \ .$$

This set of four equations is time-reversible, with solutions forward and backward in time existing in pairs :

$$\{ +q, +p, +\zeta, +\xi, +t \} \longleftrightarrow \{ +q, -p, -\zeta, -\xi, -t \} \ .$$

There are two fixed points, with $(qp) = (00)$ and $\pm\zeta = -1 = \mp\xi$. In the $(qp) = (00)$ plane these oscillator equations, though reversible, are isomorphic to the motion of a ball in a gravitational field, with one control variable (ξ) playing the rôle of friction and the other (ζ) the rôle of velocity.

No matter the initial condition, a ball subject to gravity eventually falls "down". Reversing the time and sending a ball "up" simply results in a repeat of the original trajectory. The two-dimensional flows corresponding to the Martyna-Klein-Tuckerman oscillator equations and the Nosé-Hoover gravitational problem are shown in **Figure 6.12** :

$$\{ \dot{\zeta} = -1 - \xi\zeta \ ; \ \dot{\xi} = \zeta^2 - 1 \} \longleftrightarrow \{ \dot{p} = -1 - \zeta p \ ; \ \dot{\zeta} = p^2 - 1 \} \ .$$

Returning to the oscillator problem we find that the final value of the sixth moment, 15.03 , suggests that the chain approach converges more rapidly that the moment approach, while clearly the opposite is true. The chain approach can be extended, adding even more thermostat variables. The three-thermostat version is :

$$\dot{q} = p \ ; \ \dot{p} = -q - \zeta p \ ; \ \dot{\zeta} = p^2 - 1 - \xi\zeta \ ; \ \dot{\xi} = \zeta^2 - 1 - \varsigma\xi \ ; \ \dot{\varsigma} = \xi^2 - 1 \ .$$

The sixth moment is significantly higher. Increasing the run length by a factor of ten provides a sixth moment of 15.08 , somewhat better. Chain thermostats have a defect when applied to nonequilibrium simulations. The measured kinetic temperature deviates from the target value when $\langle \ \xi\zeta \ \rangle$ is nonzero. By contrast, the Nosé-Hoover thermostat returns the correct kinetic temperature at, or away from, equilibrium.

At equilibrium the MKT Chain thermostats can be extended to arbitrary lengths while the moment method is best-suited to control of only one or two moments. The three-moment method applied to a harmonic oscillator requires an adaptive integrator as well as a much smaller average timestep (smaller by a factor of six in our exploratory calculation). The harmonic oscillator is an excellent trial problem for new thermostat

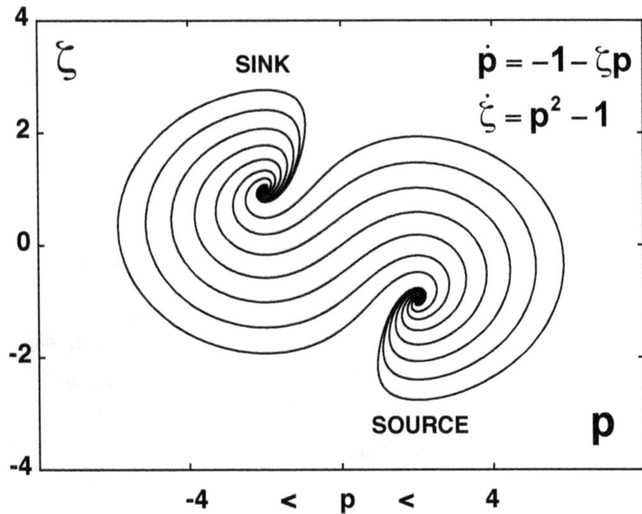

Fig. 6.12 The Martyna-Klein-Tuckerman oscillator fixed points and the Nosé-Hoover gravitational problem are isomorphic. The falling-ball problem includes the extraneous "vertical" coordinate q : $\{ \dot{q} = p \ ; \ \dot{p} = -1 - \zeta p \ ; \ \dot{\zeta} = p^2 - 1 \}$. Eventually the ball "falls" and the coordinate approaches $-\infty$. The trajectories in the figure start near the unstable fixed point $(p, \zeta) = (+1, -1)$ and flow toward the stable fixed point $(-1, +1)$. The Martyna-Klein-Tuckerman equations near the $(qp) = (00)$ plane in (q, p, ζ, ξ) space are isomorphic to the falling-ball problem with $(\xi, \zeta) \longleftrightarrow (\zeta, p)$.

ideas. We have made no attempt here to replicate the general treatment of thermostats as is so well described by Bauer, Bulgac, and Kusnezov. The variety of forms and relaxation times which could be associated with these ideas will likely reveal much more interesting physics. We recommend exploration of these problems to students.

6.10 Time-Reversible Stochastic Thermostats ?

"Stochastic" calculations are unlike all of the other thermostated approaches considered in this Chapter. They are (almost) *never* time-reversible. The prototypical stochastic motion equation, Langevin's, contains an irreversible drag force, $-(p/\tau)$, as well as a random "fluctuating force". Reversing the drag force leads immediately to divergence, which we must reject.

An approach called "Brownian Dynamics" is more promising from the standpoint of reversibility. In Brownian dynamics particle coordinates are incremented *randomly* with a mental picture of diffusion described by a diffusion coefficient D ,

$$\langle\ \delta_x^2\ \rangle = 2Ddt^2\ ;\ \delta_x \propto \mathcal{R} - (1/2)\ .$$

Most random number generators discard information and lack time reversibility. But, in the spirit of Levesque and Verlet, it *is* possible, by using integer arithmetic, to generate *time-reversible* random number generators. We give only one example here, courtesy of Professor Federico Ricci-Tersenghi (Università La Sapienza, Rome). He began with a simple and useful generator which generates $2^{22} = 4\ 194\ 304$ distinct pseudorandom numbers from the two 11-bit seeds, `intx` and `inty`, with each seed restricted to the range [0,2047] , and in the process generates the two seeds for the next pseudorandom number. This portion of the random number generator is lines 4-9 in the FORTRAN fragment below.

Federico won the 2013 Snook Prize by finding the *inverse* of this generator, going *backward* from the current seeds to the previous pair `(oldx,oldy)` as given in lines 11-14 below. This approach allows one to progress or regress through the pseudorandom numbers in either time direction, and can be used to generate time-reversible stochastic algorithms. Most stochastic simulation techniques are *intrinsically* irreversible. For example, the divergence implied by a negative friction coefficient ruled out considering a time-reversed Langevin equation.

On the other hand an equilibrium Monte Carlo simulation *could* be made time-reversible, and could conceivably provide a useful method for investigating heat flow by considering a coordinate-dependent temperature field. The time reversal of a simple random-number generator is described in a pair of arχiv papers, 1305.1805 and 1305.0961 , which detail the Ian Snook Memorial Challenge for 2013 and its solution by Ricci-Tersenghi. Here is the forward-in-time generator, lines 1-9 of the following fragment :

```
implicit real*8 (r)
read(5,*) intx,inty
write(6,*) intx,inty
i = 1029*intx + 1731
j = i + 1029*inty + 507*intx - 1731
intx = mod(i,2048)
j = j + (i - intx)/2048
inty = mod(j,2048) + 4194304
rund = (intx + 2048*inty)/4194304.0
```

```
write(6,*) intx,inty,rund
oldx = mod(205*intx + 1497,2048)
inty = inty - 1536*oldx - (1029*oldx + 1731 - intx)/2048
inty = mod(205*inty,2048)
intx = oldx
write(6,*) intx,inty
```

Lines 11-14 give the *previous* `intx` and `inty` which can then be used to compute the *preceding* pseudo random number,

$$\texttt{rund} = (\texttt{intx} + 2048 * \texttt{inty})/4194304.0 \ .$$

Let us use this approach to generate a time-reversible Brownian Dynamics.

6.10.1 *Time-Reversible Brownian Dynamics !*

As a demonstration of the utility of Ricci-Tersenghi's reversible generator, we consider an ensemble of 10,000 Brownian walkers evolving over 1000 timesteps with randomly chosen displacements in the range :

$$-(1/2) < \mathcal{R} < +(1/2) \longrightarrow \int_{-(1/2)}^{+(1/2)} \mathcal{R}^2 d\mathcal{R} = (1/12) \ .$$

$$\mathcal{R} = \texttt{rund(intx, inty)} - 0.5\texttt{d00} \ .$$

To reverse the motion (as shown in **Figure 6.13**) it is only necessary to process the ten million random numbers used in generating the walks in reversed order. Two points need attention. A single extra forward random number needs to be generated before reversal. Also, in the time-reversed version of the dynamics the sums over particles $1 \longrightarrow N$ forward in time need to be reversed in order, $N \longrightarrow 1$. With these two precautions the dynamics is precisely reversible, as shown in **Figure 6.13**. The long-time theoretical slope, $\langle \ q^2 \ \rangle/t = (1/12)$, closely describes the numerical data. The distribution of the 10 000 squared walk lengths after 1000 steps likewise closely resembles the analytic most-likely Gaussian distribution shown in the righthand panel of the figure :

$$f(q^2)dq = 10\ 000\sqrt{6/1000\pi}\exp[\ -12q^2/2000\ ]dq \ .$$

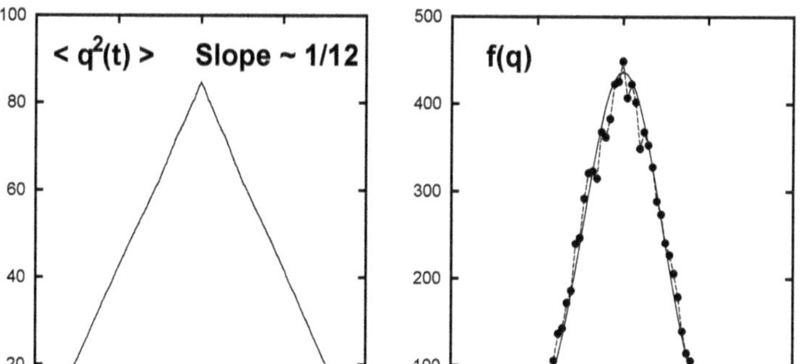

Fig. 6.13 The time-reversible random number generator was used to simulate the motion of ten thousand random walkers for one thousand timesteps. The mean-squared displacement, shown on the left, is accurately linear in time, with the theoretical slope, $\langle\ [\ \sum q_i\ ]^2\ \rangle \longrightarrow \langle\ \sum q_i^2\ \rangle = (t/12)$. After reversal the precise decrease back to $q^2 = 0$ shows the perfect bit-reversibility of the generator. The right panel shows the good agreement (at time 1000) between the numerical histogram, points for a bin-width of unity, and the analytic Gaussian distribution for a continuous walk.

6.11 Ergostats, Barostats , . . .

The energy or enthalpy or stress or heat flux of selected degrees of freedom can be controlled in much the same way as temperature. A simple problem, the harmonic oscillator, illustrates both the general idea and some of the pitfalls. Consider the simplest imaginable problem, keeping the mean energy of a harmonic oscillator equal to unity. Of course this problem in Hamiltonian mechanics has analytic solutions, all of which keep the energy constant. The simplest example is :

$$\mathcal{H} = (1/2)(q^2 + p^2) \longrightarrow q(t) = +\cos(t)\ ;\ p(t) = -\sin(t)\ ;\ E \equiv (1/2)\ .$$

But, ignoring our knowledge that Hamiltonian mechanics conserves energy we could instead blindly apply integral control with a characteristic "ergostat" (constant energy) time τ :

$$\dot{q} = +p\ ;\ \dot{p} = -q - \zeta p\ ;\ \dot{\zeta} = (q^2 + p^2 - 1)/\tau\ ,$$

and "see what happens". Varying the initial conditions and the response time τ can give a wide variety of results, none of them "wrong", but all of them unnecessarily complicated relative to the analytic solution above.

In real applications of molecular dynamics there is no analytic solution. The motion is chaotic, so that judgment is required. Simplicity, efficiency, transparency, and faithfulness to the macroscopic description are desirable features of numerical work. This said, it is feasible to construct algorithms consistent with Gibbs' "other" ensembles, such as the "isobaric" (constant-pressure) ensemble, in which the volume V is a variable, responding to differences between the instantaneous and desired values of the pressure. Unless these simulations have a straightforward connection to macroscopic physics (such as the periodic compression of a fluid by a sound wave) the resulting formalism is of little lasting interest.

To summarize, control of temperature and velocity are basic methods for generating nonequilibrium states in a useful and reproducible way. The "Gaussian isokinetic" control of kinetic temperature has proved useful in many simple nonequilibrium simulations. Nosé-Hoover temperature control, with a finite, rather than vanishing, characteristic time for thermal fluctuations is enough for most applications. Despite the lack of ergodicity for a single Nosé-Hoover harmonic oscillator, ergodicity appears *not* to be an issue for many-body problems. In any event, ergodicity is seldom worth taking seriously, due to the enormous times required for Poincaré recurrence in systems of more than a very few degrees of freedom.

6.12 Summary

Shuichi Nosé's 1984 work was revolutionary for statistical mechanics and the thermostats discussed here represent the legacy of his ideas. Nosé himself had little interest in extending his ideas to include nonequilibrium simulations. But thermostats or ergostats are essential to the simulation of nonequilibrium steady states.

It *is* interesting that the useful control algorithms, like classical mechanics, are all of them time-reversible. Thermostats based on kinetic energy follow the lead of kinetic theory, basing temperature on mechanical collisions. Although one can *define* a configurational temperature analogous to kinetic temperature (based on $\nabla\mathcal{H}$ and $\nabla^2\mathcal{H}$) there is no accompanying physical picture of a thermometer to measure it. Configurational temperature can be negative or zero.

Hybrid thermostats can control two or more characteristics of Gibbs' canonical distribution. Their construction follows the principles derived from the phase-space continuity equation, as laid down by Bauer, Bulgac, and Kusnezov. It appears that two-moment thermostats are sufficient to control a harmonic oscillator's canonical distribution. Although stochastic methods are for the most part inherently irreversible, because they incorporate an explicit damping, Brownian dynamics *can* be reversed by using a time-reversible random number generator.

6.13 References

Although his methods are unnecessarily complicated Nosé's two 1984 papers are fascinating reading: "A Unified Formulation of the Constant Temperature Molecular Dynamics Methods" in the Journal of Chemical Physics and "A Molecular Dynamics Method for Simulations in the Canonical Ensemble" in Molecular Physics. Our recent World Scientific, Singapore (2012) book, *Time Reversibility, Computer Simulations, Algorithms, Chaos* is a good source of references to the construction and use of analytic thermostats. More recent progress can be tracked on the internet.

The papers generated by Campisi's work are reviewed on the arχiv and in Sponseller and Blaisten-Barojas' article, "Failure of Logarithmic Oscillators to Thermostat Small Atomic Clusters", Physical Review E **89**, 021301R (2014). See also Marc Meléndez Schofield's 2014 PhD thesis "The Theory of Coarse-Graining Without Projection Operators" at the Universidad Nacional de Educación a Distancia in Madrid.

The two Annals of Physics papers on "Canonical Ensembles from Chaos" (1990 and 1992) by Bauer, Bulgac, and Kusnezov and the work carried out by Brańka and Wojciechowski are valuable sources for earlier work. Because it is far and away our most cited work we also mention "Canonical Dynamics: Equilibrium Phase-Space Distributions", Physical Review A **31** 1695-1697 (1985) as well as its much more detailed sequel with Franz Vesely and Harald Posch, "Canonical Dynamics of the Nosé Oscillator: Stability, Order, and Chaos", Physical Review A **33**, 4253-4265 (1986).

6.14 Problems

1. Start with a one-dimensional ideal-gas thermometer, a Maxwell-Boltzmann gas composed of particles with mass m and a Gaussian distribution of velocities $\{ \dot{x} \}$. Analyze the collision of a mass-M and velocity $\dot{X}$ particle with the thermometer. That is, consider the conservation of mo-

mentum and energy for a particle/thermometer collision. Using the facts that $\langle\ \dot{x}\ \rangle$ vanishes and that $\langle\ m\dot{x}^2\ \rangle \equiv kT$, show that when the averaged energy change of the colliding particle vanishes :

$$\langle\ M\dot{X}^2\ \rangle_{\text{after}} = \langle\ M\dot{X}^2\ \rangle_{\text{before}} \longrightarrow \langle\ M\dot{X}^2\ \rangle = \langle\ m\dot{x}^2\ \rangle = kT\ .$$

2. Show that the configurational temperature, $kT_C \equiv \langle\ F^2\ \rangle / \langle\ -F'\ \rangle$, diverges for headon velocities of $\dot{x} = \pm 18/49$ where the pair potential is $\phi(r<1) = (1-r^2)^4$.

3. Write an *eight*-equation Runge-Kutta computer program solving both Nosé's original set of four harmonic oscillator equations,

$$\{\ \dot{q} = (p/s^2)\ ;\ \dot{p} = -q\ ;\ \dot{s} = p_s\ ;\ \dot{p}_s = (p^2/s^3) - (kT/s)\ \}$$

and the four additional "scaled" oscillator equations :

$$\{\ \dot{q} = (p/s)\ ;\ \dot{p} = -sq\ ;\ \dot{s} = sp_s\ ;\ \dot{p}_s = (p/s)^2 - kT\ \}$$

Initial conditions close to periodic orbits for both sets are $(q,p,s,p_s) = (0.00, 1.55, 1.00, 0.00)$, and give periods of 7.20 and 5.58 for the two versions of the oscillator equations. Confirm that the two (q,p) trajectories are identical, apart from their dependence on time.

4. Repeat the calculation of Figure 6.4 on page 158 using $\dot{\zeta} = [\ (p^2/T) - 1\ ]$ rather than $\dot{\zeta} = [\ p^2 - T\ ]$ where $T = T(q) = 1 + \epsilon \tanh(q)$.

5. Solve the Patra-Bhattacharya oscillator equations with initial conditions $(q,p,\zeta,\xi) = (0,5,0,0)$:

$$\{\ \dot{q} = p - \xi q\ ;\ \dot{p} = -q - \zeta p\ ;\ \dot{\zeta} = [\ p^2 - 1\ ]\ ;\ \dot{\xi} = [\ q^2 - 1\ ]\ \}\ ,$$

and plot 200,000 $p(q)$ points. Add a fifth differential equation, `yyp(5) = q*q*p*p` , to the set of four so that $\langle\ q^2p^2\ \rangle \equiv$ `yy(5)/time` , confirming the strong correlation between coordinate and momentum mentioned in the text.

6. Show that the two fixed points for the Tuckerman chain equations,

$$\{\ \dot{q} = p\ ;\ \dot{p} = -q - \zeta p\ ;\ \dot{\zeta} = p^2 - 1 - \xi\zeta\ ;\ \dot{\xi} = \zeta^2 - 1\ \}\ ,$$

have opposite stabilities with one acting as a source and the other a sink for the flow in the (ζ,ξ) plane. Consider also the flow in the (q,p) plane with (ζ,ξ) at their fixed-point values. Show that the two flows act in unision to generate and maintain an overall Gaussian distribution in the four-dimensional space. This example is the subject of the Snook Prize problem.

Chapter 7

Key Results from Nonequilibrium Simulations

Topics

Nonequilibrium Simulations / Periodic Shear with Doll's and Sllod Algorithms / Boundary-Driven Shear Flows / Green-Kubo and Homogeneous Simulations of Periodic Heat Flow / Boundary-Driven Simulations of Heat Flow / The ϕ^4 Heat Flow Model / Phase-Space Attractor Dimensionality Loss / Mesoscopic Reversibility with Smooth Particles / Shockwaves with Smooth-Particle Representations / Tensor Temperature / Time Delay / Continuum Models for Shockwave Structure /

7.1 Prototypical Nonequilibrium Simulations

The great variety of nonequilibrium systems (including the Universe and everything in it!) suggests the usefulness of small-scale benchmark simulations. Both the models chosen and the numerical methods used to simulate them need to be readily reproducible. The models need to include the physical effects of interest in as simple a context as possible. In this Chapter we formulate prototypical shear flow and heat flow models for the nonequilibrium transfer of momentum and energy. We show how to analyze these simulations so as to estimate the dimensionality of their corresponding phase-space strange attractors. We also implement shockwave simulations, which include nonequilibrium flows of mass, momentum, and energy and which underline the importance of *nonlinear* transport effects. Before considering these nonequilibrium simulations in detail we can review their emergence from the equilibrium and nonequilibrium simulations pioneered a half century ago by Fermi, Pasta, and Ulam, by Alder and Wainwright, and by Gibson, Goland, Milgram, and Vineyard.

7.1.1 *Early Equilibrium Simulations*

Equilibrium many-body simulations began in the decade after the Second World War. At Los Alamos in 1952 Enrico Fermi planned a variety of interesting applications described in Stanislaw Ulam's comments on their paper with John Pasta, "Studies of Nonlinear Problems". That work showed clearly that one-dimensional anharmonic chains, at low enough energies, can easily avoid ergodic equilibration, instead returning closely, and relatively quickly, near to their initial states. One would likely say today that this is "obvious" because harmonic systems are far from ergodic. But in the early 1950s this was evidently a big surprise, even to Fermi.

Soon Berni Alder and Tom Wainwright pursued both equilibrium and nonequilibrium studies with hard disks and hard spheres at the Livermore Radiation Laboratory and found a very different behavior, opposite to the Fermi-Pasta-Ulam work. Hard disks and spheres approach equilibrium rapidly, in just a few collision times. Alder and Wainwright showed this by computing the evolution of Boltzmann's velocity-space "H function", a one-body ideal-gas approximation to Gibbs' many-body entropy :

$$H_{\rm Boltzmann} \propto S_{\rm Boltzmann} = -k\langle \ \ln f_1(p) \ \rangle \ ;$$

$$S_{\rm Gibbs} \equiv -k\langle \ \ln f_N(\ \{ \ (q,p) \ \} \ ) \ \rangle \ .$$

It is not quite so apt to characterize the rapid hard-sphere equilibration as "obvious", even though it is tempting, given the spreading of nearby trajectories for simple scattering problems, as was illustrated in **Figure 2.5** on page 36 . Sinai is credited with a "proof" of ergodicity, but the complexity of his mathematics makes Alder and Wainwright's numerical demonstration more convincing.

George Vineyard's group, at the Brookhaven National Laboratory, was interested in the propagation of radiation damage in metal crystals, looking at the details of the lattice defects formed during the slowdown of an initially high-energy particle. Although none of these three efforts (aside from the longer runs of Alder and Wainwright, which compared theoretical estimates of pressure to those from molecular dynamics simulations) was directed toward equilibrium properties, most of the work set out simply to solve Newton's equations of motion, $\{ \ F = ma \ \}$. Vineyard's simulations included special viscous damping of boundary particles' motions, tuned to absorb, rather than reflect, the maximum possible wave energy at the system boundaries (providing "quiet boundaries") .

In the 1960s, as large-scale computation grew and spread from the American National Laboratories into the Universities and to other countries, a much wider selection of problems came under investigation. In the area of statistical mechanics there were three main foci : first came checking the basic validity of Gibbs' statistical mechanics ; next Green and Kubo's Gibbs-based perturbation theory of transport coefficients ; and finally Maxwell and Boltzmann's low-density kinetic theory. Phase-space mixing and collisional "molecular chaos" resulting from the exponential growth of perturbations soon validated all three theoretical approaches.

In addition to these fundamental questions, chemists, and even biochemists, sought to understand particular materials. Detailed investigations of special ionic or molecular problems, involving water, sodium chloride, benzene, ... emerged. Materials scientists and engineers found problems useful to their specialities too : the structures of dislocations and their dynamics, as well as wave formation, propagation, and damping were all fair game for simulation. There was a parallel effort to understand the irreversible flow properties of solids subject to plastic deformation and fracture through shock and detonation waves.

A basic engineering tool, linear elastic theory, with singularities – divergent stresses and strains at dislocations, interfaces, and crack tips – was another obvious target for simulations. The paradox that the macroscopic equations describing diffusion, shear flow, and heat flow are *irreversible*, but predict rather well the results of *time-reversible* atomistic mechanics, came in for investigation too. By early Spring of 2014 Google listed fourteen million results for "molecular dynamics" searches, too much information for any single book-length survey to span. By mid-June the number of results had dropped below six million. In early September there were fewer than 900,000 results. In early October 700,000. Evidently budget cuts at Google must be influencing the outcome! Let us be content with the dozen or so problems which we can discuss in the course of a Chapter.

7.1.2 *Nonequilibrium Simulations*

Around 1970 an amazing number of scientists had the idea to use Gibbsian perturbation theory to estimate the Helmholtz free energy of liquids. The idea was to vary a "reference system" (often hard spheres, with a variable diameter) for which the distribution of pairs was known (often from an approximate fluid theory such as the Percus-Yevick integral equation for the pair distribution) , keeping track of the estimated free energy by treating

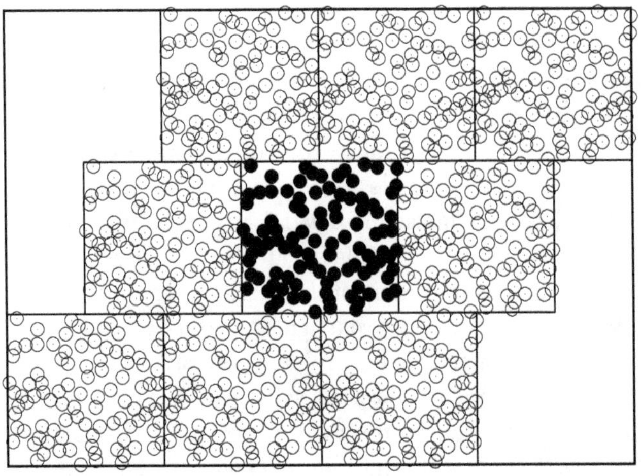

N = 100, Periodic Shear Flow

Fig. 7.1 A periodic 100-particle system with the motion in the central periodic cell driven by periodic images above and below. Eight moving periodic images of the central cell are shown for clarity. Only 400 ordinary differential equations are required to describe the motion. Either the internal energy or the kinetic temperature relative to the mean motion can be constrained by a friction coefficient. [This illustration is a copy of Figure 3 from our paper with Janka Petravic, Physical Review E **78**, 046701 (2008)] .

attractive forces as a perturbation. By choosing the *minimum* free energy (corresponding to the *maximum* phase volume) a very useful "theory" of liquids resulted. The success of this equilibrium theory was reviewed by John Barker and Doug Henderson in their 1976 Reviews of Modern Physics' "What is Liquid ? Understanding the States of Matter". That understanding caused our own research direction to change. Ever since we have concentrated on *nonequilibrium* systems, for which there is no corresponding model like Gibbs'. Instead we must rely on simulations.

Our own research interest in nonequilibrium simulations has never flagged since the 1970s. We have sought to simulate and model relatively simple flow processes using similarly simple force laws. Consequently we consider mostly two-dimensional example problems, from which algorithms for more complicated two- and three-dimensional systems follow easily. **Figure 7.1** illustrates a periodic *shear flow* driven by oppositely-moving periodic images of a 100-body unit cell. **Figure 7.2** illustrates a *heat flow* driven by the thermostated temperature difference between two chambers, one "hot" and one "cold". Both these "steady-state" problems are simpler to formulate and understand than are the many time-dependent inhomo-

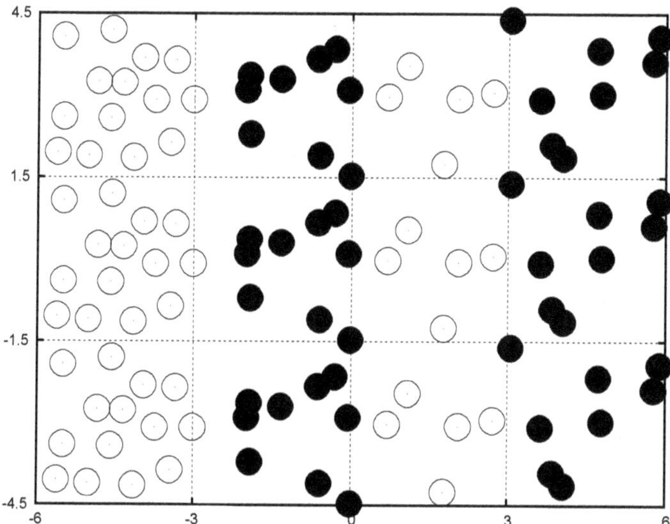

Fig. 7.2 Three images of a doubly-periodic 36-particle system with cold, Newtonian, hot, and Newtonian regions as we move from left to right. The particles are free to move throughout all four chambers. The cold and hot temperatures are 0.25 and 1.00 , controlled by two Nosé-Hoover thermostat variables, $\zeta_{\rm cold}$ and $\zeta_{\rm hot}$ with $\dot{\zeta} = \sum[\, p^2 - 2T \,]$ where the sum includes all particles currently occupying the appropriate chamber. The short-ranged pair potential is $\phi(r < 1) = (1 - r^2)^4$. For more details see "Fractal Dimension of Flows", Figure 1, Chaos **2** , 245 (1992) .

geneous problems involving moving interfaces, cracks, or shockwaves.

From a pedagogical standpoint it is interesting to note that the simplest shear flow model can be based on the dynamics of just *two* particles with the periodic shearing boundary conditions of **Figure 7.1** . The details of extracting the irreversible heat with thermostat forces while using shearing periodic boundaries to drive the flow make this two-body problem a manageable and educational exercise in numerical methods.

Three bodies are enough to treat heat flow. See, for several examples, Bill's 1983 Physica **118A** paper, "Atomistic Nonequilibrium Computer Simulations". By 1983 Denis Evans and Mike Gillan had already developed, independently, a *periodic single-chamber* heat-flow model that accelerates hot particles to the right and cooler ones to the left without any net flow of mass. Their approach was designed to generate a heat current matching the exact Green-Kubo theory in the small-current limit. Small-scale models using these ideas are ideal for comparisons with the predictions of the low-density Boltzmann equation and are useful exercises in preparing to program larger-scale simulations.

Nonlinear heat-flow simulations *with* temperature gradients can be complicated by thermal expansion, particularly where the system is solid. The fluid system shown in **Figure 7.2** has periodic boundary conditions in both directions, so that thermal expansion is not a problem. There are four interconnected chambers, with time-dependent kinetic temperatures, two of them thermostated (open circles in the figure), and two of them not (filled circles). The expansion problem is avoided by allowing particles to migrate freely among the four chambers, one hot, one cold, and two Newtonian. Although we could smear the transition from chamber to chamber by using a weight function like Lucy's to smooth the thermostating, such a precaution is unnecessary. The dynamics is well-behaved despite the occasional discontinuous temperature jumps whenever a particle changes cells.

In addition to the 144 dimensions corresponding to the particle coordinates and momenta, the phase space for this system includes two thermostat variables, $\zeta_{\rm cold}$ and $\zeta_{\rm hot}$. The results for the dimensionality of the steady-state attractor were specially interesting. They showed a reduction in the phase-space dimensionality from the equilibrium value of 146 to 125, a 14% reduction of 21 . We return to this topic of phase-space dimensionality loss in Sections 7.6 and 7.7 . Here we simply emphasize that the reduction in dimensionality (as opposed to volume) indicates the extreme rarity of nonequilibrium states. The reduction can greatly exceed the dimensionality of the thermostat variables.

7.1.3 *Green-Kubo Perturbation Theory of Transport*

Green-Kubo theory is a "first-order" perturbation theory for phase-space distribution functions. It describes the *linear* response of mechanical systems to small density, velocity, and temperature gradients. The corresponding linear transport coefficients, diffusivity, shear and bulk viscosity, and heat conductivity, all turn out to be given by time integrals of equilibrium correlation functions. Shear viscosity, for example, can be traced to the correlation function of the shear stress :

$$\eta = (V/kT)\int_0^\infty \langle\, P_{xy}(0)P_{xy}(t)\,\rangle_{\rm eq} dt \ .$$

The correlation function is readily accessible from a single long equilibrium computer run in which one or more shear stress components are measured so that their time correlations can be computed. For two-dimensional fluids the two shear-stress possibilities are $P_{xy}(t)$ and $[\ P_{xx}(t) - P_{yy}(t)\ ]/2$.

Thermal conductivity is similarly related to the decay of equilibrium fluctuations in the heat flux vector :

$$\kappa = (V/kT^2)\int_0^\infty \langle\ Q_x(0)Q_x(t)\ \rangle_{\rm eq}dt = (V/kT^2)\int_0^\infty \langle\ Q_y(0)Q_y(t)\ \rangle_{\rm eq}dt\ .$$

These integration formulæ for transport coefficients from equilibrium fluctuations are not particularly convenient routes to the transport coefficients. They do have the teaching advantage of a pedagogical background and a computational advantage of making *all* the transport coefficients accessible from a single equilibrium run. On the other hand the number-dependence and fluctuations are relatively large using this Green-Kubo formulation.

Instead we can use a more direct *nonequilibrium* approach, based on computer simulations with imposed velocity or temperature gradients. Then, we only need divide the measured flux, of momentum or of energy, by the imposed gradient :

$$\eta \equiv -P_{xy}/[\ (du_x/dy) + (du_y/dx)\ ]\ ;\ \kappa \equiv -Q_y/(dT/dy)\ .$$

To pursue another route to viscosity or heat conductivity, less direct but equally valid, we could compute the global entropy production directly from its definition as the dissipated work or energy divided by temperature. This works for either of the two transport processes. For the case that the gradients are only in the x direction we find (1) the rate at which work is done on an element V by a uniform strain rate ; and (2) the rate at which entropy accumulates within an element V due to an imposed temperature gradient :

$$T\dot{S} = \int_V[\ \eta(du_x/dy)^2 + \kappa T(d\ln T/dx)^2\ ]dV\ .$$

The transport coefficients obtained in any one of the three fundamental ways (fluctuations, fluxes, or heating) can be compared to those inferred from more complex flows, such as the prototypical shockwave problem illustrated in Figure 4.6 on page 94 and discussed in Section 4.7 .

In such shockwave problems choosing a coordinate frame centered on the shockwave reveals that a part of the kinetic energy of the cold rapidly-right-moving particles at the left appears as an additional thermal internal energy of the slower hotter fluid exiting to the right. In an alternative coordinate system, fixed on the hot fluid, this same inhomogeneous shock process can be viewed as an *inelastic* collision of fast cold fluid with a hot motionless fluid, converting *all* of the cold-fluid kinetic energy to heat. The *width* of the shockwave transition zone linking the cold and hot equilibrium states depends mainly on viscosity, as we will detail later in this Chapter.

Let us begin our description of benchmark computational models with an analysis of the simple shear flow which converts the work done by a velocity gradient into heat. Following our description of three *different* shear flow algorithms we will consider dissipation in the *absence* of velocity, due instead to the *transfer* of heat from a warmer to a cooler body. These two prototypical steady flows can be used to determine the viscosity and the thermal conductivity. We follow them up with a discussion of shockwaves, in which both transport properties, along with several nonlinearities, plus a tensor temperature, and time delay, all play a rôle.

7.2 Periodic Shear with the Doll's and Sllod Algorithms

7.2.1 *The Doll's Algorithm*

In order to eliminate surface effects of order $N^{-1/D}$ in D dimensions, periodic shear algorithms were developed in the 1970s, using moving images of a cell under shear to drive the motion. See again **Figure 7.1** . A Hamiltonian which is compatible with an averaged uniform motion in the x direction, a motion which varies linearly with y , is the "Doll's-Tensor" Hamiltonian, so called because the (qp) combination in that Hamiltonian sounds like the popular porcelain "Kewpie" doll of the early 20th century.

Consider the equations of motion following from the (yp_x) version of the Hamiltonian :

$$\mathcal{H}_{\rm Doll's} = \mathcal{H}_{\rm eq} + \dot{\epsilon}\sum(yp_x)_i \ ,$$

where the sum is over all particles :

$$\{ \ \dot{x} = p_x + \dot{\epsilon}y \ ; \ \dot{y} = p_y \ ; \ \dot{p}_x = F_x \ ; \ \dot{p}_y = F_y - \dot{\epsilon}p_x \ \} \ .$$

The "extra" terms in the equations of motion are linear in the "strain rate", which is, in the case illustrated in **Figure 7.1** , $\dot{\epsilon} \equiv (du_x/dy)$. The periodic boundaries imply that the forces be calculated according to Bill Wood's "nearest-image" rule. The rule is unambiguous for potentials with a range less than half the box length. In the y direction the rule is time-independent and straightforward :

```
yij = y(i) - y(j)
if(yij.lt.-ely/2.0d00) yij = yij + ely
if(yij.gt.+ely/2.0d00) yij = yij - ely
```

In the x direction there is an added time-dependent complexity, which we leave as a problem for the reader. The time-dependent horizontal shift in the location of the particles' periodic images depends upon the images' separation in the y direction. The time-dependence of this shift is linear in the time and the strainrate, but with time-periodic jump discontinuities.

Because the dynamics is derived from the Doll's Hamiltonian the time derivative of that Hamiltonian vanishes, $\dot{\mathcal{H}}_{\rm Doll's} = 0$. This shows directly that the work done by the two strain-rate terms is in exact agreement with the Virial Theorem :

$$\dot{\mathcal{H}}_{\rm eq} = \dot{\mathcal{H}}_{\rm Doll's} - \dot{\epsilon}(d/dt)\sum_i (yp_x)_i = 0 - \dot{\epsilon}(d/dt)\sum_i (yp_x)_i =$$

$$-\dot{\epsilon}[\ \sum_i (p_yp_x)_i + \sum_{i<j}(yF_x)_{ij}\ ] \equiv -\dot{\epsilon}P_{xy}V\ ,$$

The pressure-tensor definition follows the derivation given in Section 2.6 .

This conservation relation is the usual First Law thermodynamic identity linking the energy change $\int \dot{\mathcal{H}}_{\rm eq}dt$ to the work done by the integrated external shear stress, $-\epsilon P_{xy}V$. The isotropic version of Doll's Tensor, with $\dot{\epsilon}_{xx} = \dot{\epsilon}_{yy}$ in two dimensions, generates constant enthalpy states when $\dot{\epsilon}$ is small, and a frequency-dependent bulk viscosity, when $\dot{\epsilon} \propto \sin(\omega t)$.

The equations of motion are obtained from the Hamiltonian in the usual way, with $\dot{q} = +(\partial\mathcal{H}/\partial p)$ and $\dot{p} = -(\partial\mathcal{H}/\partial q)$ for each degree of freedom. Solving the $4N$ equations for N particles of unit mass, $\{\ \dot{x},\ \dot{y},\ \dot{p}_x,\ \dot{p}_y\ \}$, soon shows that the energy continually increases due to the work being done by the periodic boundaries through the positive shear stress, $\sigma_{xy} \equiv -P_{xy}$. To offset this irreversible heating, so as to stabilize a stationary flow state, an ergostat or a thermostat variable ζ is added to the motion equations :

$$\{\ \dot{p}_x = F_x - \zeta p_x\ ;\ \dot{p}_y = F_y - \dot{\epsilon}p_x - \zeta p_y\ \}\ .$$

To achieve constant internal energy, ($\dot{E} = \dot{\Phi} + \dot{K} \equiv 0$) we combine the definition $\dot{x} = p_x + \dot{\epsilon}y$ with $\dot{E}$, yielding :

$$\sum[\ -F_x\dot{x} - F_y\dot{y} + p_x(F_x - \zeta p_x) + p_y(F_y - \dot{\epsilon}p_x - \zeta p_y)\ ] \equiv 0\ .$$

where the sum is over all particles in the system. Equivalently :

$$\sum[\ -F_x(p_x + \dot{\epsilon}y) - F_yp_y + F_xp_x - \zeta p_x^2 + F_yp_y - \dot{\epsilon}p_yp_x - \zeta p_y^2\ ] = 0 \longrightarrow$$

$$-\dot{\epsilon}[\ \sum yF_x + p_xp_y\ ] - 2\zeta NkT = 0 \rightarrow \zeta \equiv -\dot{\epsilon}(P_{xy}V/2NkT)\ ,$$

where $T = (T_{xx} + T_{yy})/2$ is the instantaneous value of the kinetic temperature. Combining $(yF_x)_i + (yF_x)_j = (yF_x)_{ij}$, using the nearest-image distance in the y direction, gives again the usual virial-theorem expression for $P_{xy}V$.

This computational approach to steady isoenergetic shear with periodic boundaries is very successful in minimizing size effects. That is, there is very little dependence of the resulting viscosity on system size. The constancy of the internal energy is a welcome check of the programming. In our 2008 paper with Janka Petravic we studied the size dependence of simulations with both the internal energy per particle and the density set equal to unity. See Table II in that paper. It is convenient to take an initial zero-energy square lattice, with the comoving velocities $\{ p \}$ scaled to produce the desired internal energy. The pairwise-additive potential we used is :

$$\phi_{\rm soft}(r < 1) = 100(1 - r^2)^4 \ .$$

At a strain rate of $\dot{\epsilon} = 0.50$ the resulting Doll's-tensor shear viscosities for 64 disks ($\eta = 1.23_4$) and for 16 384 disks ($\eta = 1.24_4$) were nearly identical. These results are a benchmark, a good check of the programming.

7.2.2 *The Sllod Algorithm*

For the simple shear flow $\{ \dot{x} = \dot{\epsilon}y + p_x \}$ the momenta $\{ p_x \}$ correspond to velocity measurements made in the comoving (shearing) frame. The corresponding Newtonian equations of motion, expressed in the (fixed) laboratory frame are :

$$\{ \ (d/dt)(\dot{x}) = F_x = +\dot{\epsilon}p_y + \dot{p}_x \ ; \ \dot{p}_y = F_y \ \} \ .$$

The equivalent set of motion equations in the comoving frame has been termed the "Sllod" algorithm :

$$\{ \ \dot{x} = \dot{\epsilon}y + p_x \ ; \ \dot{y} = p_y \ ; \ \dot{p}_x = F_x - \dot{\epsilon}p_y \ ; \ \dot{p}_y = F_y \ \} \ .$$

At zero strain rate these motion equations give constant energy. The additional rate of change of the internal energy due to the strain rate $\dot{\epsilon}$ agrees exactly with the linear-response obtained from Doll's Tensor :

$$\dot{E} = \dot{K} + \dot{\Phi} = \dot{\epsilon}[\ \sum_{i<j} -(yF_x)_{ij} - \sum_i p_xp_y \ ] \equiv -\dot{\epsilon}P_{xy}V \ .$$

In the absence of thermostating, *both* algorithms, Doll's and Sllod, satisfy the thermodynamic relation from the adiabatic form of the First Law of Thermodynamics : $\dot{E} = -\dot{\epsilon}P_{xy}V$.

Solving the *thermostated* Sllod equations of motion at the strain rate $\dot{\epsilon} = 0.50$, for 64 disks and 16 384 disks gives viscosities of 1.23_0 and 1.24_6 , not very different to the Doll's results using the same energy, density, and pair potential, $\phi_{\rm soft}(r < 1) = 100(1 - r^2)^4$.

The nonlinear "normal-stress" effects are *even* functions of the strain rate, and are sensitive to the chosen shear-flow algorithm. It is significant that in the absence of interparticle forces to stabilize the momenta both the Doll's and the Sllod algorithms predict divergent normal stresses :

$$[\ \dot{p}_y = -\dot{\epsilon}p_x\ ] \longrightarrow [\ p_y^2 \to \pm\infty\ ]\ [\ \text{Doll's}\ ]\ ;$$

$$[\ \dot{p}_x = -\dot{\epsilon}p_y\ ] \longrightarrow [\ p_x^2 \to \pm\infty\ ]\ [\ \text{Sllod}\ ]\ .$$

The detailed results *with* forces (again for a strain rate of 0.50 and with 16 384 soft disks) bear out the direction of these limiting observations :

$$(P_{xx}V/Ne) = 0.679 + 3.193 = 3.872\ ;$$

$$(P_{yy}V/Ne) = 0.694 + 3.233 = 3.926 > (P_{xx}V/Ne)\ [\ \text{Doll's}\ ]\ ;$$

$$(P_{xx}V/Ne) = 0.692 + 3.211 = 3.903\ ;$$

$$(P_{yy}V/Ne) = 0.681 + 3.214 = 3.895 < (P_{xx}V/Ne)\ [\ \text{Sllod}\ ]\ .$$

The normal pressure components are expressed here in the form "kinetic + potential = total".

Although these differences between the horizonal and vertical stresses are "small" they are large enough to account for the Weissenberg Effect, the climbing of paint or other viscous fluids, up a rotating stirring rod. In that case it is P_{zz} which causes the climbing. In our own three-dimensional periodic simulations P_{zz} turned out to be *less* than both P_{xx} and P_{yy} , suggesting a depression rather than a climbing in response to shear.

Because both these two shear algorithms, Doll's and Sllod, are consistent with energy conservation and with linear-response theory either one of them can be used to determine an accurate shear viscosity. But the nonlinear thermostated normal-stress effects are inaccurate and require more realistic boundary conditions, to which we turn next.

7.3 Boundary-Driven Shear Flows

Our familiarity with boundary-driven shear flows dates back to the early 1970s, when Bill Ashurst developed both boundary-driven and periodic stationary flows. In those days it was of interest to confirm the validity of the Green-Kubo relations for the transport coefficients, particularly since Verlet, Levesque, and Kürkijarvi's early simulations included factor-of-two errors in two of their four Green-Kubo analyses.

We revisited the shear-flow problem with Janka Petravic about thirty-five years later. By then several algorithms had been developed for shear flow and two rather different algorithms had been developed for heat flow. Because the linear transport coefficients were understood well our focus was on nonlinear effects, both the bulk effects intrinsic to a fluid, and the somewhat larger boundary effects dependent on the simulation algorithm.

Straightforward shear flow and heat flow algorithms can be based on the dynamics of four different ($L \times L$) chambers, two of them Newtonian and two thermostated or ergostated. In both the constrained chambers each of the particles "driving" the velocity gradient or the temperature gradient is tethered to its individual lattice site. In the shear-flow case these sites move, with all of those in a constrained chamber moving steadily at the same speed. The driving particles interact with their individual sites through a time-dependent quartic potential :

$$\phi_{\rm tether} = (\kappa/4)(r_i - r_{\rm tether})^4 \ ; \ x_{\rm tether}(t) = x_{\rm tether}(0) \pm (\dot\epsilon/2)tL \ .$$

The summed-up constant kinetic energy of each driver chamber, relative to the sites' motion, is thermostated by a "Gaussian" thermostat (based on Gauss' Principle of Least Constraint) :

$$\{ \ \dot p = F - \zeta p \ \} \ ; \ \zeta = \sum (F\cdot p)/(2mK)$$

$$\longrightarrow \dot K = \sum (p\cdot \dot p/m(1/2)) \equiv 0 \ .$$

Thus the velocities and kinetic temperatures of the two heat reservoirs are constants of the motion.

In the two Newtonian chambers our interest is focused on the strain-rate dependence of the pressure tensor. The contribution of each Newtonian Particle i to the pressure tensor is given by the Virial Theorem :

$$(P_{xx}V)_i = (p_x^2/m)_i + (1/2)\sum_j x_{ij}^2 F/|r_{ij}| \ ;$$

$$(P_{xy}V)_i = (p_x p_y/m)_i + (1/2)\sum_j x_{ij}y_{ij} F/|r_{ij}| \ .$$

We minimize integration errors by choosing smooth pair-potential forces. See again **Figure 2.8** on page 47 for a comparison of two very similar choices. Either of the potentials shown there, one proportional to Lucy's smooth-particle weight function, the other a soft-disk pair potential, is suitable :

$$\phi_{\rm Lucy}(r < 1) = (1 + 3r)(1 - r)^3 = 1 - 6r^2 + 8r^3 - 3r^4 \ ,$$

$$\phi_{\rm soft}(r < 1) = (1 - r^2)^4 \ ;$$

The maximum difference between these two functions is less than 0.035 .

A careful investigation of boundary-driven shear flows, in both two- and three-dimensional systems, shows that the differences in the normal stresses, such as $[\ P_{xx} - P_{yy}\ ]$, are too small to measure for four-cell simulations containing 20×20 or $20 \times 20 \times 20$ particles per cell. This experience confirms Liem, Brown, and Clarke's 1992 shear flow work (with 37 500 "bulk" particles and 5610 "driver" particles) in which the authors concluded, no doubt with great regret, that in the end their normal-stress effects were too small to measure. Increasing the strain rate was to no avail because higher strain rates caused the bulk fluid to separate from the driving walls.

By reducing the system size to 10×10 or $10 \times 10 \times 10$ particles per cell it *is* possible to estimate the signs and magnitudes of the normal-stress differences. Our work with Janka Petravic established that these differences were quite different to those obtained with periodic Doll's or Sllod shear simulations. The (tensor) temperature results in three dimensions were relatively precise and interesting :

$$T_{yy} > T_{xx} > T_{zz} \ [\ \text{Periodic Doll's}\ ] \ ;$$

$$T_{xx} > T_{yy} > T_{zz} \ [\ \text{Periodic Sllod}\ ] \ ;$$

$$T_{xx} > T_{zz} > T_{yy} \ [\ \text{Boundary Driven}\ ] \ .$$

Our explorations of periodic shear flow show that *either* algorithm, Doll's or Sllod, can give an accurate shear viscosity but that *neither* is trustworthy for normal-stress effects. No doubt convincing normal-stress simulations could simultaneously model gravity, a free surface, and a rotating rod, requiring relatively long times as well as hundreds of thousands of particles. Such simulations are quite feasible today, and worth exploring.

In 1995 it was thought that transport coefficients would diverge for two-dimensional systems, exhibiting a logarithmic dependence on system size. We set out to check on that. We found that the shear viscosity varies only a little with system size, and in a regular way, just as do also

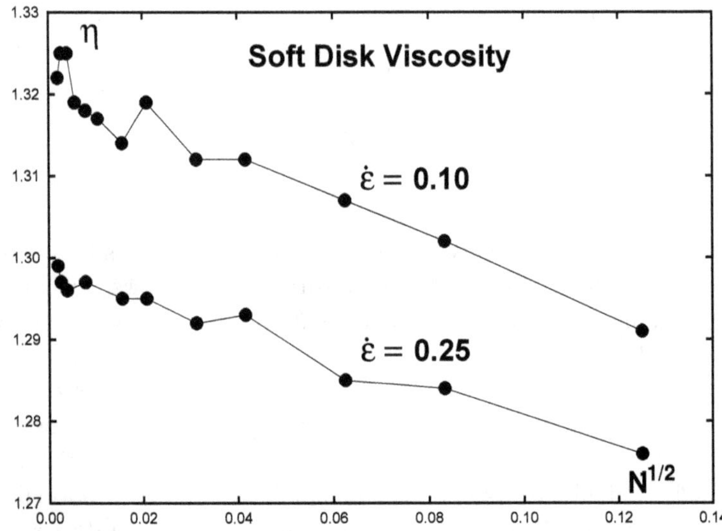

Fig. 7.3 Shear viscosity as a function of system size using the soft-disk pair potential $\phi = 100(1 - r^2)^4$ at unity density and unit energy per particle. Results are shown for two shear strain rates, 0.10 and 0.25 . The *decrease* of viscosity with strain rate is "shear thinning". In shockwaves viscosity typically *increases* with strain rate.

equilibrium equation of state measurements. In our January 1995 Physical Review E paper with Harald Posch we carried out a series of periodic Sllod simulations. The soft-disk system varied in size from $L^2 = 64$ up to $L^2 = 264\,196$. Deviations from the large-system limit varied accurately as $1/L$ as is shown in **Figure 7.3** .

7.3.1 *High-Strain-Rate Deformation of Solids*

Exactly these same Doll's and Sllod shear flow algorithms can be, and have been, applied to the irreversible shear of solids. Solids can be idealized as resisting any flow until a critical value of the shear stress is applied. That threshhold shear stress is called the "plastic yield strength". In the simplest models for the flow of solids the yield strength is considered to be a material property. Simulation results indicate an order of magnitude increase in the shear stress (or yield strength) with increasing strain rate. At constant volume this rate dependence decreases with increasing temperature.

A semiquantitative model for solid-phase plastic deformation, consistent with the simulations, can be based on dislocation motion. Rather than having the entire lattice slip under shear, parallel to a crystal plane, dislocations allow the relative motion to occur locally, involving just a few core atoms at a time. Detailed models of interacting dislocations can be built up and solved, treating the dislocations as point defects (in two dimensions) or line defects (in three) which move in response to stress. In addition to the boundary stresses dislocation motion is affected by the fields of nearby dislocations. The combination of molecular dynamics with dislocation dynamics provides a consistent picture in which stress varies as a temperature-dependent power of the strain rate. The power is close to linear in the temperature indicating that cold solids can be approximated by perfect plasticity, with a well-defined rate-independent yield strength, while solids approaching the melting point behave more like viscous fluids, with a shear stress proportional to the strain rate.

7.4 Green-Kubo and Homogeneous Heat-Flow Simulations

Heat flow is a more complicated process than shear flow. This is apparent from the fact that *three* particles are required to simulate heat flow, while two are enough for shear. With only two particles and a fixed center of mass there can be no heat flow, by symmetry. With *three* particles the "hotter" one can be accelerated to the right with the "colder" ones accelerated to the left, leading to a net flow of heat without any net mass flow. The simplest form of the Boltzmann equation shows that the nonequilibrium distribution is cubic in the velocity components for heat flow, $\propto v_x v^2$, and quadratic for shear flow, $\propto v_x v_y$.

In the laboratory, heat flow at constant pressure implies a density gradient; density gradients are *not* a good fit with periodic boundaries. Gillan and Evans developed an artificial means for generating homogeneous heat flow, *without any gradients*, by using an energy-sensitive external field. The field, by pushing more-energetic and less-energetic particles in opposite directions, generates heat flow in the absence of temperature gradients. The field needs to be chosen so as to agree with linear-response theory. Green and Kubo's linear-response theory, a perturbation theory based on Gibbs' canonical ensemble, provides an equilibrium recipe for the conductivity κ :

$$\kappa = (V/kT^2)\int_0^\infty \langle\, Q_x(0)Q_x(t)\,\rangle_{\rm eq} dt \ .$$

The heat flux, $Q = (Q_x, Q_y)$ in two dimensions and (Q_x, Q_y, Q_z) in three, is the flow of energy per unit length or area in those two cases, and time.

Even in the absence of an imposed temperature gradient and at equilibrium, small heat currents arise and decay. Their fluctuations make it possible to determine the conductivity from equilibrium fluctuations. There are two separate contributions to heat flux, a convective contribution from the energy carried by particles as they move, and an action-at-a-distance contribution in which neighboring particles transfer energy through the mechanism of interparticle forces. We consider both contributions next.

The x component of heat flux includes the summed-up convective contributions describing the rate at which Particle i , for instance, transports its own energy in the x direction ,

$$Q_xV \supset \sum_i(\dot{x}_ie_i) \ ; \ e_i \equiv (mv_i^2/2) + \sum_j(\phi_{ij}/2) \ .$$

The probability, per unit area and time, that Particle (i) crosses a plane perpendicular to the x axis is $\dot{x}_i/V$. The potential energy ϕ_{ij} is divided evenly between Particles i and j, accounting for the factor of one half in the convected potential energy.

There is also an action-at-a-distance contribution of the pair forces to the heat flux. The need for this flux contribution can be seen by considering the one-dimensional head-on collision of a moving mass i with a motionless, but otherwise identical, mass j . The interparticle force F_{ij} transmits both momentum $(m\dot{x}_i)$ and energy $(m\dot{x}_i^2/2)$ from one mass to the other in the course of the collision.

As a numerical example of this collisional energy transfer we take two one-dimensional particles of unit mass, with $\phi(r < 1) = 10(1 - r^2)^4$ and initial conditions :

$$\{ \ x_1, x_2, \dot{x}_1, \dot{x}_2 \ \} = \{ \ -0.5, +0.5, +1.0, +0.0 \ \} \ .$$

The separation and accelerations are given by

```
dx = x(2) - x(1)
f(1) = -8.0d00*dx*10.0d00*(1.0d00 - dx*dx)**3
f(2) = +8.0d00*dx*10.0d00*(1.0d00 - dx*dx)**3
```

so that righthand sides of the four equations of motion for $\{ \ \dot{x}_1, \dot{x}_2, \ddot{x}_1, \ddot{x}_2 \ \}$ are :

```
yyp(1) = p(1)
yyp(2) = p(2)
yyp(3) = f(1)
yyp(4) = f(2)
```

At the left **Figure 7.4** shows the changing velocities of the two particles during the collision, as well as the rate of change of $e_2 - e_1$. Because the lefthand Particle "1" is always left of the origin, while the righthand Particle "2"is to the right the rate of energy transfer from 1 to 2 is necessarily equal to the energy flux at the origin. Particle 1 soon comes to rest at $x_1 = -0.1984$, having transferred all of its momentum and energy to Particle 2 during the collisional period $\Delta t = 0.600 = 600dt$ where the Runge-Kutta timestep is 0.001 . An analytic expression for the heat flux includes $\dot{e}_2(x_2 - x_1)/V$ where $(x_2 - x_1)/V$ is the probability that a sampling plane in the volume V intersects the flow of energy from Particle 1 to Particle 2 . The rate of energy transfer is :

$\dot{e}_2 = -\dot{e}_1 = (1/2)\dot{\phi} + p_2\dot{p}_2 = (-1/2)F_2(p_2 - p_1) + p_2F_2 = (1/2)(p_1 + p_2)F_2$. The two expressions for the heat flux (including the factor $x_2 - x_1$) are both plotted in the right panel of **Figure 7.4** and agree within the width of the line. The measured energy transfer multiplied by $x_2 - x_1$ *is* the solid line. The points are the force-based heat flux.

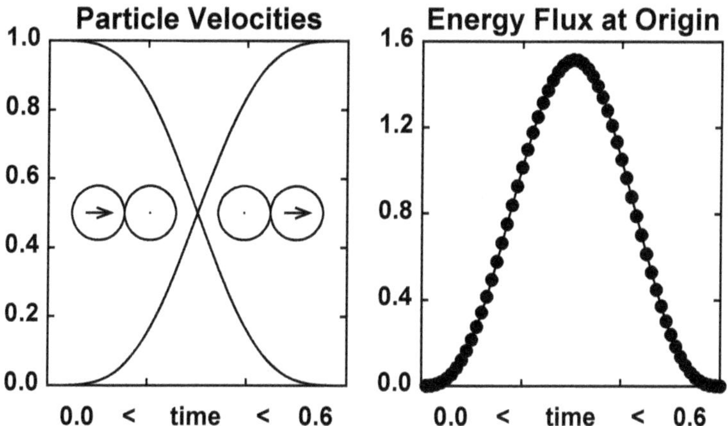

Fig. 7.4 The velocities of two colliding particles are shown on the left. On the right we compare two heat-flux expressions, $\dot{e}_2(x_2 - x_1)$ and $F_2(p_1 + p_2)(x_2 - x_1)/2$, the collisional transfer from Particle 1 to Particle 2 using the potential $\phi(r < 1) = 10(1 - r^2)^4$.

Including the contributions from all such pair interactions in the x component of the atomistic energy flux gives :

$$\sum_i (\dot{x}_i e_i) + \sum_{i<j} P^{\phi}_{xx} V(\dot{x}_i + \dot{x}_j)/2 ,$$

where P^{ϕ} is the "potential" part of the pressure tensor. e is the internal energy per unit mass, including the *total* kinetic energy $v^2/2$. Provided that we use the kinetic definition of temperature the velocity products $\dot{x}e$ can be expressed in terms of the hydrodynamic velocity $\langle \dot{x} \rangle = u_x$ and the temperature tensor :

$$(1/2)\sum^{N} \dot{x}v^2 = (N/2)[\ u_x^3 + 3u_xkT_{xx} + u_xkT_{yy}\ ] ,$$

where the "3" comes from averaging :

$$\langle \dot{x}^3 \rangle = \langle (u_x + p_x)^3 \rangle = \langle u_x^3 + 3u_x^2p_x + 3u_xp_x^2 + p_x^3 \rangle \simeq u_x^3 + 3u_xkT_{xx} ,$$

where the mean value of $p_x^3 \equiv (\dot{x} - u_x)^3$ has been assumed to vanish. This assumption is an indicator of the need for kinetic-theory analyses of strong shockwaves.

Now, for comparison, recall from Section 4.7 the stationary continuum fluxes of mass, momentum, and energy for a steady-state shockwave with a mass velocity x component u :

$$\{\ \rho u,\ P_{xx} + \rho u^2,\ \rho u[\ e + (P_{xx}/\rho) + (u^2/2)\ ] + Q_x\ \} .$$

Comparing the two microscopic and macroscopic energy fluxes it is interesting to see that the "missing" kinetic part of the pressure tensor is actually included in the "internal energy" e above, and comes from $\dot{x}^3$:

$$\sum_{N\subset V} \dot{x}(\dot{x}^2 + \dot{y}^2)/2 = u[\ VP_{xx}^K + (Nk/2)(T_{xx} + T_{yy})\ ] .$$

It is satisfying to see that the microscopic and macroscopic approaches agree (at least near equilibrium) .

7.5 Boundary-Driven Simulations of Periodic Heat Flow

In solids the flow of heat can be described in terms of phonon scattering. With conventional few-body interaction forces simulations need propagation lengths longer than the phonon mean free path. The phonon approach is relatively messy, as it involves averages over the "k space" described by a "Brillouin zone". Vectors in that zone, (k_x, k_y) or (k_x, k_y, k_z) describe propagating waves. Interactions among waves provide scattering which in turn provides a phonon mean-free-path computation of the thermal conductivity. Close to the melting point, or in fluids, the simpler atomistic description is adequate.

Two thermostated chambers can be maintained at different temperatures by controlling their kinetic energy :

$$\{\ \dot{p}_{\rm cold} = F - \zeta_{\rm cold}p_{\rm cold}\ ;\ \dot{p}_{\rm hot} = F - \zeta_{\rm hot}p_{\rm hot}\ \} .$$

Newtonian dynamics in connecting chambers provides not only a heat flow but also a temperature gradient, from which the conductivity can be measured :

$$\kappa \equiv -Q_x/(dT/dx) \ ,$$

Just as in viscosity simulations there is good agreement between such nonequilibrium simulations and the Green-Kubo equilibrium recipe. The main (æsthetic) fly in this ointment is the competition between the periodic boundaries and the mechanical equilibrium of heat-sensitive pressure forces.

Although one might expect that one-dimensional heat-flow simulations could avoid the need for dealing with thermal expansion, one-dimensional solids with harmonic forces transmit energy from one end to the other at the speed of sound, unhindered by any (x to y) scattering mechanism. A way to avoid this difficulty, making it possible to study the size-dependence of thermal conductivity in more "realistic" solids, is the ϕ^4 model, to which we turn next. This model has the fringe benefit of shedding light on the mechanism by which nonequilibrium steady-state simulations come to satisfy the Second Law of Thermodynamics.

7.6 The ϕ^4 Model and Phase-Space Dimensionality Loss

The ballistic transfer of energy by sound waves in solids is a nuisance because the scattering of such waves, essential for a size-independent heat conductivity, becomes weak at low temperature, even vanishing for a perfect lattice because the vibrational amplitude, responsible for scattering, varies as $\sqrt{kT/m\omega^2}$. The resulting weak scattering makes straightforward numerical simulations difficult at low temperatures. In a *one*-dimensional system with nearest-neighbor interactions the contribution of low-frequency phonons to the conductivity gives divergent results – the apparent heat conductivity increases with system size and never approaches a large-system limit. For example, a harmonic chain has no finite conductivity at all because the lattice vibrations are neither scattered nor damped.

Nevertheless, a very clever modification of the crystal-lattice potential does provide scattering and a limiting large-system heat conductivity. Adding a potential guaranteeing phonon scattering provides a good model for Fourier heat conductivity in one, two, or three dimensions. Such a model has been carefully studied and characterized by Ken Aoki and Dimitri Kusnezov. This "ϕ^4" model gives a convergent large-system conductivity rather than diverging. The details are described next.

To stabilize a crystal lattice, even in one dimension, a special cubic *tethering* force from the quartic "ϕ^4" potential, linking each particle to its own lattice site, can be added. Because the tethering force varies as x^3 rather than x the harmonic distribution of frequencies is undisturbed. The tethering force prevents the large-system divergence of $\langle\ [\ x - \langle x\rangle\ ]^2\ \rangle$. The simplest ϕ^4 problem links two bodies to their lattice sites at ± 0.5 and to each other :

```
x1dot = p1
x2dot = p2
p1dot = (-0.5d00 - x1)**3 - z1*p1
p2dot = (+0.5d00 - x2)**3 - z2*p1
p1dot = p1dot + x2 - 1.0d00 - x1
p2dot = p2dot - x2 + 1.0d00 + x1
z1dot = p1*p1 - Tcold
z2dot = p2*p2 - Thot
```

Here there is no purely-Newtonian mechanics. Such a two-body run, with `Tcold = 1` and `Thot = 2` gives, after twenty million steps of $dt = 0.01$:

$$\langle\ z_1\ \rangle = 0.0595\ ;\ \langle\ z_2 = -0.0298\ \rangle \quad \longrightarrow Q = -0.0595\ ;\ \kappa = 0.0595\ .$$

for a flux of -0.0595 and a conductivity of 0.0595 . Evidently this relatively simple problem generates a *fractal* (fractional dimensional) phase-space attractor. Directly from the equations of motion the change of phase volume with time is :

$$(\dot{\otimes}/\otimes) = \sum_i [\ (\partial\dot{x}/\partial x) + (\partial\dot{p}/\partial p)\ ] = -z_{\rm cold} - z_{\rm hot} = -0.0298 < 0\ .$$

The steady-state continuous loss of phase volume signals the formation of a strange attractor, with the separation between two nearby trajectories growing exponentially in time while the phase volume simultaneously shrinks exponentially in time ! The "strange" attractor is well-named. Even this simplest possible two-particle version of the ϕ^4 model provides interesting features. Varying the temperature of Particle 2 while leaving Particle 1 at unit temperature reveals some interesting fractal attractors as well as a limit cycle for $T_2 = 5$.

This limit cycle points out some interesting physics, and in the simplest possible way. Just one of the six phase-space dimensions is directly involved in a non-Newtonian loss of phase volume. The (time-averaged) values of all six terms, with $(T_1, T_2) = (1, 2)$, are as follows,

$$\otimes(t) \simeq \otimes(0) \exp[\ (\ 0.0 + 0.0 + 0.298 + 0.0 + 0.0 - 0.0595\ )t\ ]\ ,$$

indicating the contributions of the dynamics to expansion or contraction in the $\{\ (q, p, \zeta)_{\text{cold}}$ and $(q, p, \zeta)_{\text{hot}}\ \}$ phase-space directions, in that order. With $(T_1, T_2) = (1, 5)$ *any* initial phase-space distribution collapses onto a one-dimensional limit cycle [!] . This is possible because the phase volume not only translates, while expanding in the ζ_{cold} direction and contracting in the ζ_{hot} direction, but it also *rotates* and *shears* due to the influence of momenta on coordinates, coordinates plus friction coefficients on momenta, and momenta on friction coefficients :

$$\dot{q} = \dot{q}(p)\ ;\ \dot{p} = \dot{p}(q, \zeta)\ ;\ \dot{\zeta} = \dot{\zeta}(p)\ .$$

The geometric and physical effects of this phase-volume collapse are both interesting and profound. Nonequilibrium stationary states have not just a *small* volume in phase space. That volume actually *vanishes.* In this particular case a six-dimensional volume $\prod dqdpd\zeta$ vanishes exponentially fast so that the trajectory soon becomes a one-dimensional track. We will see that this loss of phase-space dimensionality is *typical* of nonequilibrium steady states and that the loss often, as in this two-body case, exceeds the dimensionality associated with the friction coefficients or other constraints on the motion.

See **Figure 7.5** for typical (qp) phase-plane projections of the two particles' motions. The equal-temperatures case is interesting too, though it is not shown here. That equilibrium situation, with no net heat flow, reveals a noticeable correlation of the two interacting particles' displacements and momenta.

In the interest of developing simple models of nonequilibrium systems which include a Newtonian portion, the ϕ^4 model stands out. With just three degrees of freedom (one "hot", one "cold", and one Newtonian) it is quite feasible to generate a nonequilibrium steady state, with the flow of heat controlled by "cold" and "hot" friction coefficients and with cold and hot temperatures of 1 and 2. There are eight differential equations to solve, which include the cubic forces from the ϕ^4 potential and the linear forces from the nearest-neighbor Hooke's-Law potential :

$$\phi_{\text{tether}} = (1/4)(x - x_0)^4\ ;\ \phi_{\text{Hooke}} = (1/2)(x_{ij} - x_{0ij})^2\ .$$

The righthand sides of the eight differential equations are as follows :

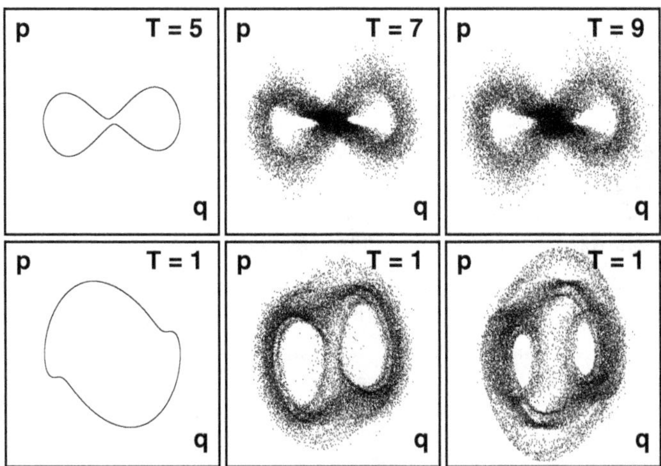

Fig. 7.5 Phase plane projections for the cold particle (lower row) and the hot particle (upper row) for hot-particle temperatures of 5 (limit cycle), 7, and 9 (the latter two give strange attractors). Evidently the six-dimensional phase space for this two-particle problem includes distributions varying from a one-dimensional limit cycle to a six-dimensional space-filling object (when the two temperatures match) . The range of coordinates is ± 4 for the hot particle and ± 2 for the cold one. The range of momenta is ± 15 for the hot particle and ± 2.5 for the cold one.

```
x1dot = p1
x2dot = p2
x3dot = p3
p1dot = (-1.0d00 - x1)**3 - z1*p1
p2dot = ( 0.0d00 - x2)**3
p3dot = (+1.0d00 - x3)**3 - z3*p3
p1dot = p1dot + x2 - 1.0d00 - x1
p2dot = p2dot - x2 + 1.0d00 + x1
p2dot = p2dot + x3 - 1.0d00 - x2
p3dot = p3dot - x3 + 1.0d00 + x2
z1dot = p1*p1 - 1.0d00
z3dot = p3*p3 - 2.0d00
```

Solving the eight equations for a billion timesteps with `dt = 0.001d00` gives the phase plots shown in **Figure 7.6** . The time-averaged friction coefficients for the simulation are :

$$\langle\ z_1\ \rangle = 0.086\ ;\ \langle\ z_3 = -0.043\ \rangle\ \longrightarrow Q = -0.086\ ;\ \kappa = 0.172\ .$$

The rate at which cold-particle heat is extracted, $0.086 p_1^2\ =\ 0.086$,

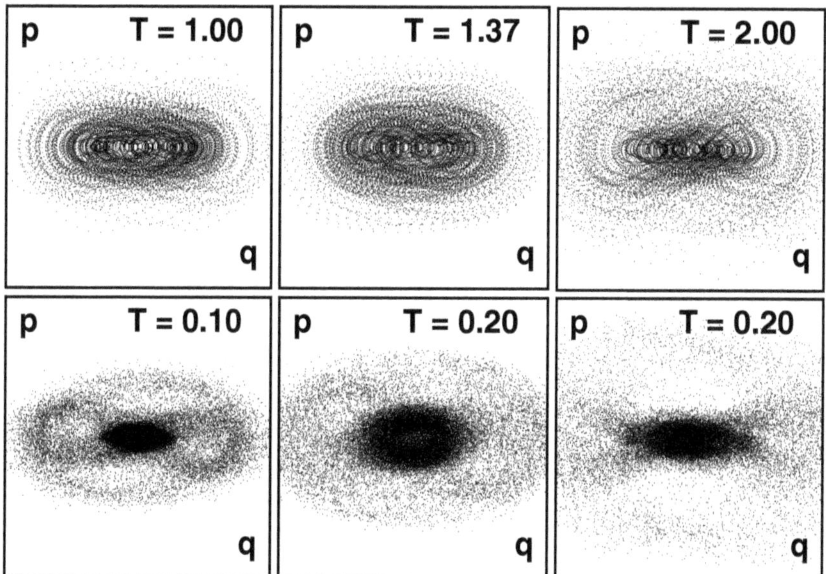

Fig. 7.6 Phase portraits for two three-body heat flow problems with eight phase-space dimensions. The right and leftmost particles' temperatures are controlled by Nosé-Hoover thermostats with $\tau = 1$. The lower row, with temperatures 0.1 and 0.2 , has a coordinate range of 2 ; the upper row, with temperatures of 1.0 and 2.0 , has a coordinate range 4. The lower row has a momentum range of 5 ; the upper row 12 .

is exactly the same as the rate at which hot-particle heat is inserted, $0.043p_3^2 = 0.086$. Because the temperature gradient is 0.5 the conductivity is $-Q_x/(dT/dx) = 0.086/0.500 = 0.172$.

Using temperatures of 0.10 and 0.20 gives instead :

$$\langle\ z_1\ \rangle = 0.80\ ;\ \langle\ z_3 = -0.40\ \rangle\ \longrightarrow Q = -0.080\ ;\ \kappa = 1.60\ ,$$

showing that these problems are very far from the linear-response regime and deserve a more detailed explanation than we can afford to give here.

Nevertheless, these provocative results, for simple three-body systems, illustrate the same apparent paradox as did the two-body thermostated problem above. Liouville's Theorem can be applied to the change in comoving phase volume within the eight-dimensional state space :

$$(d\ln\otimes/dt) = \sum^{3}(\partial\dot{x}/\partial x) + \sum^{3}(\partial\dot{p}/\partial p) + \sum^{2}(\partial\dot{\zeta}/\partial\zeta) =$$

$$0 + [-\zeta_1 - \zeta_3\ ] + 0 = -0.043 \text{ for } T = (1.00, 1.37, 2.00)\ ;$$

$$0 + [-\zeta_1 - \zeta_3\ ] + 0 = -0.40 \text{ for } T = (0.10, 0.20, 0.20)\ .$$

$$\longrightarrow$$

$$\otimes(t) = \otimes(0)e^{-0.043t} \text{ for } T = (1.00, 1.37, 2.00) ;$$

$$\otimes(t) = \otimes(0)e^{-0.40t} \text{ for } T = (0.10, 0.20, 0.20) .$$

How is it possible in "steady states" or "stationary states" like these that the phase volume can decrease forever? The answer is through the attractor's loss of not only phase volume, but also *dimensionality*, just as in the extreme two-particle limit cycle shown in **Figure 7.5** .

7.7 Phase-Space Attractor Dimensionality Loss

The ϕ^4 model provides the clearest possible many-body example of the rarity of nonequilibrium states. Boltzmann and Gibbs thought in terms of the smooth microcanonical and canonical phase-space distributions, no doubt imagining that nonequilibrium states occupied a relatively small volume in the phase space. This same point of view persists today, in intellectually-isolated pockets, although nonequilibrium simulations have made the "small-volume" (as opposed to zero-volume) point of view obsolete. A new mechanism for understanding irreversibility away from equilibrium is provided by the Lyapunov spectrum.

The Lyapunov spectrum for the ϕ^4 model can provide a quantitative estimate for the "smallness" or "rarity" of nonequilibrium states. A 24-atom chain is long enough for a convincing demonstration. If the temperatures of the two ends of the chain are imposed by thermostating, so that $T_1 = 0.003$ and $T_{24} = 0.027$ the Lyapunov spectrum can be determined by following the motion of a reference trajectory and 50 satellite trajectories. The phase-space dimensionality of 50 includes 24 particle coordinates, 24 momenta, and the two friction coefficients used to thermostat the first and last particles' temperatures.

Figure 7.7 shows the summed-up spectrum computed in this way. By summing up n of the local Lyapunov exponents, starting with the largest, we can compute the instantaneous comoving rate of volume change in an n-dimensional volume. The *maximum* growth rate describes an eight-dimensional phase-space hypervolume. For an n-dimensional object the time-averaged rate is given by the corresponding sum of global Lyapunov exponents :

$$(\dot{\otimes}/\otimes)_n \equiv \sum_1^n \lambda_i(t) \longrightarrow \langle\ (\dot{\otimes}/\otimes)_n\ \rangle \equiv \sum_1^n \lambda_i\ .$$

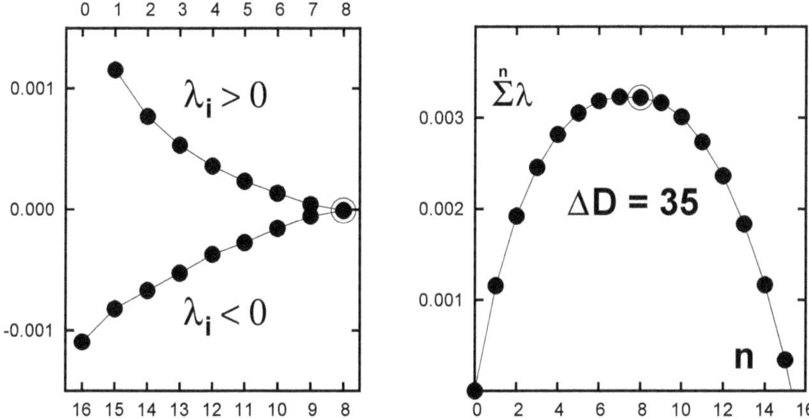

Fig. 7.7 Lyapunov exponents (left), and their sums (right), for a 24-particle ϕ^4 chain with "cold" temperature 0.003 and "hot" temperature 0.027 . The dimension of the phase-space attractor is only 15 while the dimension of the entire phase space is 50 .

The same identity applies to both the instantaneous and the time-averaged phase-volume changes. Now consider the 24-particle ϕ^4 chain. For the 24 chain particles the first 15 of the time-averaged sums are positive, indicating that a 15-dimensional volume element will grow exponentially in the time. Higher-order sums, with 16 or more Lyapunov exponents time-averaged, are negative, so that a (16 or more)-dimensional volume will forever shrink. The attractor corresponding to the motion is between 15 and 16 in its dimensionality, indicating a *dimensionality loss* of 35, relative to an ergodic equilibrium state.

To confirm that this (originally surprising) result is no fluke we can extend the chain length for our simulation from 24 to 36, so that the state space has grown from 50-dimensional to 74-dimensional. We find that the first 30 of the 74 sums are positive, indicating that a 30-dimensional volume element will grow exponentially in the time. Adding the 31st Lyapunov exponent so as to describe the growth rate of a 31-dimensional volume element gives a *negative* sum, so that the steady-state attractor fills only a 31-dimensional region in the 74-dimensional space. The dimensionality loss in this larger system is 43. The physical interpretation of this loss is both simple and profound. The probability of finding a typical phase point associated with nonequilibrium heat flow for this system is not just small or infinitesimal. *It is zero!* Nonequilibrium states occupy less than a finite volume in the many-body state space. Even the *dimensionality* of these states is far less than that of the phase space itself.

7.7.1 *Fractal Information Dimension*

Fig. 7.8 Construction of the triangular Sierpinski Carpet (at the top) showing the construction of this 1.58496-dimensional fractal object. The 2.72683-dimensional Menger Sponge shown below is constructed similarly. Both objects are self-similar fractals. Those encountered in nonequilibrium simulations have fractal dimensionalities that vary from point to point within their attractors. Such more-complex objects are "multifractal".

From the standpoint of fractals and dynamical-systems theory the *dimensionality* of the attractor can be defined in a variety of ways, all of which give the same result for homogeneous self-similar fractals like the Cantor Set, Sierpinski Carpet, and Menger sponge suggested by **Figure 7.8**. Because the thermodynamic Gibbs' entropy is given by the logarithm of the number of states (or minus the logarithm of the probability for those states, if their probabilities are equal) the "information dimension" is of primary interest in statistical mechanics. Its dimension deficit, relative to equilibrium, describes the way in which Gibbs' entropy and $\sum_{\rm boxes}[\,\text{prob} \ln \text{prob}\,]$ diverge as the box size approaches zero.

Because a D-dimensional object (like a line segment, a square, or a cube for $D = 1$, 2, and 3) can be divided into $\simeq d^{-D}$ similar objects of scale length d , the simplest definition of a "fractal dimension" is based on this scaling of volume with length. For a general object the dimension is defined as the logarithm of the number of similar objects divided by the logarithm of the inverse scale length, $\ln(1/d)$.

The simplest model illustrating this construction is the Cantor Set, the limiting zero-length set of line segments obtained by repeatedly discarding the "middle third". Discarding the middle third of the line segment between zero and one leaves two segments of length $(1/3)$. A second discard operation leaves four segments of length $(1/3)^2$. After n such operations there are 2^n segments of length $(1/3)^n$ so that the dimensionality of the Cantor set is $D_C = \ln(2)/\ln(3) = 0.63093$. The dimensionality of the Sierpinski carpet, which is composed of 3^n triangular regions of scale length $(1/2)^n$, is $D_S = \ln(3)/\ln(2) = 1.58496$. The dimensionality of the Menger Sponge shown in **Figure 7.8**, with 20^n cubes of scale length $(1/3)^n$, is $D_M = \ln(20)/\ln(3) = 2.72683$.

The *nonequilibrium* fractal distributions obtained with thermostats and ergostats are different to these idealized Cantor-Sierpinski-Menger distributions in that *nonequilibrium* fractals typically (at least, not too far from equilibrium) have no "holes". Their dimensionality is defined in terms of probability density rather than volume. This lack of holes can best be understood as a consequence of time reversibility along with ergodicity. Because the equations of motion are time-reversible *any* state in the phase space could be used as an initial condition. Motion started there will reach the attractor. Now consider the consequences of time reversibility. Motion from the time-reversed attractor state (a "repellor" state) will eventually reach the reversed initial condition. Thus in the (typical, at least close to equilibrium) case that there are *no* holes in the nonequilibrium distributions the entire phase space participates in an ergodic flow *from* an unstable repellor *to* its reflected image, a somewhat less unstable strange attractor.

The dissipative Galton Board problem mentioned in Section 1.6 and detailed in Section 8.5 eventually *does* produce a hole (with a set of conservative tori filling the hole). The limit cycles found for short cold ϕ^4 chains likewise suggest that sufficiently far from equilibrium those systems can cease to be ergodic and start to display the relatively complicated structure of equilibrium systems like the Nosé-Hoover oscillator. That oscillator has an *infinite* number of periodic or toroidal solutions embedded in a surrounding "chaotic sea".

So far we have considered simple models for the viscous transport of momentum and the thermal transport of energy. The common features of all of these models are time-reversible determinstic thermostats, or ergostats, which lead to dissipation with the consequent formation of strange attractors.

The stationary shockwave problem includes a step up in complexity, with many reproducible nonlinear features. In order to compare the necessarily-atomistic simulations of strong shockwaves to their continuum analogs we need a *mesoscopic description* capable of converting atomistic coordinates and momenta to the field variables of continuum mechanics–density, stress, heat flux, ... Smooth Particle Applied Mechanics furnishes such an interpolation method. Let us first describe it and then apply it to shockwave problems with detailed examples.

7.8 Mesoscopic Reversibility through SPAM Interpolation

Smooth Particle Applied Mechanics provides not only a handy means for solving the continuum equations, but also a useful and flexible interpolation technique, just the "right thing" for a continuum description of atomistic shockwaves. SPAM converts digital particle data (temperatures, pressures, energies, . . .) defined at discrete Particle locations to twice-differentiable continuum functions defined *throughout* space. This indispensable interpolation tool can link a macroscopic continuum description to the microscopic particulate one. The basic idea is to spread smoothly the contributions of individual particles so that the superposition of the discrete particle contributions provides a smooth continuum field.

In Chapter 4 we demonstrated the power of this interpolation technique by solving the continuum equations using *discrete* particles with *ordinary* differential equations of motion. SPAM solutions of the continuum equations satisfy the continuity equation exactly and can also be formulated so as to conserve linear momentum and energy. *Angular* momentum remains an active research area. Here we focus on using SPAM weight functions to obtain *continuum* representations of many-body *molecular dynamics* solutions. Rather than solving the continuum equations directly we wish instead to *measure* constitutive properties (such as the stress tensor and heat flux vector) directly from underlying particle data.

Contributions of *pairs* of particles to the energy, pressure tensor, and heat flux vector can be idealized as located *at* the particle points (the "Irving and Kirkwood choice") or distributed along a line (or curve) linking the two particles ("Hardy's choice"). Converting the localized pair-potential contributions to a corresponding continuum potential-energy field

$\phi(r)$, using a SPAM weight function, requires summing up (integrating in Hardy's case) nearby pair-potential contributions :

$$\phi(r) = \sum_{r_i} w(r - r_i)(1/2) \sum_j \phi_{ij} \ ;$$

$$w(r < h) = C[\ 1 + 3(r/h)\ ][\ 1 - (r/h)\ ]^3 \ .$$

When the weight-function range h is a few particle diameters there is no significant difference between the fields obtained by summing or by integrating. For simplicity's sake we invariably adopt summation.

Although there *are* other weight functions than Lucy's, his is the simplest finite-range polynomial which has two everywhere-continuous derivatives w' and w'' . The proportionality constant C in $w(r)$ is fixed by normalization :

$$C_{1D} = (5/4h) \ ; \ C_{2D} = (5/\pi h^2) \ ; \ C_{3D} = (105/16\pi h^3) \ .$$

The range h is a free parameter which can be chosen to optimize the correspondence between the continuum and particle descriptions. **Figure 7.9** shows the extreme values of the density of a one-dimensional lattice with a lattice spacing of unity for weight-function ranges of $1.5 < h < 5$. The density error is less than a percent for any h greater than 2.1 . Based solely on these numerical data a suitable choice of h likely lies in the range from 2.5 to 3.5 where a typical nearest-neighbor particle spacing is unity. Let us apply these ideas to a challenging problem, the microstructure of shockwaves as was briefly sketched in Chapter 4 .

7.9 Shockwaves and Their Smooth-Particle Representation

In addition to homogeneous steady states there are many kinds of steady *inhomogeneous* flows. Nonequilibrium flows can be steady, or transient, and close to, or far from, equilibrium. Closeness or distance depends on the size of the velocity and temperature gradients. To simplify simulations the gradients need to be localized. It is possible, for instance, to convert one steadily moving equilibrium fluid flow to another within a tiny localized region in space. Suppose the flow is "one-dimensional", with the entire conversion process occuring within a small "black box". What conditions must be obeyed to make such a transformation possible? Conservation of mass, momentum, and energy all apply. In addition the flow must also obey the Second Law of Thermodynamics, with entropy increasing over time.

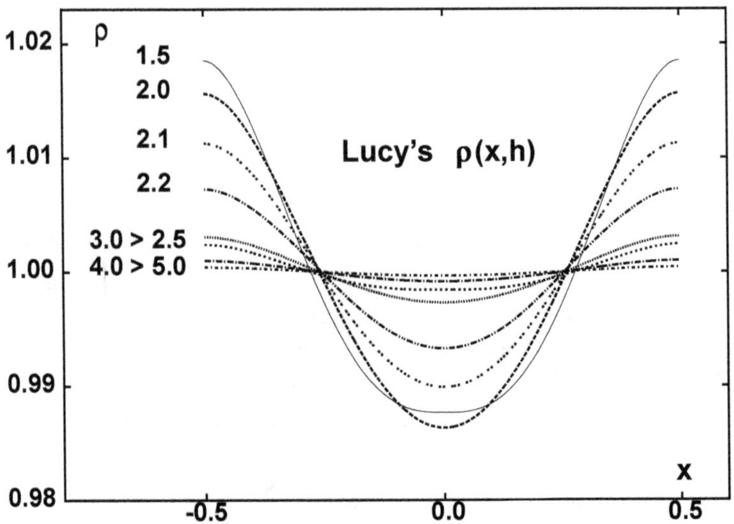

Fig. 7.9 Maximum deviations for one-dimensional chain density for $1.5 < h < 5.0$.

Evidently the mass, momentum, and energy entering and leaving the box must match, as described by the conservation laws of Chapter 4 :

$$\rho u \ ; \ P + \rho u^2 \ ; \ \rho u[\ e + (P/\rho) + (u^2/2)\ ] \text{ All Match .}$$

In addition to these conservation relations the *entropy* leaving the box must exceed that entering it. We will describe an interesting exception to the momentum-flux matchup in the Joule-Thomson flow after our discussion of shockwaves.

The flow source at $x = -\infty$ and the sink at $x = +\infty$ are necessarily in different equilibrium (ρ, u, e) states. A relation between the two can be obtained by eliminating the two velocities, $u_{\rm left}$ and $u_{\rm right}$, from the three conservation laws. The result is the "Hugoniot Relation" :

$$E_{\rm right} - E_{\rm left} = (P_{\rm right} + P_{\rm left})(V_{\rm left} - V_{\rm right})/2 \ .$$

If the "right" and "left" states are similar, the Hugoniot relation approaches the First Law of Thermodynamics for adiabatic compression or expansion, $\dot{E} = -P\dot{V}$.

Given the Hugoniot relation, the equilibrium equation of state gives the exit state in terms of the initial state or "*vice versa*". For instance, the two-dimensional ideal-gas equation of state, $PV = E = NkT$, gives :

$$\frac{P_{\rm right}}{P_{\rm left}} = \frac{3V_{\rm left} - V_{\rm right}}{3V_{\rm right} - V_{\rm left}} \ .$$

For the two-dimensional ideal gas the localized black-box compression is limited to be less than three-fold. For three-fold compression the pressure ratio diverges. In the one- and three-dimensional cases, respectively, two-fold and four-fold compressions correspond to divergent values of $P_{\rm right}$.

The *two*-dimensional application of smooth-particle weighting functions to shockwave density profiles illustrates the dependence of continuum field variables on the the range of the weight function h . The width of continuum shockwaves is governed by the viscosity, through the strain-rate (du_x/dx) , and also depends, roughly linearly, on the range of the weight function.

As a warmup demonstration example problem consider the two-fold shockwave compression of a two-dimensional square-lattice soft-disk fluid, with the short-ranged very smooth pair potential :

$$\phi(r < 1) = (10/\pi)(1 - r)^3 \ ; \quad \rho \ : 1 \longrightarrow 2 \ .$$

We use the results of equilibrium molecular dynamics simulations to compute the velocities, pressures, and energies corresponding to the two-fold compression of the zero-pressure "fluid". Further details of this specific simulation can be found in our paper "Tensor Temperature and Shockwave Stability in a Strong Two-Dimensional Shockwave" arχiv 0905.1913: Physical Review E **80** , 011128 (2009) . The "shock velocity" u_s for two-fold compression is necessarily exactly twice the "particle velocity" u_p :

$$u_s = 1.750 \ ; \ u_p = 0.875 \ ; \rightarrow \{ \ \rho : 1 \longrightarrow 2 \ \} \ .$$

The equilibrium simulations give the conservation relations for the fluxes of mass, momentum, and energy, based on an initial zero-pressure state at a square-lattice density of unity and a final pressure of 1.531 at a density of 2.0 :

$$\rho u : 1.0 \times 1.75 = 2.0 \times 0.875 = 1.75 \ ;$$
$$P + \rho u^2 : 0.0 + 1.0 \times 1.75^2 = 1.531 + 2.0 \times 0.875^2 = 3.062 \ ;$$
$$\rho u[\ e + (P/\rho) + (u^2/2) \ ] = 1.75[\ 0.383 + 0.766 + 0.383 \ ] = 1.75 \times 1.531 \ .$$

Here P is the equilibrium pressure and e is the equilibrium internal energy (per unit mass). Of course mass, momentum, and energy are conserved *away* from equilibrium too. In the steady nonequilibrium state the constant mass flux is also ρu . The momentum flux requires replacing P by P_{xx} , accounting for the additional momentum flux resulting from the velocity gradient [or density gradient, as $(d\ln\rho/dx) = -(d\ln u/dx)$] . The

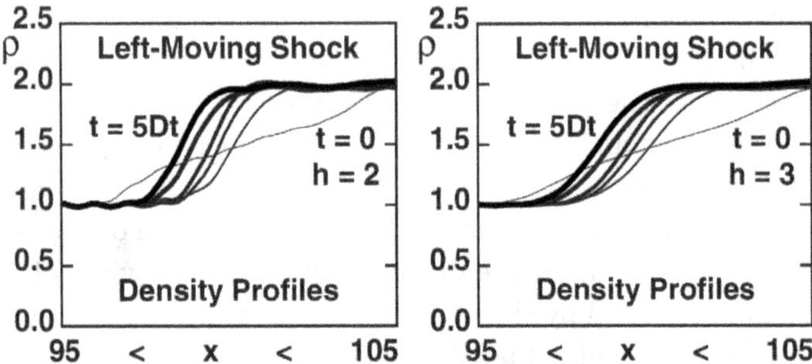

Fig. 7.10 Density profiles for two-fold compression beginning with a stress-free square lattice at unit density. The profiles at five successive multiples of the time $Dt = (40/u_s) = 2000dt$ are shown, with $h = 2$ on the left and $h = 3$ on the right. Increasing time corresponds to increasing line width as the shockwave moves slowly to the left.

energy flux too has a gradient-dependent part, the heat flux Q . Likewise, e is not necessarily the equilibrium internal energy. Conventionally viscosity is used to describe the nonequilibrium part of the pressure tensor :

$$P_{xx} = P_{\rm eq} - (\lambda + 2\eta)(du/dx) \ ; \ P_{yy} = P_{\rm eq} - \lambda(du/dx) \ .$$

Varying the range of the weight function from 2 to 4 provides a velocity gradient which varies by roughly a factor of two while the shear stress, $(P_{yy} - P_{xx})/2$ is relatively insensitive to h .

Recall that the smooth-particle density is given by the sum of nearby particle contributions :

$$\rho(x) \equiv \sum_j mw(\ |x_j - x| \ ) \ ; \ m \equiv 1 \ ;$$

$$w = (5/4h)[\ 1 + 3(z/h)\ ][\ 1 - (z/h)\ ]^3 \ ; \ z \equiv |\ x_j - x\ | \ .$$

The density profiles shown in **Figure 7.10** indicate that a reasonable choice for the range is $h = 3$. The profiles with $h = 2$ are undesirably "wiggly" and would defeat our goal of generating reasonable continuum representations of atomistic shockwaves. In the next Section we implement a detailed analysis, starting out with a triangular lattice with initial density $\rho = \sqrt{4/3}$.

7.10 Shockwaves, Tensor Temperature, and Time Delay

7.10.1 *Evolution of a Sinusoidal Shockwave*

Let us now consider in more detail the most important example of a localized adiabatic deformation, a stationary shockwave. In the coordinate

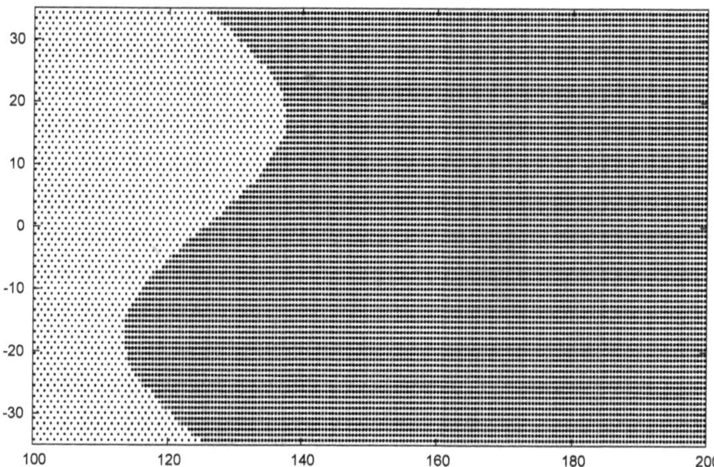

Fig. 7.11 The initial conditions, with time $t = 0$, corresponding to a strong sinusoidal shockwave where the flow is from left to right. Here the short-ranged pair potential is $\phi(r < 1) = (10/\pi)(1-r)^3$. The boundary conditions correspond to two-fold compression of a stress-free triangular lattice with shock velocity $u_s = 1.930$ and particle velocity $u_p = 0.965$. The next three Figures illustrate the rapid stabilisation of a planar wave from this initial condition with the density changing from $\sqrt{4/3}$ to $2\sqrt{4/3}$.

system of **Figures 7.11-7.14** proper choices of the mass, momentum, and energy fluxes leave the shockwave structure (a stationary irreversible compression process) fixed in space, with cold material flowing in at the left and hot fluid exiting at the right. From the computational standpoint it is relatively simple to introduce a new column or a new plane of particles at regularly spaced times, giving the mass flux directly (number of particles introduced per unit time and per unit length or per unit area, for two-dimensional or three-dimensional problems).

The righthand boundary can likewise be treated as a computational treadmill, with any particle closer than the range of the forces to the boundary given the fixed exit speed until it does exit. If desired, the entrance and/or the exit speed can be adjusted until the shockwave location is stabilised at its desired value of the x coordinate. Because it is not very time consuming there is nothing wrong with an alternative approach, measuring the equilibrium equation of state and then applying it to the Hugoniot equation and the conservation laws to find the exit states compatible with any specified entrance state. The figures show four stages in the evolution from a sinusoidal shockwave to a planar one suitable for analysis. The analysis is carried out using the smooth-particle ideas from Chapter 4 .

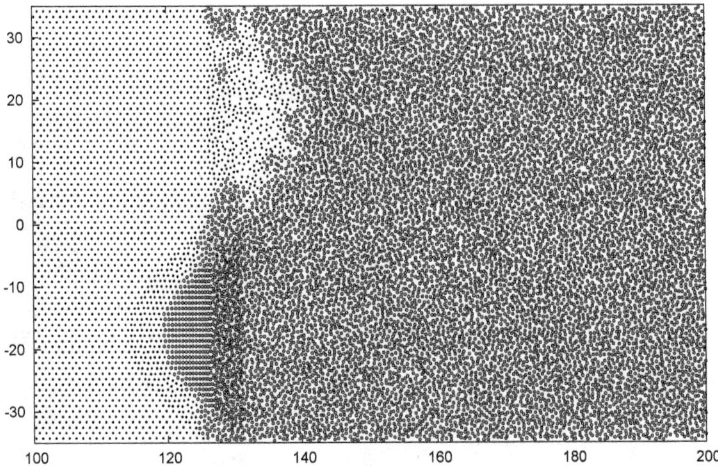

Fig. 7.12 Time $t = 2$ in the evolving sinusoidal shock problem. The system height is $80\sqrt{3/4}$ with an initial density (from a close-packed triangular lattice) of $\sqrt{4/3}$.

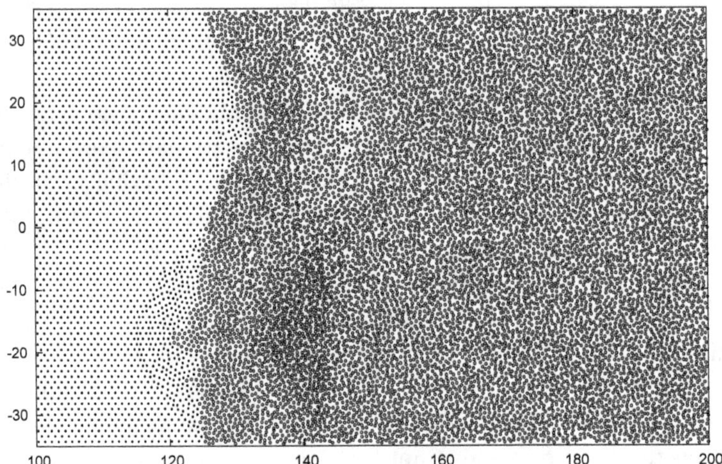

Fig. 7.13 Time $t = 4$ in the evolution from a strong sinusoidal shockwave to a stable planar wave. The cold lattice is converted to a hot fluid with temperature $T_{\rm eq} = 0.12$.

7.10.2 *Analysis of a Planar Shockwave Profile*

Let us use smooth-particle techniques to analyze the long-time limiting shockwave of **Figure 7.14**. We compute the mechanical and thermal av-

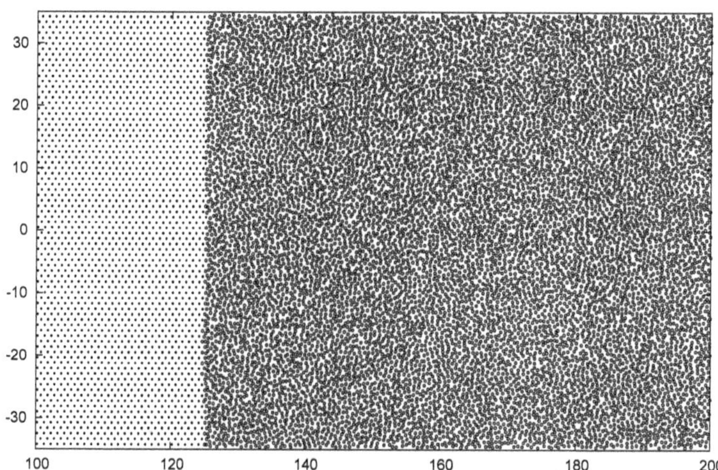

Fig. 7.14 Time $t = 120$ snapshot confirming the long-time stability of the planar shockwave evolving from the initial sinusoidal shockwave of **Figure 7.11** . The corresponding smooth-particle thermal and mechanical profiles appear in **Figure 7.15** .

erages using Lucy's weight function, as described in Chapter 4 :

$$\langle\ f(x_g)\ \rangle = \sum_{x-h}^{x+h} f_j w_{gj}\ ;\ w \propto [\ 1 + 3(|x|/h)\ ][\ 1 - (|x|/h)\ ]^3\ .$$

They appear in **Figure 7.15** . There is no *a priori* uniquely correct choice of the range of the weight-function h . Ideally h should be chosen so as to simplify the macroscopic description of the microscopic averages. In order that the chosen averaging procedure be applicable to transient as well as stationary states a circle of radius h needs to include about twenty particles.

To summarize this shockwave work — we observed several interesting features in the series of shockwave simulations. These features can all be seen in **Figures 7.11-7.15** .

[i] If the initial location of the shockwave is sinusoidal, $\propto \sin(2\pi y/H)$, where H is the height of the simulation cell, the wave soon becomes planar. This indicates that one-dimensional shockwaves *are* stable in two space dimensions.

[ii] If the directional dependence of the kinetic temperature is measured the longitudinal temperature typically exceeds the transverse one, often by an order of magnitude, $T_{xx} >> T_{yy}$.

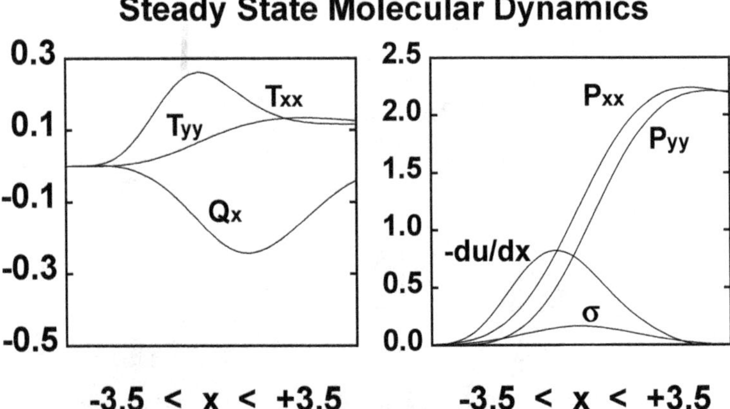

Fig. 7.15 Tensor temperature and heat flux (on the left) along with the pressure tensor, strain rate, and shear stress (on the right) as computed from the shockwave profile of **Figure 7.14** . All of the averages were calculated using Lucy's twice-differentiable one-dimensional weight function, with $h = 3$. Notice that the temperature gradients precede (are to the left of) the heat flux and that the velocity gradient precedes the shear stress. In the usual irreversible constitutive equations no such delays are included.

[iii] If the strain-rate is measured—(du/dx) can readily be determined by a numerical differentiation of $u(x)$—typically the maximum strainrate precedes the maximum anisotropy in the pressure tensor, $(P_{xx} - P_{yy})$. Likewise, the maximum temperature gradient precedes the maximum amplitude of the heat flux vector component Q_x . These examples show that shockwaves can exhibit *time delay*, with the "cause" (a gradient) preceding the "effect" (a flux) by a time of order the collision time.

[iiii] It is not unusual to find that T_{xx} , or $(1/2)(T_{xx}+T_{yy})$, has a maximum *within* the shockwave. If Fourier's Law held in this situation there would have to be a *negative heat conductivity* on the hot side of the shockwave.

[v] The strain-rate dependence of the viscosity in shockwaves (shear thickening rather than shear thinning) is opposite to that found in simple shear. Data for moderately-strong shockwaves suggest that the effective viscosity, $\eta \equiv -(P_{xx} - P_{yy})(dx/du)/2$ can exceed the low-strain-rate Newtonian viscosity by as much as thirty percent.

The stationary shockwave, in particular the time delays between "causes" (velocity and temperature gradients) and "effects" (shear stress and heat flux) , are themselves challenging projects in the mesoscopic analysis of atomistic flows. Even more challenging are descriptions of the transient geometry seen in **Figures 7.12 and 7.13** . We are confident that the pursuit of accurate descriptions will enhance future understanding of real-world systems far from equilibrium.

7.11 The Simplest Continuum Models for a Shockwave

The sophistication of continuum models for one-dimensional stationary shockwaves varies from the simplistic to the baroque. A relatively thorough discussion emphasizing models on the simplistic side occupies Chapter 6 of our 2012 book, *Time Reversibility, Computer Simulation, Algorithms, Chaos*. Of the many models discussed there let us review the simplest constitutive model leading to a stationary shockwave in a two-dimensional fluid, a constant shear viscosity η . To avoid a negative tangential pressure component it is convenient to choose the bulk viscosity equal to the shear. This means that the viscous resistance associated with rapid compression is comparable to that associated with rapid shear. In two space dimensions this choice gives $\eta_V = \lambda + \eta = \eta \rightarrow \lambda = 0$. The corresponding pressure tensor is :

$$P_{xx} = P_{\rm eq} - (\lambda + 2\eta)(du_x/dx) = P_{\rm eq} - 2\eta(du_x/dx) \ ;$$

$$P_{yy} = P_{\rm eq} - \lambda(du_x/dx) = P_{\rm eq} \ .$$

Although choosing the bulk viscosity equal to zero might *seem* simpler that choice leads to *tension* in the y direction, which is completely implausible for a simple fluid with short-ranged repulsive forces.

To illustrate this simplest of shock structures consider the two-fold compression of a fluid with an equilibrium equation of state chosen to simplify the numerical work :

$$P = \rho e \ ; \ e = (\rho/2) + 2T \ .$$

This model is suggested by Grüneisen's idea of separating pressure into a "cold" part and a "thermal" part.

For an initial cold state ($T = 0$) and with the density changing from 1 to 2 the Hugoniot relation, $\Delta e = -\langle \ P \ \rangle \Delta V$, can be solved analytically :

$$e_{\rm hot} - e_{\rm cold} = (1/2)e_{\rm hot} + (1/4)e_{\rm cold} \rightarrow e_{\rm hot} = (5/2)e_{\rm cold} = (5/4) \ .$$

Evidently the changes of the state variables within the shockwave are as follows :

$$\rho \ : 1 \rightarrow 2 \ ; \quad P \ : (1/2) \rightarrow (5/2) \ ; \quad u \ : 2 \rightarrow 1 \ ;$$

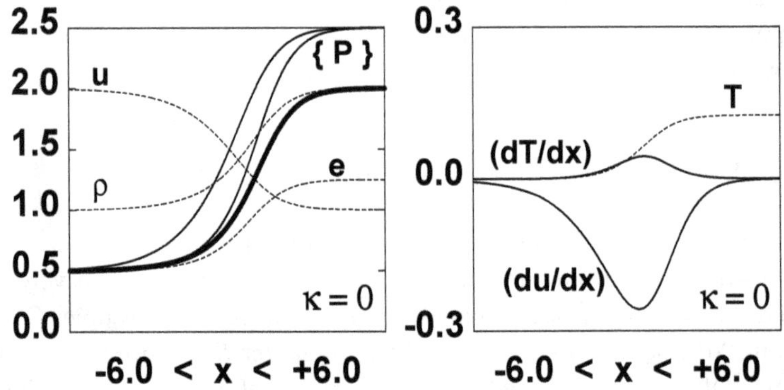

Fig. 7.16 Stationary shockwave profiles given $P_{xx} = \rho e - 2(du/dx)$; $e = (\rho/2) + 2T$. The mass, momentum, and energy fluxes, $\{ 2, (9/2), 6 \}$, are constant throughout. The heavy line is the "cold-curve" pressure, $(\rho^2/2)$.

$$e \ : (1/2) \to (5/4) \ \ ; \ \ T \ : 0 \to (1/8) \ ,$$

from which the constant fluxes of mass, energy, and momentum follow :

$$\text{Flux}_{\text{mass}} = \rho u = 2 \ ;$$

$$\text{Flux}_{\text{energy}} = \rho u[\ e + (P_{xx}/\rho) + (u^2/2) \ ] + Q_x = 6 \ ;$$

$$\text{Flux}_{\text{momentum}} = P(\rho, e) - 2\eta(du/dx) + \rho u^2 = P_{xx} + \rho u^2 = (9/2) \ .$$

In the interest of simplicity we choose the heat flux $Q_x \equiv 0$. Then these three flux equations can be solved for $\{ \ \rho, e, (du/dx) \ \}$. Eliminating the density and energy gives a simple differential equation for the velocity profile (where the shear viscosity is equal to unity) :

$$(du/dx) = (3/2u)[\ 2 - u \ ][\ 1 - u \ ] < 0 \ ; \ 1 < u < 2 \ .$$

A single-equation Runge-Kutta solution, started near the "hot" state, provides the velocity profile shown in **Figure 7.16** . With the velocity determined, the flux equations provide the other mechanical and thermal variables. The sequence of steps is as follows :

$$u \longrightarrow \rho \longrightarrow P_{xx} \longrightarrow e \longrightarrow P(\rho, e), T(\rho, e) \longrightarrow (du/dx) \ .$$

Mechanical and thermal variables corresponding to this simple model are shown in **Figure 7.16** . Similar steps, but with *two* differential equations rather than just *one*, can be carried out if $Q_x = -\kappa(dT/dx) \neq 0$.

7.12 More Complex Continuum Models for a Shockwave

Given the results from molecular dynamics, shown in **Figure 7.15** , it is desirable to include additional ideas in a shocked fluid's constitutive equations relating heat flux, shear stress, tensor temperature, and energy. Of course it is essential to confirm that the improved constitutive equations conform to the constancy of the mass, momentum, and energy fluxes in the shockwave. We illustrate the use of all the new ideas by modifying our Grüneisen model to allow for differences between the longitudinal and transverse tempereatures while still separating the equilibrium pressure and energy into "cold" density-dependent and "thermal" temperature-dependent parts :

$$P_{\rm eq} = \rho e \ ; \ e = (\rho/2) + T_{xx} + T_{yy} \ .$$

This equilibrium equation of state (where $T_{xx} = T_{yy} = T$) supports exactly the same changes in equilibrium properties as the simplest model of the previous Section :

$$1 < \rho < 2 \ ; \ \rho u \equiv 2 \ ;$$

$$(1/2) < e < (5/4) \ ; \ P_{xx} + \rho u^2 = (9/2)$$

We have successfully augmented this simplest two-temperature model with all *three* modifications of the previous Section's simple model suggested by molecular dynamics simulations :

First, given that temperature can be anisotropic, temporarily, the longitudinal and transverse temperatures must necessarily *relax* toward a common value :

$$\dot{T}_{xx} \supset (T_{yy} - T_{xx})/\tau \ ; \ \dot{T}_{yy} \supset (T_{xx} - T_{yy})/\tau \ .$$

Second, the thermal parts of the compressive rate of work and the conductive flow of heat (call these $\dot{W}$ and $\dot{Q}$) can contribute differently to the longitudinal and transverse components of temperature :

$$\rho\dot{T}_{xx} \supset -\alpha \times \dot{W} - \beta \times \dot{Q} \ ;$$

$$\rho\dot{T}_{yy} \supset -(1-\alpha) \times \dot{W} - (1-\beta) \times \dot{Q} \ .$$

Third, the heat flux itself responds differently to the longitudinal and transverse temperature gradients, and, like shear stress $[\ P_{xx} - P_{yy}\ ]/2 = \sigma$ grows or decays with a characteristic relaxation time :

$$\sigma + \tau_\sigma\dot{\sigma} = -\eta(du/dx) \ ; \ Q + \tau_Q\dot{Q} = -[\ \kappa_{xx}(dT_{xx}/dx) + \kappa_{yy}(dT_{yy}/dx)\ ]/2 \ .$$

The details required to implement all of these ideas can be found in our paper with Paco Uribe, "Maxwell and Cattaneo's Time-Delay Ideas Applied to Shockwaves and the Rayleigh-Bénard Problem", arχiv 1102.9560 : Communications in Science and Technology **19** , 5-12 (2013) .

Two points discovered in that implementation are of particular interest. If the response times are too long the relaxation equations become unstable and have no solution. Also, the presence of the relaxation times destabilizes the usual practice, integrating the constant-flux equations from "hot" to "cold". This hot-to-cold instability can be avoided by solving the complete transient set of continuum equations beginning with an approximate model profile, such as that generated in Section 7.11 . Provided that the relaxation times are not too long this algorithm produces stable stationary solutions including both thermal relaxation and time delay. A "staggered grid" of cells, with the density, stress, and heat flux defined "in" the cells, the velocity, temperatures, and energy defined at the two nodes bounding each cell, and the gradients approximated by finite differences provides a stable algorithm. Such problems are ideally suited to thesis work.

7.13 Summary

Nonequilibrium simulations require driving forces to maintain their distance from equilibrium. This driving can be imposed by [i] boundary conditions (as in shear flow or heat flow) or by [ii] moving particles (as in the shock wave). Because nonequilibrium states are dissipative, producing heat and entropy, some ongoing mechanism for extracting the heat must be supplied. Thermostat or ergostat forces and outgoing flows (as in shockwaves) can all of them remove the extra entropy produced by dissipation within the system. Green-Kubo perturbation theory can provide a guide to designing nonequilibrium systems in a way consistent with linear transport theory.

A general finding from the nonequilibrium simulations, quite consistent with Boltzmann's equation, is that *nonlinear* effects (such as quadratic or cubic constributions to nonequilibrium fluxes) are relatively small and difficult to observe. The normal stresses in a shear flow and the increase (or decrease) of viscosity with strain rate are examples. Fractal nonlinearity for heat flow *can* be observed in the ϕ^4 model of heat conduction. This model provides the simplest possible illustration of phase-space dimensionality loss away from equilibrium. With this model it is feasible to simulate dimensionality losses which are a significant fraction of the total number of degrees of freedom. This loss provides an understanding of the Second Law of Thermodynamics, to which we return in the next Chapter.

Shockwaves are a particularly rich source of nonequilibrium information. The boundary conditions are simple: cold and hot equilibrium states. Among the many interesting features of shockwaves are the anisotropicity of the temperature, with $T_{xx} >> T_{yy}$ in strong shockwaves, the maximum of the temperature within the shock, signalling the failure of Fourer's Law to provide even the direction of heat flow, and the delayed response of the stress and heat flux to the velocity and temperature gradients which drive them. Understanding and modelling these effects by incorporating Lucy's weight-function analysis into the description of nonequilibrium states will remain an active research area for some time. Such problems are particularly well-suited to graduate and post-graduate research.

7.14 References

The shear flows are described in our paper with Janka Petravic, "Simulation of Two- and Three-Dimensional Dense-Fluid Shear Flows *via* Nonequilibrium Molecular Dynamics: Comparison of Time-and-Space-Averaged Stresses from Homogeneous Doll's and Sllod Shear Algorithms with Those from Boundary-Driven Shear", Physical Review E **78**, 046701 (2008). See also the normal-stress search : "Investigation of the Homogeneous-Shear Nonequilibrium-Molecular-Dynamics Method", Physical Review A **45**, 3706-3713 (1992), by S. Y. Liem, D. Brown, and J. H. R. Clarke.

The shear viscosity data showing that the two-dimensional transport coefficient is well-defined for fluids is taken from Wm. G. Hoover and Harald Posch, "Large-System Hydrodynamic Limit", Molecular Physics Reports **10**, 70-85 (1995). For some well-executed three-dimensional Rayleigh-Bénard simulations see Denis Rapaport's "Hexagonal Convection Patterns in Atomistically Simulated Fluids", Physical Review E **73**, 025301R (2006).

The high-strain-rate shear of solids is a measurement of the dynamic yield strength. For a summary of the early simulations see William G. Hoover, Anthony J. C. Ladd, and Bill Moran, "High-Strain-Rate Plastic Flow Studied *via* Nonequilibrium Molecular Dynamics", Physical Review Letters **48**, 1818-1820 (1982). Tony Ladd and Bill also undertook a quantitative study of the utility of dislocation dynamics in explaining the shear of solids : "Plastic Flow in Close-Packed Crystals *via* Nonequilibrium Molecular Dynamics, Physical Review B **28**, 1756-1762 (1983).

7.15 Problems

1. Develop thermostated Doll's Tensor motion equations to measure bulk viscosity in a periodic system of variable sidelength $L(t)$.

2. For a periodic shear-flow simulation work out the equations describing the "nearest-image" treatment of the x coordinate for a particle leaving its periodic box. Work out also the nearest-image distance for a pair of particles in neighboring horizontal rows of periodic cells.

3. Show that the Cantor set on the line segment $[\ 0 \ldots 1\ ]$ consists of all the base-3 fractions composed of 0s and 2s (without any 1s).

4. Show that the Hugoniot relation of Section 7.9 ,

$$E_{\rm right} - E_{\rm left} = (P_{\rm right} + P_{\rm left})(V_{\rm left} - V_{\rm right})/2 \ ,$$

follows from the conservation relations for a steady shockwave which converts one equilibrium (P,V,E) state to another.

5. Find the ratio $[\ P_{\rm right}/P_{\rm left}\ ]$ as a function of $[\ V_{\rm right}/V_{\rm left}\ ]$ for the shockwave compression of a *one*-dimensional ideal gas with the equation of state $PV = 2E = NkT$.

6. Use Runge-Kutta integration to solve the Maxwell-Cattaneo relaxation equation for relaxation times $\tau = 10,\ 1,$ and 0.1 :

$$\sigma + \tau\dot{\sigma} = \dot{\epsilon} \text{ with } \dot{\epsilon} = [\ e^{-t} + e^{+t}\ ]^{-1} \ .$$

Chapter 8

Second Law, Reversibility, Instability

Topics

Time Reversibility, Boltzmann, Loschmidt, Zermélo / Second Law of Thermodynamics for Hamiltonian Systems / Second Law of Thermodynamics for Thermostated Systems / Averaging, Ergodicity, Recurrence, Constraints / Thermostated Ergodicity for the Galton Board / Thermostated Ergodicity for Harmonic Oscillators / Nonequilibrium Harmonic Oscillators / Hamiltonian Exponent Pairing for Small Systems / Lyapunov Instability for Many-body Systems / Antique Dogs and Stochastic Fleas /

8.1 Time Reversibility: Boltzmann, Loschmidt, Zermélo

The time-reversibility of microscopic classical mechanics is apparent from the motion equations themselves. The classical equations of Newton, Lagrange, and Hamilton, are not only "time-reversible". They also conserve phase volume. Today's *nonHamiltonian* "dissipative" motion equations, with *decreasing* phase volume on average, leading to fractals, are still time-reversible. The "new" approach has evolved over a 40-year period dating for us back to Bill Ashurst's thesis work in the 1970s. New discoveries, both conservative and dissipative, are still coming to light today. Levesque and Verlet's bit-reversible version of the Leapfrog algorithm exhibits precise time reversibility. By contrast, no such algorithm is possible for dissipative motion equations, despite their formal time reversibility, as their overall many-to-one phase-space mappings can only be reversed by storing their past history.

Microscopic time-reversible motion equations, either conservative or dissipative, are relatively easy to "solve" with a variety of numerical algorithms. Chaos, through Lyapunov instability, necessarily renders any longtime "solution" open to question and, in the dissipative case, any time-reversed longtime solution open to the same incredulity provoked by time-reversed movies.

Microscopic mechanics as well as the bit-reversible algorithm pass the fundamental reversibility test: reversed movies of an isolated Newtonian system can be played in either direction and obey the same evolution equations in both directions. Time-reversed movies of continuum problems, like time-reversed movies of real life, look unphysical, wrong, and do not obey the equations used to produce the forward movie. Time-reversed movies of microscopic simulations of free expansion and shockwave propagation look unphysical too, despite their close relationship to Newton's equations of motion and to viscous, heat-conducting continuum mechanics. In all our simulations, microscopic and macroscopic, our conclusions based on specific example problems are not sensitive to our choice of numerical methods, despite ubiquitous Lyapunov instability. We believe that by promoting mixing and ergodicity this instability is actually a help rather than a hindrance.

In contrast to the microscopic view, macroscopic mechanics is typically *irreversible*. The underlying evolution equations are also harder to solve because they are *partial* differential equations in which both the space and the time dependence of the dependent variables must be found. Solutions of these macroscopic problems are more challenging because they include *gradients* on the righthand sides, often include singularities, and are likely also to be subject to Lyapunov instability. Newtonian viscosity and Fourier heat conduction provide clear examples of the irreversible nature of the macroscopic approach. Newton's viscosities *always* act to reduce velocity gradients. Likewise Fourier's model for heat flow is *invariably* in the direction from hot to cold, regardless of the direction of macroscopic velocities.

How could it be that the *irreversible* processes we see all around us stem from time-reversible laws? Of course this is a "Good Question" and a longstanding one. Boltzmann's statistical treatment of low-density gas dynamics is often accepted as providing "The Answer". His kinetic equation for the evolution of a gas' density, velocity, energy, and entropy predicts that an isolated Newtonian system would never lose entropy. Evidently the averaging process he used in order to motivate faith in the irreversible Boltzmann equation contradicts mechanics and is simply an approximation. On just these grounds Boltzmann's irreversible equation was criticized by Loschmidt. Loschmidt reasoned that because any solution of Newton's equations leading to an entropy *increase* could be reversed, giving a *de-*

crease, Newton's, Lagrange's, and Hamilton's evolution equations *cannot* obey the Second Law. From this reversibility viewpoint an isolated Newtonian system can only have a *constant* entropy. This same conclusion follows from Liouville's Theorem.

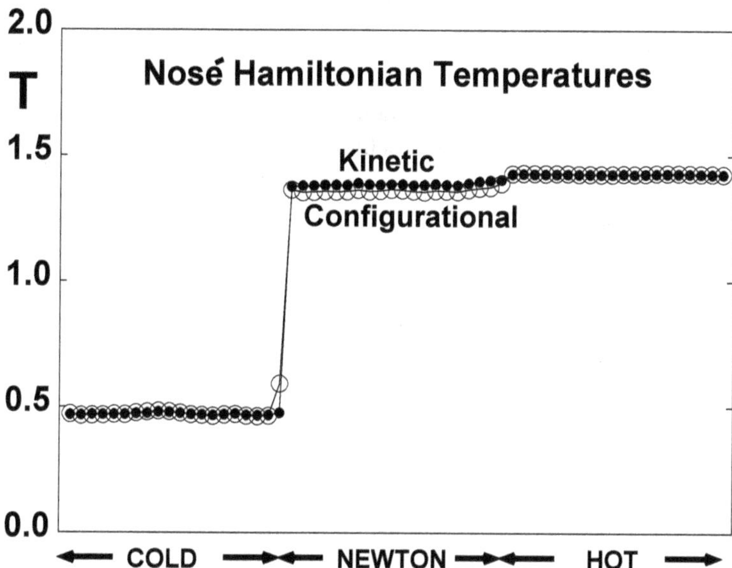

Fig. 8.1 Temperature profiles for a ϕ^4 chain of 60 one-dimensional particles. In the two 20-particle reservoir regions as well as the Newtonian region the configurational and kinetic temperatures agree. At the same time the Hamiltonian nature of the two heat reservoirs prevents the transfer of heat despite the the tremendous temperature gradient.

Loschmidt would have applauded our **Figure 6.9** on page 167 . That figure was generated using two Hamiltonian (Hoover-Leete) isokinetic thermostats, with fixed kinetic energy, $K_0 = K_{\rm cold}$ or $K_{\rm hot}$:

$$\mathcal{H}_{\rm HL}(T) = 2\sqrt{K_pK_0} - K_0 + \Phi \ ;$$

$$K_p = \sum (p^2/2m) \ ; \ K_0 = (DNkT/2) \ ,$$

Though these thermostats *do* simultaneously constrain the kinetic temperatures in the two heat reservoirs, the resulting system *does not* promote heat flow.

We display a second such example above, based on Nosé's original idea for temperature control :

$$\mathcal{H}_{\rm Nosé}(T) = (K_p/s^2) + \Phi + (p_s^2/2M) + K_0 \ln(s^2) \ ,$$

in **Figure 8.1**. Two different canonical temperatures were imposed on hot and cold reservoir regions (20 ϕ^4 particles each) using two independent values of Nosé's time-scaling variable, $s_{\rm cold}$ and $s_{\rm hot}$. In the two Nosé thermostat regions the kinetic and configurational temperatures agree, while in the Hoover-Leete case they do not. But in both these Hamiltonian examples the 20-particle Newtonian regions coupled to the two thermostats showed no tendency to transport heat from the hot region to the cold or to display a reasonable temperature gradient. These two examples, both of them fully time-reversible, vindicate Loschmidt's view that classical mechanics cannot "explain" the Second Law, despite Boltzmann's approximate work. *Any entropy increase in a a purely-classical-mechanical evolution would imply the possibility of an (illegal) decrease, by time reversal.* We emphasize that Loschmidt's and Zermélo's objections properly apply to *individual mechanical systems* rather than to hypothetical ensemble averages.

A tempting rejoinder to these Loschmidt-Zermélo objections, or paradoxes, is that continuum evolutions and classical trajectories are simply two different *models* of behavior. In order for them to conform some form of ensemble averaging over trajectories must be carried out. Proper averages *will* obey the Second Law and will not obey Liouville's incompressible Theorem. The fractal nature of nonequilibrium steady states' distributions indicates that the required average will not be simple as it must extend over all distance scales. It seems implausible that *any* averaged incompressible flow can lead to an *increased* phase volume.

Boltzmann's 1872 point of view, that a low-density gas evolves in the direction of more available states, is often accepted today. But in fact Boltzmann's equation, like the Navier-Stokes equations, is just a "model", clearly a model unlike classical mechanics and even less like continuum mechanics. It has the virtue of simplicity and utility, as it provides a quantitative basis for dilute-gas dynamics. In developing models Thoreau's message, "Simplify, simplify, simplify", is good advice. Roberts and Quispel, in their 1992 Physics Reports paper, sought to generalize time-reversibility by including "involution" mappings of phase space into itself. Their definition of "time-reversible" dynamical systems, with time t and with phase-space variables $\{ x \}$, includes motion equations invariant under the combination of operations :

$$+t \to -t \ ; \ x \to G \cdot x \text{ with } G \cdot G = \text{Identity} \ .$$

Carl Dettmann showed, in his 2014 paper "Diffusion in the Lorentz Gas", Communications in Theoretical Physics **62** , 521-540, that the underdamped and fully dissipative simple harmonic oscillator,

$$\dot{q} = p \; ; \; \dot{p} = -q - p \; ;$$

$$G(q,p) = (q, -q - p)/[\ q^2 + (4/3)(p + (q/2))^2\ ] \; ,$$

satisfies the not-so-simple Roberts-Quispel reversibility definition. A likely moral is that arguments based on arbitrary combinations of coordinates and momenta (such as $q + p$) are more mathematical than physical in nature. Let us accept Hamiltonian mechanics as a model and see to what extent it is consistent with the Second Law.

8.2 Second Law for Hamiltonian Systems

Despite Loschmidt's and Zermélo's objections it is apparent that some Hamiltonian systems *can* obey the Second Law if they have so many states that covering all of them would require the Age of the Universe, around 14 billion years. Assuming that we can shuffle through 10^{12} operations per second the Age allows perhaps 10^{30} operations :

$$10^{12}[\text{ ops/sec }] \times 3 \cdot 10^7[\text{ sec/year }] \times 14 \times 10^9[\text{ years }] \simeq 10^{30} \simeq 2^{100} \; .$$

This means that shuffling two decks of cards together is enough to create more states than one can conceivably examine.

The constant-energy free expansion of a Hamiltonian gas (with more than 100 particles) into a larger container appears to correspond to an increase in thermodynamic entropy, as more states are available in the larger container. Evidently the Second Law and ensemble-averaged Hamiltonian mechanics disagree, as Liouville's Theorem allows for no increase in Gibbs' entropy for an isolated Hamiltonian system. Zermélo identified and objected to a related difficulty. Liouville's Theorem of incompressible flow for phase-space probability density implies that representative systems from a tiny volume element in phase space will *eventually* return to that element's initial location. This recurrence of the initial state implied by Hamiltonian phase-volume-conserving classical mechanics is counter to macroscopic continuum mechanics, which is, like life itself, irreversible.

It has gradually become clear that Hamiltonian mechanics by itself is inadequate to the task of describing steady nonequilibrium flows. Because phase-space flows that generate heat necessarily expand to larger and larger hypervolumes, a *nonHamiltonian shrinkage* of phase volume, corresponding to a controlled extraction of heat, is an integral part of any steady-state nonequilibrium simulation. We will soon consider the details of two simple example problems, [i] the field-driven Galton Board, in which a particle wandering through a periodic lattice undergoes impulsive discontinuous collisional jumps in velocity, and [ii] the heat conducting Doubly-Thermostated Oscillator. In both these *nonequilibrium* cases *any* equilibrium initial condition, chosen from Gibbs' ergodic microcanonical Board or his canonical Oscillator distribution, leads to the *same* steady-state nonequilibrium fractal result. The ergodic and reversible oscillator dynamics is smoothly controlled by *two* time-reversible thermostat variables. Dissipation, driven by a temperature gradient, leads to dissipative phase-space structures. These distributions can be either fractal or limit-cycle, depending on the size of the temperature gradient. The Galton Board, like the conducting oscillator, also leads to dissipation, through the conversion of gravitational field-energy into heat. The corresponding phase-space distribution is likewise represented by fractal or limit-cycle distributions. We will explore both these nonequilibrium problems after a consideration of conservative time-reversible Hamiltonian mechanics.

For conservative Hamiltonian systems Boltzmann and Gibbs' views of a digital phase space geometry, "coarse graining", provide accurate pictures of "irreversibility", in the overwhelming tendency of systems to occupy portions of phase space closer to "equilibrium" than is their initial condition. For this reason it is tempting to start with Hamiltonian mechanics in our efforts to "understand" the Second Law. Let us specialize to conservative Cartesian systems, with a digital dynamics sharing both its (Poincaré) recurrence and its (Loschmidt) reversibility features with the corresponding properties of the Levesque-Verlet bit-reversible algorithm :

$$\{ \ q_{t\pm dt} - 2q_t + q_{t\mp dt} = [\ F(\{ \ q_t \ \})(dt^2/m) \ ]_{\rm Integer} \ \}$$

$$[\text{ Integer Coordinates }] \ .$$

Such a coarse-grained integer-coordinate description, in a bounded space, is *the* natural representation of conservative mechanics, patently reversible [Loschmidt], as well as recurrent [Zermélo]. Notice that the *only* trajectories possible with these motion equations are (if we are patient) periodic cycles of states. Because the momenta are not a part of the coordinate-based "leapfrog" equations of motion it is natural to think of two successive coordinate sets as defining a single integer-space "state".

With just a finite number of these paired-coordinate states $\{ \ q_t, q_{t\pm dt} \ \}$ any given pair has a unique successor set as well as a unique predecessor set ,

$$\{ \ q_{t-dt} \text{ and } q_t \longrightarrow q_{t+dt} \ ; \ q_{t+dt} \text{ and } q_t \longrightarrow q_{t-dt} \ \} \ .$$

If the space is bounded, eventually the motion must recur, giving a closed cyclic periodic orbit which can be followed in *either* time direction. The entire cycle can equally well be played either forward or backward. It is *time-reversible.* Beginning with any "unlikely" initial condition guarantees in principle that the very same condition will recur in the future. Likewise the reversed trajectory, stumbling backward onto that same "unlikely" state, is only unlikely because its information content, that is, the difference from a typical equilibrium state, is so great. For such a system any apparent irreversibility is "temporary", though it will likely outlast the Universe.

Forward

1	6	11	16	21	26	31	36
41	46	51	56	61	2	7	12
17	22	27	32	37	42	47	52
57	62	3	8	13	18	23	28
33	38	43	48	53	58	63	4
9	14	19	24	29	34	39	44
49	54	59	64	5	10	15	20
25	30	35	40	45	50	55	60

Backward

64	59	54	49	44	39	34	29
24	19	14	9	4	63	58	53
48	43	38	33	28	23	18	13
8	3	62	57	52	47	42	37
32	27	22	17	12	7	2	61
56	51	46	41	36	31	26	21
16	11	6	1	60	55	50	45
40	35	30	25	20	15	10	5

Fig. 8.2 Checkerboard dynamics. The forward algorithm begins in the upper lefthand corner and has visited every square after 63 iterations, ending on the square marked "64". The backward algorithm retraces these steps precisely ending where the forward algorithm began. The sum of the integers in corresponding squares is 65 .

For a small-scale clarification of the recurrence and reversibility features of conservative dynamics imagine a standard 64-square configurational checkerboard with a rule specifying the "next" forward square and the "previous" backward one. Given an initial square q_0 the two sequences $q_0, \ q_{\pm dt}, \ q_{\pm 2dt}, \ \ldots$ will sooner or later (with at most 63 jumps) reach q_0 once more. So long as the rule is reversible (so that the sequence can be played backward to time 0) it is inevitable that the cycle reaches q_0 again.

Let us look at a simple example using an increment of 13 and with periodic boundaries imposed by the modulo function :

`inew = mod(i + 12,64) + 1` $\longrightarrow$ 1, 14, 27, 40, 53, 2, 15, 28, ...

[Iterating Forward in time]

This algorithm is easy to reverse. Because *subtracting* 13, modulo 64 (in order to go back) is the same as *adding* 51, the time-reversed equivalent of this "pseudorandom" generator has a similar form :

`iold = mod(i + 50,64) + 1` $\longrightarrow$ ..., 28, 15, 2, 53, 40, 27, 14, 1 .

[Iterating Backward in time]

Figure 8.2 shows the equivalent sequence of checkerboard squares covered by the forward and backward versions of this reversible algorithm. Starting at square 1 (also labelled "1") the result of a single forward iteration moves to the 14th square (second row, sixth column) which we label "2" . After the 63rd iteration we arrive at the 52nd square (seventh row, fourth column) which we label "64" . To reverse the algorithm starting at the 52nd square (now labelled "1", we apply the reversed algorithm which takes us to the 39th square (fifth row, seventh column) which we label "2". Another 62 iterations get back to the upper lefthand corner of the checkerboard, labelled "64". Notice that the algorithms forward and backward are perfectly reversible, deterministic, and ergodic (though with their constant stride of ± 13 from one square to its successor or from its predecessor they are not very good pseudorandom number generators) . Although the results of the forward and reversed algorithms "look" reversible the algebraic form are certainly not identical.

The ergodicity of the two forward and backward jump rules just given is exceptional. Jumps of 2, 4, 8, 16, 32, or 64 squares return after 32, 16, 8, 4, 2, or 1 jumps, respectively. Ordinarily simulated motion in a discrete state space recurs long before all states are visited. Good random number generators behave in the opposite fashion, visiting every single state before recurring. Because the atomistic equations of motion are second-order in time dynamical "states" are represented by *pairs* of numbers (two coordinates or one coordinate and one velocity). To make a rough estimate of the recurrence time for finite-precision dynamics (the time required for the integer algorithm to reach the starting pair of integers once more) suppose the integers have eight digits (corresponding to single-precision arithmetic) and that a "state" is represented by the 32 digits necessary to specify two succeeding pairs of x and y coordinates. A simulated N-body problem then charts its path through 10^{32N} states. If the jumps from one state to the next were random then the number of jumps required for recurrence is of order $\sqrt{10^{32N}} = 10^{16N}$. If we imagine a generous trillion jumps per second, $\simeq 10^{19}$ per year, we see that it is unrealistic ever to expect to see a repeat. And of course double precision would take much longer ! Boltzmann was well aware of this argument and used it as a defense against Zermélo's recurrence paradox.

To sum up, isolated conservative systems obey the Second Law of Thermodynamics through a statistical seeking out of the *many* closer-to-equilibrium states available to them. This type of understanding applies to (for instance) the collision of macroscopic bodies, the fracture of solids under tension and the thermalization of sound waves. The last of these problems is typical of the Green-Kubo description of the linear dissipative response of deviations from equilibrium. Simpler situations, including the one often used as a synonym for Second-Law behavior, steady heat flow from hot to cold, cannot be described by conservative Hamiltonian mechanics. A development from the 1970s and 1980s, "thermostated dynamics" makes possible not only just a new dynamics, but a new interpretation of the Second Law. Let us consider "thermostated" systems next.

8.3 Second Law for Thermostated Systems

By thermostating the *surroundings* of a "controlled" or "managed" mechanical system irreversibility comes about in a way which avoids Loschmidt's and Zermélo's objections. By introducing time-reversible reservoir variables to constrain local energies or temperatures we produce a time-reversible dynamics that reduces the phase-space *dimension* of the Hamiltonian systems to which they are coupled. The Second Law for such "open systems" is a consequence. The system's phase-space probability density shrinks in dimension and quickly converges to a multifractal flow, usually ergodic and connecting a repellor source to an attractor sink. This approach to implementing and analyzing thermostated irreversibility makes use of chaotic dynamical-systems analysis, providing tools more powerful and æsthetic than those available to Boltzmann and Gibbs.

Boltzmann's approximate treatment of dilute gases suggests that *conservative* Hamiltonian systems obey the Second Law by seeking out the abundance of states which are closer to "equilibrium". The mixing and phase-shifting effects of collisions combine microscopic time-reversible events so as to yield dissipative macroscopic behavior. But Boltzmann's approach is very special, and not readily generalized to liquids or solids. By taking advantage of thermostated ergodic simulations we can turn to a much more general *nonHamiltonian* explanation of the Second Law.

Thermostating makes it possible to model nonequilibrium steady states, in which a driven thermostated system soon comes to occupy a very limited (fractal) portion of phase space. Bit-reversible dynamics cannot be applied to such systems. Thermostated motion equations contain time-reversible friction coefficients which act, on the average, to *decrease* phase volume. This decrease is necessary for stability. Bit-reversible dynamics on the other hand, like Hamiltonian dynamics, necessarily conserves phase volume.

In a "coarse-grained" version of phase space changing volume means either a one-to-many or a many-to-one mapping, depending upon the direction of heat flow. A rough analog of Second-Law behavior *can* be formulated and analyzed :

$$\{ \, q, p, \zeta \, \}_t \stackrel{\rm Runge-Kutta}{\Longrightarrow} \{ \, q, p, \zeta \, \}_{t+dt} \quad [\text{ Cell} - \text{Centered Integers }] .$$

In this approach an integer phase-space, perhaps with only a few digits per integer, can be used to describe a "state", while the propagation from state to state, indicated by the arrow, is carried out precisely, with floating-point arithmetic. The variables at time $t + dt$ are then placed in the center of their integer phase-space cell before the next iteration. This type of "coarse-graining" closely corresponds to the usual double-precision simulations when 16-digit integers are used.

For a thermostated steady-state system, with one or more friction coefficients controlling the temperature (by extracting the irreversibly generated heat) of selected degrees of freedom, we can express the time average of the time-dependent comoving change of phase volume $\otimes$ in terms of the time-averaged friction coefficients $\{ \, \zeta \, \}$ controlling the flow :

$$\langle \, (\dot{\otimes}/\otimes) \, \rangle = \langle \, \sum_{\{ \, p \, \}} (\partial \dot{p}/\partial p) \, \rangle = \langle \, \sum_{\{ \, p \, \}} -\zeta \, \rangle < 0 \, .$$

Here the direction of the Second-Law inequality is clear. For any *steady* process the phase volume *cannot* grow; unlimited growth corresponds to instability and divergence. Collapse of phase volume, though it may seem peculiar, is an essential part of the mechanism for generating strange attractors, the fractional-dimensional phase-space objects with a fractal dimension less than that of the embedding space. For very simple fractal objects see again the classical examples of **Figure 7.8** on page 206 .

Phase-volume collapse is accompanied by Lyapunov-unstable stretching and folding, which together provide mixing. Without mixing, macroscopic results would depend upon the initial conditions. It is surprising that the equations of motion generating *attractors* are typically *time-reversible*. The resulting nonequilibrium steady states provide answers to [i] Loschmidt's paradox with Lyapunov instability and to [ii] Zermélo's paradox with an indifference to recurrence while tucked away in that fractal part of phase space representing the steady state. From the standpoint of continuum mechanics, where the constitutive equations are typically irreversible, there is no Loschmidt paradox, and Zermélo's paradox is likewise no problem because continuum simulations are necessarily restricted to a strange-attractor portion of the continuum phase space. Our fractal understanding of the Second Law, based on findings from computation, is quite satisfactory. Let us explore the topology of both Hamiltonian and thermostated phase-space flows from this standpoint of computation.

8.4 Computational Averaging, Ergodicity, and Recurrence

We have just considered two views of irreversibility, conservative and dissipative. Both of them lead to a common conclusion: the number (or density, or probability) of "nonequilibrium" states is small (in the limit, negligible) relative to their equilibrium neighbors. For the philosophers this conclusion is perhaps enough. For those of us that prefer or demand a more quantitative and detailed understanding it is necessary to consider the same question from the presentday computational standpoint.

Accordingly, consider two computational approaches, Hamiltonian leapfrog mechanics in the conservative case, Nosé-Hoover Runge-Kutta mechanics in the dissipative case. Hamiltonian mechanics obeys Liouville's Theorem. This might appear to disqualify it from agreement with the Second Law. But if the Second Law discusses only equilibrium states then the periodic orbit of a leapfrog simulation *is* a good analog for a thermodynamic equilibrium state. If pressed we could select the *longest* of the periodic orbits as the one to use in characterizing equilibrium.

Thermostated mechanics allows for changes of phase volume, with thermostats providing a many-to-one loss of volume and the formation of fractals. The thermostated motion equations can certainly be solved with Runge-Kutta integration, choosing the timestep dt to guarantee that the accuracy of the algorithm is limited by roundoff. For fans of centered differences an alternative scheme, a bit like leapfrog, can be developed for steady-state simulations. But such a scheme, mapping the past toward future invariably gives a many-to-one average. An example of such an algorithm, with a thermostated temperature T , is :

$$\{ q_{t+dt} = 2q_t - q_{t-dt} + (F_t dt^2/m) - \zeta_t[\ q_{t+dt} - q_{t-dt}\ ](dt/2) \}\ ;$$

$$\zeta_{t+dt} = \zeta_t + (dt/\tau^2)\left[\sum \frac{m[\ q_{t+dt} - q_t\ ]^2}{(NDkTdt^2)} - 1\right] .$$

The top set of equations can be solved for "new" coordinates, $\{ q_{t+dt} \}$, which can in turn be used to find the new friction coefficient(s) ζ_{t+dt} .

What are the common features of these two approaches to irreversibility and can they be successful? We believe the answer to the second question to be "Yes". Let us consider the first in more detail, in order to justify our answer to the second. The argument follows Bill's 2000 Physical Review E work with Christoph Dellago, "Finite-Precision Stationary States At and Away from Equilibrium". That work showed by example that steady-state finite-precision dissipative simulations consist of two parts, [i] a transient, followed by [ii] a periodic orbit. See **Figure 8.3** , which illustrates both of them. In either equilibrium or nonequilibrium situations the "accurate" integration is rounded to the center of a phase-space cell at the conclusion of each timestep. In the case of integer leapfrog no rounding is necessary and the entire orbit is periodic. To make the picture more concrete let us consider an example problem of Second-Law Hamiltonian dynamics, the four-fold free expansion problem of Section 4.5, page 89 . We imagine an integer phase space with $2L$ the largest integer and with the system, initially confined to an $L \times L$ square, suddenly released into the larger $2L \times 2L$ square.

There are $N = 16\,384$ particles so that the number of "accessible" states increases from L^{2N} to $(2L)^{2N}$. The Gibbs' entropy increase reflecting this change is $Nk\ln(4)$. Suppose we simulate this problem accurately, so that the post-increase dynamics corresponds, roughly, to random jumps among the $(2L \times 2L)^N$ configurational states, where L might in practice be of the order of 10^{16} . How many of the "accessible" states are *actually* accessible? This is a version of the Birthday problem (How many should be in a room to make it likely that two have the same Birthday?). So long as the integrator is deterministic and reproducible (the leapfrog integrator would be a good choice) the number of states visited in the periodic orbit generated by the integrator is of the order of $\sqrt{365(\pi/8)}$ in the Birthday problem and $\sqrt{\Omega(\pi/8)}$ in statistical mechanics, where Ω is the actual number of states and greatly exceeds those in a computational periodic orbit. It is necesssary to imagine that as the density of visited points (t/Ω) builds up to the recurrence point with $t \times t \simeq \Omega$ that it does so uniformly. Evidently all of the usual statistical mechanics would go through for this computational model with a "minor" adjustment. Boltzmann's constant would need to be adjusted by a factor of 2 because the "number of states" used in defining the entropy would be (apart from a negligible added constant) the square root of the pre-computational number of states. Our conclusion from this thought experiment is that the foundations of statistical mechanics are safe and secure in the new era of computation. It is only necessary to adjust Boltzmann's constant by a factor of two to bring the "new" point of view, with $\sqrt{\Omega}$ states into consonance with the "old" Ω-state point of view, $k_{\rm new} \ln \sqrt{\Omega} = k_{\rm old} \ln \Omega$.

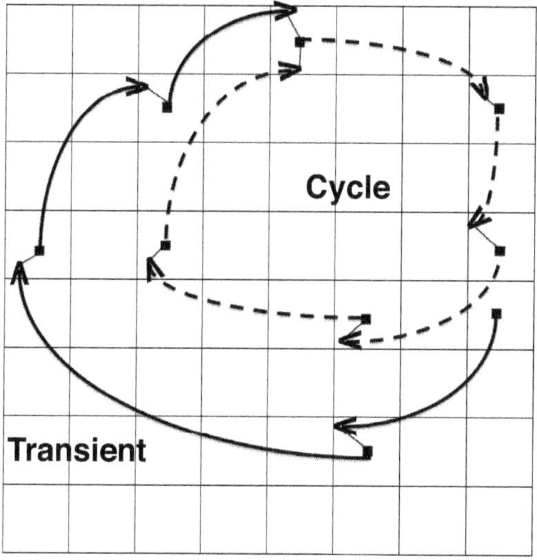

Fig. 8.3 Small-scale model illustrating the "transient" approach to a coarse-grained periodic "cycle". Because each of the accurate integrations begins at a cell center and ends with a jump to another cell center this algorithm is intrinsically irreversible unless Levesque and Verlet's bit-reversible algorithm is used for all the jumps.

What happens in the case of *nonequilibrium* steady states? Here we imagine a definite algorithm for solving the thermostated equations of motion, once again settling onto a periodic orbit, this time with a *fractal* dependence of orbit length on computer precision. Evidently the phase-space structure becomes multifractal, down to the resolution scale of the phase space. Celso Grebogi, Edward Ott, and James Yorke pointed out that the correlation dimension D_C of the attractor (the variation of the number of points in the neighborhood of a given point is proportional to r^{D_C}) describes the number of states. Dellago and Hoover give many detailed examples in their 2000 paper. We can approximate the dimensionality loss ΔD, and the resulting correlation dimension of the finite-precision attractor with an estimate based on entropy production :

$$(\Delta D/D) \simeq \dot{S}/(Nk\lambda_1) = \langle\ \zeta\ \rangle/\lambda_1\ ,$$

where the average is over both time and all the thermostated degrees of freedom. Evidently the *dimensionality of* the finite-precision analog of a strange attractor lies below that of the equilibrium case. Any attempt to compute a nonequilibrium entropy will necessarily be unsuccessful (giving a divergent $S_{\rm noneq} \rightarrow -\infty$) . The conclusion that entropy is not a viable concept away from equilibrium is not at all new, but is probably controversial in some places. We will not attempt a review of the literature here, being content to point out that there need be no contradiction between the continuous space of mathematicians and the countable integer space of computation. We refer the interested reader to Wikipedia for more elaborate discussions.

Let us return to the main idea of this Chapter on the Second Law, concrete computational examples of Second Law behavior. Let us first consider a chaotic and dissipative one-body problem, known variously as the "Quincunx", the "Periodic Lorentz Gas", and the "Galton Board".

8.5 Thermostated Ergodic Galton Board

The Galton Board is a one-body caricature of Alder and Wainwright's hard-sphere studies. The dynamics of a single particle is much like the dynamics of two particles, which already has a melting transition [!], Lyapunov instability, and ergodicity. Adding a field and a constraint produce the simplest possible nonequilibrium steady state, a two-dimensional discontinuous but time-reversible map. The dissipative Baker Map is the same order of complexity but arguably less physical and less interesting. The Galton Board has both conservative and dissipative behavior including fractals and limit cycles. Thus a complicated 2D map is enough to establish chaos and its close relationship to the repellor-attractor version of the Second Law.

Perhaps the *simplest* time-reversible model exhibiting *irreversible* behavior *is* the Galton Board, which contains a particle moving, or falling, through a fixed periodic array of scatterers. Two other plausible time-reversible candidates for "simplest" are the dissipative Baker Map [see

Chapter 2 of our 2012 *Time Reversibility, Computer Simulation, Algorithms, Chaos* book] and Section 6.3's heat-conducting Nosé-Hoover oscillator. The periodic Galton Board geometry has a fringe benefit. It is ideally suited to demonstrating the binomial distribution of random-walk steps enabled by the scattering of particles to the right or to the left. Rather than following many inelastic particles for a short time we choose instead to follow a single wandering perfectly-elastic and thermostated hard-disk particle for a longer period, typically on the order of a million collisions, resulting in a downward current with overall fluctuations of order 0.001 . We provide a unit-cell description of an infinite triangular lattice within which the dynamics evolves. Although phase space for a moving isokinetic particle is three-dimensional [$(x, y, \arctan(p_y/p_x)$)] we can reduce the description to two dimensions by focusing on the Poincaré section describing the collisions. Each collision is described in terms of its position along the scatterer's perimeter, and the direction, relative to the outward normal, of the post-collision velocity.

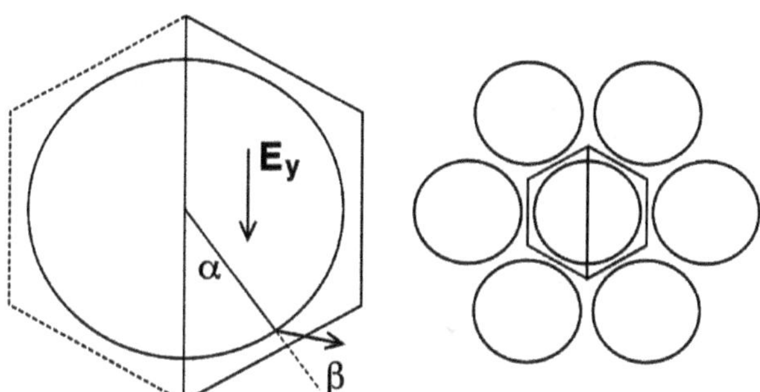

Fig. 8.4 Definition of $0 < \alpha < \pi$, which gives the location of a collision, and β , which gives the direction of the post-collision velocity. Collisions in the left half of the triangular-lattice unit cell are mapped into the right half using $(+x, +y) \longrightarrow (-x, +y)$. The field E_y is nonzero in the nonequilibrium case.

Figure 8.4 shows corresponding definitions of two angles, α and β . The first gives the *location* of a collision along the scatterer's perimeter. The second gives the (post-collision direction of the wanderer's) *velocity* relative to the outward normal of the scatterer surface. The collisions which punctuate both ends of each free-flight trajectory segment are represented by points. Each representative point is plotted within the segment [$\alpha, \sin(\beta)$] which describes all possible collisions :

$$[\, 0 < \alpha < \pi \ , \ -1 < \sin(\beta) < +1 \,] \ .$$

Points in this rectangular segment are an analog to a Poincaré section and record the phase flow at the instant of each collision. The points in this sectional record represent complete phase-space locations for each collision. By using periodic boundaries the entire wanderer trajectory in an infinite lattice can be mapped into a unit cell of the lattice of scatterers. At equilibrium this catalog of states could be further reduced, by a factor of twelve, by using the 12-fold symmetry of the lattice. **Figure 8.5** shows an equilibrium sequence of 100 Newtonian collisions. They are a typical portion of a trajectory embedded in an infinitely-extended lattice structure. The trajectory could equally well be displayed as a series of straight-line segments within a single unit cell of the lattice.

We wish to go beyond equilibrium collisions to explore *nonequilibrium* situations. We choose a vertical external driving field $g \equiv E_y$ (reminiscent of gravity) capable of stimulating a dissipative current. The best we can do to retain some symmetry beyond periodicity is to orient the field parallel to the long axis of a crystallographic cell, retaining a two-fold mirror-reflection symmetry.

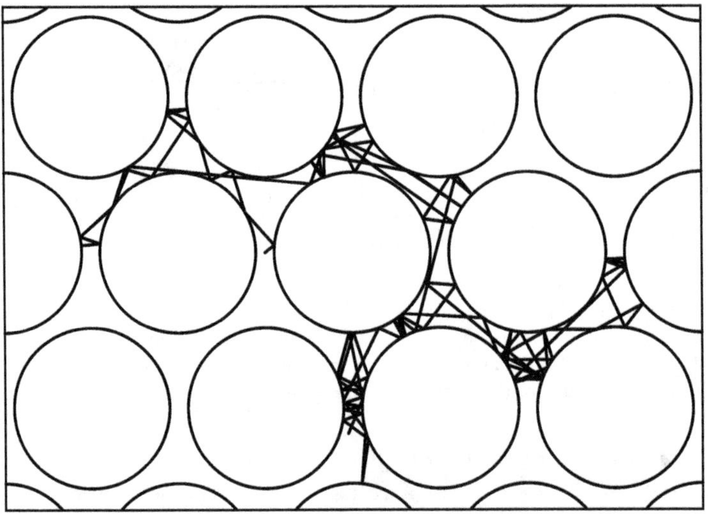

Fig. 8.5 100 equilibrium collisions in the field-free Galton Board at a density four-fifths of the closest-packed. The figure was constructed by changing the sign of the radial velocity whenever the wandering Newtonian particle contacted a scatterer. Sir Francis Galton (1822-1911) used a manybody mechanical analog of this computational model to demonstrate the binomial distribution.

A convenient simulation method for this model, either at or away from equilibrium, and using either the leapfrog algorithm or Runge-Kutta integration, evolves the Cartesian motion { $(x(t), y(t)$ } until the coordinates lie just *within* a scatterer. The integration can then be continued by following a three-step collision process: first, return to the most recent set of pre-collision (x, y) coordinates; next, convert the Cartesian velocity $(\dot{x}, \dot{y})$ to its polar-coordinate form $(\dot{r}, r\dot{\theta})$, measured relative to the scatterer; third, change the sign of the radial velocity and convert it back to the Cartesian form $(\dot{x}, \dot{y})$. Then, given the post-collision coordinates and velocities, the free-flight trajectory can be continued until the next collision occurs.

If the scatterer is fixed at the origin of a single-particle unit cell and has a radius of 0.50 (so that the center-to-center separation of two nearest-neighbor scatterers is $\sqrt{(5/4)} = 1.118034$) the post-collision coordinates and momenta can be computed from the pre-collision coordinates `xold, yold` giving the new momenta `px, py` as follows :

```
if(x*x + y*y.le.0.25d00) then
x = xold
y = yold
r = dsqrt(x*x + y*y)
pr = -(x/r)*px - (y/r)*py
pt = +(x/r)*py - (y/r)*px
px = +(x/r)*pr - (y/r)*pt
py = +(y/r)*pr + (x/r)*pt
endif
```

Here `pr` and `pt` are respectively the radial and tangential velocity components of the wandering particle measured relative to the fixed scatterer. The two data (`x11, y11`) for the Poincaré collision plot in a 2×2 square come from the old x coordinate and tangential momentum as follows :

```
pi = 3.141592653589793d00
alpha = dacos(xold/r)
sinbe = pt
x11 = ((alpha + alpha)/pi) - 1.0d00
y11 = pt
```

Figure 8.6 shows 300,000 successive equilibrium (zero field) collisions in the [$\alpha, \sin(\beta)$] plane generated in this way. It is quite evident (and easily verifiable) from the **Figure** that the motion is ergodic, and that all

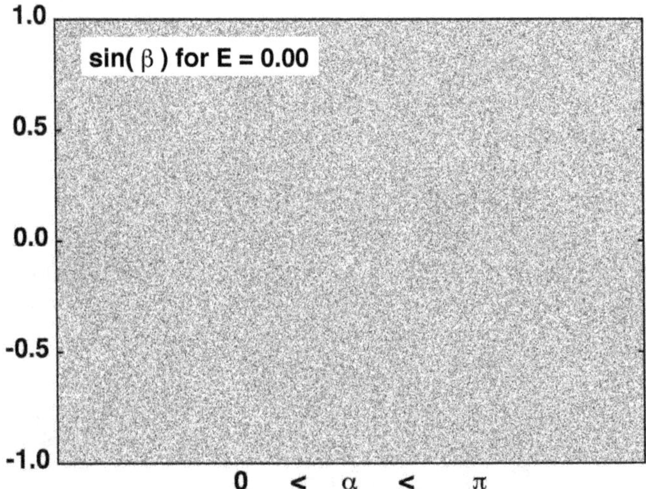

Fig. 8.6 300,000 successive equilibrium collisions in the [$\alpha, \sin(\beta)$] plane.

values of the angle α and the radial velocity component $\sin(\beta)$ are equally likely. Although the behavior of the model "looks" irreversible, in the sense that a localized ensemble of initial conditions, $dxdyd\dot{x}d\dot{y}$ will soon lead to a uniform covering of the space, it satisfies Liouville's Theorem for Hamiltonian systems. Through repeated shearings a uniformly-occupied small square retains its fine-grained area while its coarse-grained area expands to fill the space uniformly, as in **Figure 8.6** . This coarse-grained approach to equilibrium is the mixing property familiar from the Boltzmann-Gibbs picture of equilibration.

The Galton Board problem provides us with a well-equipped guide to *nonequilibrium* simulations. It is natural to introduce a constant accelerating field (Galton's classroom demonstrator used gravity) to induce a nonzero current. Away from equilibrium the downward field E_y provides an energy source, $mgh = E_y h$. The increased energy due to the field is entirely kinetic. If energy were to be conserved, with the field "on", its dependence on the y coordinate would destroy the periodic simplicity of the cell-to-cell lattice symmetry. The additional gravitational energy could be spirited away by friction, though unless the friction were to act only in the y direction the simple symmetry of the problem would be lost.

To retain the cell-to-cell symmetry of the problem most simply we *constrain* the kinetic energy of the wandering particle to its initial value. To do this we use a time-reversible frictional force, proportional to the velocity, choosing its magnitude so as to maintain a fixed kinetic energy during the motion from one collision to the next :

$$\dot{x} = p_x \; ; \; \dot{p}_x = -\zeta p_x \; ; \; \dot{p}_y = -E_y - \zeta p_y \; ; \; \zeta \equiv (-E_y p_y / p^2) \longrightarrow \dot{K} \equiv 0 \; .$$

This choice for the friction coefficient ζ constrains the kinetic energy to its initial value. Exactly the same formulæ result if we apply Gauss' Principle of Least Constraint to the problem of keeping the kinetic energy constant.

Gauss' Principle states that in enforcing a constraint one should use the least-possible constraint force. Because it is apparent that tangential, as opposed to longitudinal, forces perpendicular to the trajectory unnecessarily increase the magnitude of the constraint force, Gauss' Principle agrees with the simple longitudinal friction force we chose above, $F_{\rm Gauss} = -\zeta p$.

For simplicity in our numerical work we choose the constant value of $p_x^2 + p_y^2$ equal to unity and we choose a density of hard-disk scatterers equal to four-fifths the close-packed density. A constant friction leads to more complexity, with energy fluctuations. For an up-to-date summary of results for generalizations of the Galton Board problem see Dettmann's "Diffusion in the Lorentz Gas" arχiv 1402.7010 = [Communications in Theoretical Physics **62** , 521-540 (2014)] .

It is important to note the time reversibility of Gauss' motion equations. In the time-reversed motion both the momenta and the friction coefficient ζ change signs. Just as in the equilibrium situation it is convenient to integrate the equations of motion until a scatterer is encountered. There is an analytic form for the trajectory between scatterers but the programming is simpler if we pass over that analysis and integrate numerically. **Figure 8.7** shows how the zero-field probability density is modified for four values of the vertical field strength, $\{ \; 1, 2, 3, 4 \; \}$.

Figure 8.7 displays two features which surprised us (in 1987) . First, the probability densities shown there are all *singular multifractals* rather than smooth Gibbsian distributions. A numerical investigation shows that the probability in the (circular) vicinity of a point varies as a *fractional* power of the distance from that point. The dependence on direction is smooth in some directions (parallel to the striped regions) and is singular in others (perpendicular). The edges of the striped regions signify a change of scatterers from one neighboring particle to another. For details see our 1987 Journal of Statistical Physics article, "Diffusion in a Periodic Lorentz Gas" .

Notice that the motion is *still* ergodic for field strengths of 1, 2, and 3 in the sense that tiling the [$\alpha, \sin(\beta)$] space with rectangular cells leaves no

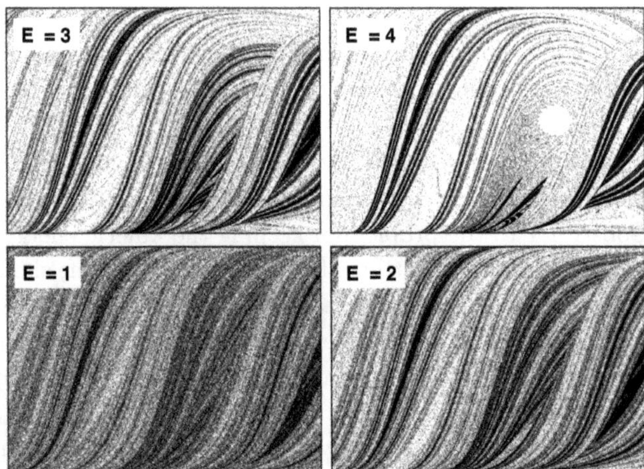

Fig. 8.7 300,000 nonequilibrium collisions with field strengths of 1.00, 2.00, 3.00 and 4.00 . The abscissa is [$0 < \alpha < \pi$] and the ordinate is [$-1 < \sin(\beta) < +1$] . The relationship of { α, β } to collisions follows the definitions in **Figure 8.4** . For simplicity's sake we have generated these figures with a time step of 0.00001 rather than using the analytic form for the trajectory given in our 1987 reference with Moran and Bestiale.

cell forever empty. A reason for this ergodicity lies in the time reversibility of the equations of motion. It is evident that a trajectory can start anywhere in the space. This means that the same trajectory, followed backward in time [which changes the sign of $\sin(\beta)$] will come back to the starting point, which could have been chosen anywhere.

The probability distribution itself (sometimes called the "natural measure") is reproducible and robust. The fractal nature of the measure varies from place to place. Though there is a superficial resemblance to the Cantor set, the spaces between the stripes in the Galton-Board probability are not empty. Careful work shows that the nonequilibrium motion is typically ergodic so long as the field strength is less than 3.69 . Near that field strength the distribution of collisions collapses to a one-dimensional "limit cycle", an attractive regularly repeating series of twenty collisions. Overall, the wanderer on this limit cycle moves steadily downward at an overall angle of 30 degrees relative to the field direction. At higher field strengths we identified another dozen limit cycles.

A second topologically interesting feature can be seen in the data for a field strength of 4.00 . There is a large near-circular hole in the dissipative chaotic sea. Investigation shows that solutions within the hole stay there, and are conservative rather than dissipative. Within the hole the

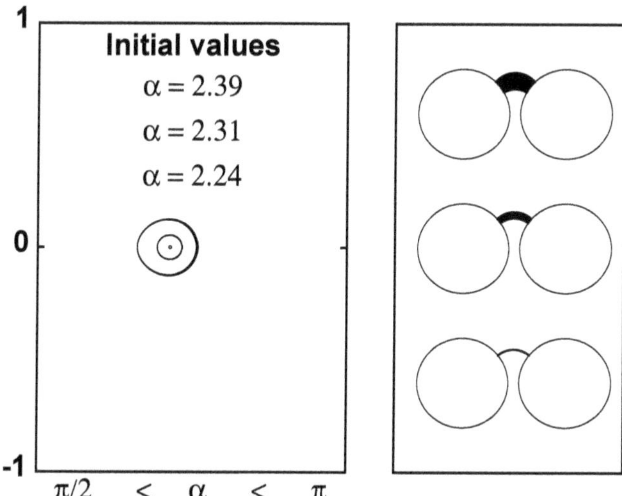

Fig. 8.8 An island of *conservative* solutions of the Galton Board problem embedded in the chaotic dissipative sea. The field strength is 4.00 . The segment of the [$\alpha, \sin(\beta)$] plane (shown at left) contains the sample [x, y] periodic orbits (shown at right).

time-averaged friction coefficient $\langle\ \zeta(t)\ \rangle$ necessarily vanishes. Outside, its mean value is positive, corresponding to dissipation. **Figure 8.8** shows three of the conservative solutions that populate the hole visible in **Figure 8.7** . All of them are stable tori in phase space, and because the time-averaged dynamics corresponds to stable horizontal bouncing rather than vertical dissipation this model illustrates an interesting coexistence of the two qualitatively different behaviors, conservative and dissipative.

This same combination of conservative and dissipative regions occurs in our 2014 work with Clint Sprott, "Heat Conduction, and the Lack Thereof, in Time-Reversible Dynamical Systems: Generalized Nosé-Hoover Oscillators with a Temperature Gradient", in Sprott's own 2014 "A Dynamical System with a Strange Attractor and Invariant Tori", and in Politi, Oppo, and Badii's 1986 study of three chaotic laser equations : "Coexistence of Conservative and Dissipative Behavior in Reversible Dynamical Systems".

The Galton Board is an extremely interesting nonequilibrium steady state which has deservedly generated a considerable historical and technical literature. The ergodicity of its fractal (zero-volume) strange-attractor phase-space distribution provides a simple geometric explanation for the Second Law of Thermodynamics. The time-reversed trajectories which follow from the same equations of motion are unlikely to be observed by chance because their probability is *zero*, not just small. The "core" of the attractor, those boxes that contain half of the probability density, requires a smaller and smaller fraction of the boxes as the box size decreases.

The Galton Board work illustrates several pedagogically desirable properties for the understanding of the Second Law for systems with time-reversible equations of motion. The space-filling ergodicity and Lyapunov instability of the dynamics guarantees typical distributions and currents independent of the initial conditions' details. Reversed trajectories (which would violate the Second Law, steadily converting kinetic energy to stored potential energy, myE_y) are unobservable, of zero probability, due to the fractal nature of the strange-attractor distribution. The instability of reversed trajectories exceeds that of the observable forward-in-time trajectories because the summed-up reversed Lyapunov spectrum is necessarily *positive*, $\sum \lambda > 0$, indicative of the repulsion characterizing the "repellor". The details of the Lyapunov spectrum for the Galton Board and for hard disks are given in Christoph Dellago's 1996 Universität Wien Dissertation as well as in two of his publications in Physical Review E, with Glatz, Posch, and Bill, "Lyapunov Spectrum of the Driven Lorentz Gas" (1995) and "Lyapunov Instability in a System of Hard Disks in Equilibrium and Nonequilibrium Steady States" (1996) .

We will see that these same paradoxical features (ergodicity with a vanishing fraction of the accessible states and a negative Lyapunov-exponent sum forward-in-time) show up in continuous-flow example problems based on the harmonic oscillator as well as with continuous potentials and with time-reversible two-dimensional maps. To emphasize the similarities of the *discontinuous* Galton Board problem with the fully *continuous* doubly-thermostated oscillator problem, let us consider their common features. As a preparatory refresher to the following section discussing ergodic thermostated oscillators, look once again at **Figure 5.7** on page 121 , and **Figure 6.11** on page 172 , which both show $(q, p, \zeta{=}0, \xi{=}0)$ cross sections for typical multifractal structures resulting from *nonequilibrium doubly-thermostated* oscillators.

Although large-system ergodicity is ruled out by the clock, small-system ergodicity is both desirable and possible, as the Galton Board example shows. Ergodicity allows time-averaged results to approach a limiting distribution independent of the initial conditions (provided that these initial conditions correspond to the same fixed macroscopic variables, such as the Galton Board's driving field or the oscillator's temperature profile). For equilibrium Hamiltonian systems at the same macroscopic energy or thermostated systems at the same temperature ergodicity implies the agreement of molecular dynamics with Gibbs' statistical mechanics.

Testing for ergodicity is very much a matter of trial and error. The ergodicity of a particular model's equations of motion is hard to assess, as the three kinds of Galton Board solutions (strange attractor, conservative tori, and limit cycles) clearly show. A systematic approach to diagnosing ergodicity includes two steps: first, find a smooth desirable distribution, perhaps Gibbs', together with a corresponding set of motion equations, perhaps Nosé-Hoover, consistent with it; second, carry out diagnostic simulations designed to search for discrepancies with the chosen desirable distribution. Let us make these ideas more concrete by considering problems that do have ergodic distributions throughout their four-dimensional unbounded phase spaces, doubly-thermostated harmonic oscillators.

8.6 Thermostated Ergodic Harmonic Oscillators

The Hamiltonian harmonic oscillator, with its elliptical (qp) phase-space orbit, is linear and ergodic, which is good, but it is too simple for the study of ergodicity and chaos problems. The problem can be generalized by adding a control variable pair (s, p_s) to the Hamiltonian as Nosé did,

$$2\mathcal{H} = (p/s)^2 + q^2 + 2\ln(s^2) + p_s^2 \ ,$$

resulting in a four-dimensional phase space with a great variety of orbits embedded in a chaotic sea. This thermostating approach is not at all ergodic, instead giving the complicated structures characteristic of Hamiltonian chaos. But the slightly more complex approaches of Martyna, Klein, and Tuckerman and of Hoover and Holian made the oscillator problem ergodic, following Gibbs' canonical ensemble. They did this by including contributions from *two* independent "control variables" (ζ, ξ) so as to manage the oscillator's (q, p). This multi-variable-control approach, greatly generalized by Bauer, Bulgac, and Kusnezov, *was* tremendously successful in simplifying isothermal simulations and ultimately heat flows.

The harmonic oscillator is the most obvious starting point. Most low-temperature problems with continuous forces approach a minimum-energy oscillator limit. Gibbs' canonical distribution for the oscillator is a Gaussian in (q, p) . Three motion equations describe the Nosé-Hoover oscillator ,

$$\{ \ \dot{q} = p \ ; \ \dot{p} = -q - \zeta p \ ; \ \dot{\zeta} = p^2 - 1 \ \} \longrightarrow \langle \ p^2 \ \rangle \equiv 1 \ ,$$

and are certainly *consistent* with the (extended) canonical distribution :

$$f(q, p, \zeta) \propto e^{-q^2/2} e^{-p^2/2} e^{-\zeta^2/2} \longrightarrow (\partial f/\partial t) \equiv 0 \ .$$

But all of the many different numerical solutions of these three motion equations are very far from ergodic. Their Poincaré sections in the [$\zeta = 0$] plane reveal an infinite variety of disjoint periodic or quasiperiodic solutions as well as a single chaotic solution. See again the cross sections of **Figure 1.6** on page 13 .

Accordingly, it was and is a complete and continuing surprise that "complicating" matters, adding just one more control variable, can actually provide a smooth space-filling four-dimensional Gaussian distribution. In fact, there are *many*, even an infinite number, of four-variable sets of equations compatible with Gibbs' canonical distribution. Some of them are evidently ergodic, with a multi-variable Gaussian distribution reaching to infinity in all four dimensions. See again the doubly-thermostated *nonequilibrium* oscillator sections shown on page 172 . The best known examples are the Hoover-Holian and Martyna-Klein-Tuckerman oscillators :

$$\dot{q} = p \; ; \; \dot{p} = -q - \zeta p - \xi p^3 \; ; \; \dot{\zeta} = p^2 - 1 \; ; \; \dot{\xi} = p^4 - 3p^2 \; ; \; [\text{ HH }]$$

$$\dot{q} = p \; ; \; \dot{p} = -q - \zeta p \; ; \; \dot{\zeta} = p^2 - 1 - \xi\zeta \; ; \; \dot{\xi} = \zeta^2 - 1 \; . \; [\text{ MKT }]$$

The Hoover-Holian equations control both the second and the fourth velocity moments while the Martyna-Klein-Tuckerman equations use ξ as a means to control the other control variable ζ .

Both sets of motion equations are fully consistent with the four-dimensional Gaussian distribution function :

$$f(q,p,\zeta,\xi) \propto e^{-q^2/2} e^{-p^2/2} e^{-\zeta^2/2} e^{-\xi^2/2} \longrightarrow (\partial f/\partial t) \equiv 0 \; .$$

To avoid giving the impression that four equations are always enough for ergodicity consider again the Patra-Bhattacharya oscillator equations given on page 161 :

$$\dot{q} = p - \xi q \; ; \; \dot{p} = -q - \zeta p \; ; \; \dot{\zeta} = p^2 - 1 \; ; \; \dot{\xi} = q^2 - 1 \; . \; [\text{ PB }]$$

Although this clever model controls both $\langle \, q^2 \, \rangle$ and $\langle \, p^2 \, \rangle$ and also satisfies the four-dimensional Gaussian distribution it is definitely not at all ergodic.

Folklore suggests that knowledge of (all) the moments of a distribution is enough to unambiguously identify that distribution. At the same time, a glance at **Figure 1.9** on page 16 shows that fixing only the second moment $\langle \, p^2 \, \rangle$ or only the fourth $\langle \, p^4 \, \rangle$ (along with *all* the odd moments) is not enough to generate the complete Gaussian. That **Figure** demonstrates that the three differential equations for ($\dot{q}, \dot{p}, \dot{\zeta}$) produce regular trajectories coexisting with chaos, rather than Gibbs' smooth Gaussian distribution, which is the properly weighted sum of *all* the various solutions.

On the other hand *any* function of energy is stationary for the Hamiltonian *pairs* of equations :

$$\{ \dot{q} = p \; ; \; \dot{p} = F(q) \} \longrightarrow \mathcal{H} = \Phi + \sum (p^2/2) \ \text{Constant} \ .$$

To get the "right" function of energy, Gibbs' canonical distribution, requires additional work to modulate the phase-space density, with controls that manage at least two moments. It is easy to fix any two of the velocity moments by choosing the appropriate control equations. It is not so appealing to go beyond the fourth moment because higher moments give stiffer equations.

Rather than undertaking a comprehensive investigation of various probability densities coming from the two successful doubly-thermostated approaches we studied the convergence rates of the time-averaged second, fourth, and sixth moments according to the Hoover-Holian and Martyna-Klein-Tuckerman schemes as applied to the harmonic oscillator. The convergence of the velocity moments favors the "moments" approach fixing $\langle \, p^2, p^4 \, \rangle$. The convergence of the coordinate moments favors Martyna's "chain" approach, fixing $\langle \, p^2, \zeta^2 \, \rangle$. Because the moments are still fluctuating after a billion oscillator timesteps their exact value is irrelevant for applications to many-body problems.

Because even slightly more complicated systems (the cell model is a good example) are sufficiently ergodic for accurate studies, the Nosé-Hoover, Hoover-Holian, or Martyna-Klein-Tuckerman thermostats are readily adaptable to large-scale simulations. The characteristic time scales of the thermostat variables can be managed so as to enhance, restrain, or optimize the control exerted by the thermostats.

An interesting aspect of the nonequilibrium simulations controlled by these thermostats is the stability of their nonequilibrium currents. Despite Lyapunov *instability* the flows invariably find and follow the dissipative direction obeying the Second Law of Thermodynamics. To emphasize this stability we consider next a one-thermostat nonequilibrium oscillator problem so "far from equilibrium" that it generates a stable one-dimensional limit cycle in the phase space. The equations of motion *are* time-reversible. Even so, the stability of this cycle, when time-reversed, is converted to Lyapunov *instability*. The time-reversed cycle, which violates the Second Law, is a thoroughly-unstable repellor cycle. It quickly finds its way back to the attractor cycle and the stable and comfortable life of dissipation.

8.7 Nonequilibrium Oscillator Limit-Cycle Stability

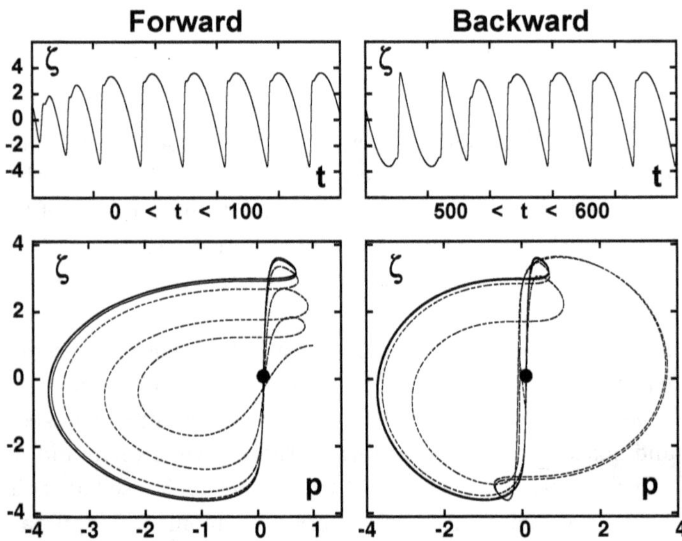

Fig. 8.9 Forward and backward solutions of the *nonequilibrium* Nosé-Hoover oscillator. The initial conditions, $(q, p, \zeta) = (0, 1, 1)$ soon lead to the limit cycles shown in the lower panels. If the dynamics is then reversed (either by changing the signs of both p and ζ or, alternatively by changing only the sign of dt) the limit cycle becomes an unstable repellor cycle. The lower left panel shows the forward trajectory from the initial condition through several periods of the limit cycle. The initial forward transient path is shown as a dotted line. The lower right panel shows the details of the *reversed* trajectory, from the reversal time ($t = 500$) until the strongly attractive limit cycle is regained. The transient path is dotted and the reversal state near $(0, 0)$ is indicated by a filled circle. The upper panels show the time history of the trajectory leading to the limit cycles in each case. These details are sensitive to the exact time of reversal as well as to the choice of integrator, here a fifth-order Runge-Kutta integrator with a timestep $dt = 0.001$.

The singly-thermostated Nosé-Hoover oscillator, with an imposed temperature profile $T(q)$, can react in a dissipative way or not, depending on the initial conditions and the profile. With two control variables the nonequilibrium oscillator phase-space distributions are typically fractals and do not show conservative tori. The cross sections of **Figure 5.7**, page 121 , are so intricate that words cannot yet describe them. But they do invariably lead to entropy production, with an overall decreasing fractional-dimensional extension in phase. Rather than consider the stability of complex attractors within four-dimensional phase spaces let us take up the nonequilibrium Nosé-Hoover oscillator under conditions so far from equilibrium that the phase-space distribution is no longer complicated.

We consider the "time-reversible" Nosé-Hoover *nonequilibrium* oscillator with a temperature gradient large enough to *guarantee* a limit-cycle solution no matter what the initial condition :

$$\{ \dot{q} = p \; ; \; \dot{p} = -q - \zeta p \; ; \; \dot{\zeta} = p^2 - T(q) \} \; ;$$
$$0 \leq T(q) = 1 + \tanh(q) \leq 2 \; .$$

Take a solution of the three equations, $\{ q(t), p(t), \zeta(t) \}$ with initial values $(q, p, \zeta) = (0, 1, 1)$. Then reverse the motion, by changing the time ordering as well as the signs of the momentum p and friction coefficient ζ. The three equations are satisfied by this new data set, with both sides of the $\dot{q}$ equation changing signs and both sides of the $\dot{p}$ and $\dot{\zeta}$ equations unchanged. The upper portion of **Figure 8.9** shows this reversal property clearly for 100,000 timesteps forward as well as 100,000 timesteps more after changing the signs of p and ζ. From the visual standpoint the numerical reversal is perfect, just as in the Hamiltonian case. But this reversal is relatively short-lived. A dissipative limit cycle is the sole stationary solution possible with this temperature profile.

The lower dissipative part of the **Figure** tells a different story on a longer timescale. From the same initial condition it is about 50,000 transient timesteps before the inevitable limit cycle is reached. The solution continues forward along the limit cycle until the reversal time, $500,000dt = 500$. The reversal does much more than just changing the signs of the momentum and friction coefficient. The signs of the heat flux and of the entropy production change too. These changed signs *violate* the Second Law of Thermodynamics : the heat flux becomes cold-to-hot and the entropy "production" becomes negative :

$$\langle \, (p^3/2) \, \rangle : -0.76 \longrightarrow +0.76 \; ; \; \langle \, -(\dot{\otimes}/\otimes) = \zeta \, \rangle : +1.2 \longrightarrow -1.2 \; .$$

For a relatively short time (two or three limit cycle times) the comoving phase volume *increases* rather than decreasing !

It is interesting, as well as reassuring, to see that after about 50,000 timesteps the reversed thermodynamically and mechanically-unstable limit cycle is attracted back to join its stable twin, once again obeying the Second Law, with a negative heat current and a positive entropy production.

Because this dynamics is then both stable and dissipative, not chaotic, we see that attractors need not be "strange" (chaotic) in order to obey the thermodynamics expressed by the Second Law. Only access to a dissipative part of phase space is necessary. Shortly after reversal, because the dynamics *is* Lyapunov unstable, the oscillator departs from the repellor and joins the attractive limit cycle exponentially fast.

8.8 Illustration of Hamiltonian Exponent Pairing

The local Lyapunov instability analysis can be applied to *any* system including systems which are not even chaotic, a harmonic oscillator or a diatomic molecule or a pair of colliding molecules. These examples, as well as the cell model, and many-body systems, can show exponent "pairing", with a Lyapunov spectrum composed of pairs of local Lyapunov exponents, $\{ \pm\lambda(t) \}$. In some cases reversing the time simply changes the signs of the exponents. In other cases the spectrum of reversed exponents is unstable and a completely different set of pairs forms. The symmetry and symmetry-breaking of the exponents appears to be related to thermodynamic stability. We will consider it here, for both few-body and many-body examples, to see what light it might shed on nonequilibrium flows.

For a more quantitative understanding and appreciation of conservative irreversibility it is useful to consider more deeply the Lyapunov instability of Cartesian Hamiltonian motion. It is typical of such systems that the phase-space offset vectors linking a reference trajectory to its Gram-Schmidt satellites occur in orthogonal *pairs* with equal-magnitude but oppositely-signed local exponents. The directions of these Gram-Schmidt offset vectors become well-defined after a sufficiently long incubation or annealing or equilibration period. After this sufficiently long time the two members of each Gram-Schmidt pair become "paired". Typically the symmetry of the paired vectors is then maintained with an accuracy of about one part per thousand.

In a Cartesian phase space (we have only a limited experience with others) the *components* of the paired vectors can be simply related, with the coordinate displacements $\{ dq \}$ of each member equal in magnitude to the momentum displacements $\{ dp \}$ of its partner, and *vice versa* :

$$\{ \delta_j = (+dq_j, +dp_j) = (+dp_{N+1-j}, -dq_{N+1-j}) \longrightarrow \delta_j \cdot \delta_{N+1-j} \equiv 0 \} \ .$$

This symmetry is quite different to the time-reversal symmetry $(+dq, +dp) \longleftrightarrow (+dq, -dp)$ linking two separating trajectories, forward in time, to the corresponding two time-reversed trajectories. The *signs* of the vectors themselves are immaterial because they parallel principal axes of a many-dimensional hyperellipsoid. The stretching and rotation rates of the two vectors parallel and antiparallel to a given principal axis are identical. Let us consider "pairing" for some simple examples.

8.8.1 *Single Oscillator and Single Diatomic*

Because an oscillator is the prototypical illustration of stability, we might well predict a pair of vanishing exponents for the oscillator. Consider perturbations $(\delta q, \delta p)$ about the simplest oscillator's circular (q, p) trajectory. Include also a Lagrange multiplier λ assigned to maintain the offset $\sqrt{\delta q^2 + \delta p^2}$ constant :

$$\{ \dot{q} = +p \ ; \ \dot{p} = -q \} \longrightarrow \{ \dot{\delta q} = +\delta p - \lambda \delta q \ ; \ \dot{\delta p} = -\delta q - \lambda \delta p \} \ .$$

Constraining the offset vector to have constant length means that $\dot{\delta q}\delta q + \dot{\delta p}\delta p$ must vanish. It does so automatically, which shows that the largest and smallest Lagrange multipliers, which are also the local Lyapunov exponents $\{ \lambda(t) \}$, both vanish, in accord with our intuition.

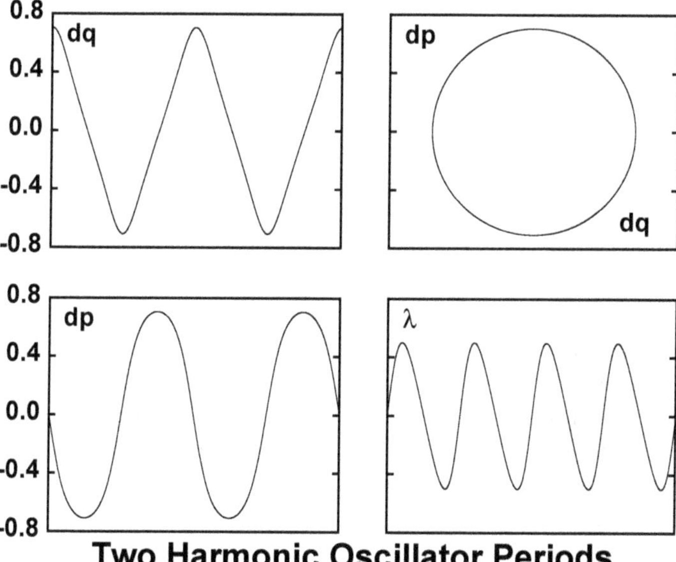

Fig. 8.10 Time dependence of the diatomic Lyapunov exponent and $\delta q(t)$ and $\delta p(t)$. The underlying motion equation is $\ddot{q} = -2q \rightarrow q = \cos(\sqrt{2}t)$; period $= \sqrt{2}\pi = 4.44288$.

For a single diatomic harmonic molecule, and a four-dimensional phase space, with unit rest length, masses, and force constant, the same approach gives the following Hamiltonian and motion equations :

$$\mathcal{H} = (1/2)[\ (q_2 - 1 - q_1)^2 + p_1^2 + p_2^2 \ ] \longrightarrow$$

$$\dot{\delta q_i} = \delta p_i - \lambda \delta q_i \ ; \ \dot{\delta p_i} = \delta q_j - \delta q_i - \lambda \delta p_i = -2\delta q_i - \lambda \delta p_i \ .$$

For simplicity, the center of mass is fixed, $\delta q_1 + \delta q_2 = 0$; $\delta p_1 + \delta p_2 = 0$. Choosing an offset vector of unity, $\delta q_1^2 + \delta p_1^2 + \delta q_2^2 + \delta p_2^2 \equiv 1$ implies that $\delta q_1^2 + \delta p_1^2 = (1/2)$.

$$\delta q_1^2 + \delta p_1^2 = (1/2) \longrightarrow \lambda = -2\delta q_1 \delta p_1 \ .$$

Knowing the Lagrange multiplier as a function of δq and δp makes the motion equations explicit cubic functions of the perturbation $(\delta q, \delta p)$. The two equations we need to solve are :

$$\dot{\delta q} = \delta p + 2\delta q^2 \delta p \ ; \ \dot{\delta p} = -2\delta q + 2\delta q \delta p^2 \ .$$

If we let $\delta q = \sqrt{1/2}\cos(\theta)$ and $\delta p = -\sqrt{1/2}\sin(\theta)$ we find

$$\dot{\theta} = 1 + \cos^2(\theta) \longrightarrow t = \sqrt{1/2}\arctan[\sqrt{1/2}\tan(\theta)] \ .$$

This analytic result (see **Figure 8.10**) agrees perfectly with the numerical solution found for the initial condition $\delta_1 \equiv (dq_1, dq_2, dp_1, dp_2) = (+\sqrt{1/2}, -\sqrt{1/2}, 0, 0)$. The relatively complicated behavior of this diatomic-molecule problem is equivalent to the dynamics of the single "scaled oscillator" with Hamiltonian

$$2\mathcal{H} = s^{+2}q^2 + s^{-2}p^2 \ ,$$

treated in Section 5.8.5 where we found the mean-squared Lyapunov exponent to be

$$\langle \ \lambda_1^2 \ \rangle = (1/2)(s^{+1} - s^{-1})^2 \ .$$

To maintain the right ratio of potential to kinetic energy we would have to choose $s = 2^{1/4}$. To maintain the correct oscillation frequency we would also need to scale the time. It is *much* simpler to choose the Hooke's-law force constant equal to (1/2) rather than unity, which causes the constant-separation Lagrange multiplier to *vanish*. In the next Section we consider the Lyapunov exponents for two *colliding* diatomics, and show that the time-dependence of the entire spectrum, not just the largest exponent, depends, in a similar way, upon the choice of units used to describe the problem.

8.8.2 *Time-Reversible Collision of Two Diatomics*

To illustrate exponent pairing and time-reversal with an educational example consider a collision of two diatomic molecules in one dimension :

$$\mathcal{H} = (1/2)(q_4 - 1 - q_3)^2 + (1/2)(q_2 - 1 - q_1)^2 + \sum_{i=1}^{4}(1/2)p_i^2 +$$

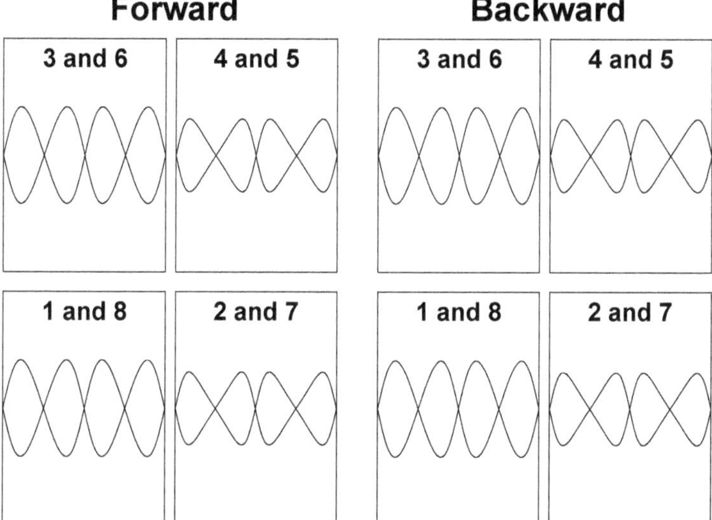

Fig. 8.11 Oscillation of the eight perfectly-paired "manybody" equilibrium exponents for two one-dimensional diatomic molecules. The two molecules are separated and do not interact either forward or backward in time. Compare this simple situation with the collisional dependence of the exponents shown in **Figure 8.12** .

$$[\ (1/4)(q_3 - 1 - q_2)^4\ ]_{\text{Repulsive}}^{\text{But only if}}$$

Notice that the quartic repulsive potential only acts when the distance between Particles 2 and 3 is less than unity. Unless those two particles interact the motions of the 12 and 34 molecules are both simple harmonic oscillations, with a frequency of $\sqrt{2}$.

To explore a very typical source of Lyapunov instability let us begin with the two molecules motionless, separated, and with both harmonic springs at their rest length of unity. We choose the eight offset vectors so as to satisfy the pairing relationship – that is, we choose the first four vectors parallel to the four coordinate directions and the last four are parallel to the momenta – paired with the corresponding coordinates :

$$\begin{aligned}
&\delta_1 \propto (1,0,0,0,0,0,0,0)\ ;\ \delta_8 \propto (0,0,0,0,1,0,0,0)\ : \\
&\delta_2 \propto (0,1,0,0,0,0,0,0)\ ;\ \delta_7 \propto (0,0,0,0,0,1,0,0)\ ; \\
&\delta_3 \propto (0,0,1,0,0,0,0,0)\ ;\ \delta_6 \propto (0,0,0,0,0,0,1,0)\ ; \\
&\delta_4 \propto (0,0,0,1,0,0,0,0)\ ;\ \delta_5 \propto (0,0,0,0,0,0,0,1)\ .
\end{aligned}$$

Figure 8.11 shows that the instantaneous Lyapunov exponents oscillate at *twice* the frequency of the molecular oscillation. When the two diatomics collide the exponents remain paired, but with each of the four pairs behaving differently. The details of the amplitudes and phases show an extreme sensitivity to the time of first contact, as one would guess from the fact that the chosen phase in **Figure 8.12** is quite arbitrary. The collision time, corresponding to the leftmost filled circles in the four forward panels can be chosen arbitrarily (by choosing the initial precollision coordinates of the two molecules). The reversed "backward" panels go through *exactly the same collision* in reversed time order. In all cases the collision occupies the interval between the two filled circles in each panel.

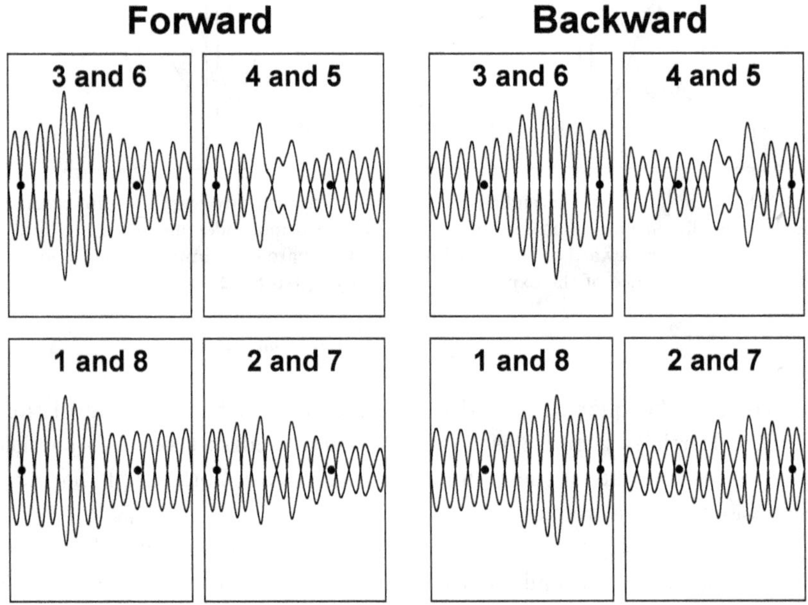

Fig. 8.12 Collision of two noisy diatomics, where the "noise" is sensitive to the moment of collision. The exponents are all perfectly paired both before and after collision. The time interval occupied by the collision is indicated by the two filled circles in each panel.

Because the motion of the offset vectors (which occurs whether or not the atoms move) is fully time-reversible as well as stable, all the eight local exponents reverse perfectly if time is reversed ($+dt \longrightarrow -dt$) in the Runge-Kutta integrator. Changing the sign of dt is the simplest way to demonstrate this reversibiity. Although the perfect pairing and reversal may seem "obvious" we will see that more complicated forces can destroy both the pairing and the forward-backward symmetry. In the meantime let us simplify the colliding diatomics a bit more, by choosing the propitious force constant (1/2) for both molecules.

8.8.3 *Spectra of Colliding Quiet Diatomics*

With the oscillator force constant chosen equal to (1/2) rather than unity, and with the centers of mass, $(1/2)(q_1 + q_2)$ and $(1/2)(q_3 + q_4)$, fixed, the equations of motion for molecules "12" and "34" require no additional constraint to keep the offset vectors constant :

$$\begin{aligned} \dot{q}_1 = +p_1 \ ; \ \dot{p}_1 = (1/2)(-q_1 + q_2) \ ; \\ \dot{q}_2 = +p_2 \ ; \ \dot{p}_2 = (1/2)(-q_2 + q_1) \ ; \\ \dot{q}_3 = +p_3 \ ; \ \dot{p}_3 = (1/2)(-q_3 + q_4) \ ; \\ \dot{q}_4 = +p_4 \ ; \ \dot{p}_4 = (1/2)(-q_4 + q_3) \ . \end{aligned}$$

In this case the perturbed equations of motion for all four particles are *exactly the same* :

$$\{ \ \dot{q} = +p \ ; \ \dot{p} = -q \longrightarrow \dot{\delta q}\delta q + \dot{\delta p}\delta p = +\delta p\delta q - \delta q\delta p \equiv 0 \ \} \ .$$

Accordingly **Figure 8.13** shows that *all eight* local Lyapunov exponents vanish before *and eventually* after the collision. In the absence of collisions the entire local Lyapunov spectrum consists of zeros. If collisions are included brief nonzero contributions to the spectrum occur. This example is most interesting. It indicates that the concept of "local Lyapunov exponents" needs to be taken with a "grain of salt". Unless special precautions are taken (and they would have to vary from case to case) the local exponents fluctuate wildly with time as well as with the chosen initial conditions.

This diatomic collision (it could equally well be viewed as the collision of one molecule with its mirror image) reveals a source of chaos and Lyapunov instability in conservative Hamiltonian systems. The motion in this example is simplicity itself, harmonic oscillations, except during the brief collision. That collision changes the phase of the harmonic oscillations as well as the apparent instability of the system. The *timing* of the collision is crucial. Evidently chaos itself, as well as the Second Law of Thermodynamics, which stems from it, can result from essentially random time-reversible events. This mechanism for opening up new phase-space regions for exploration provides a more detailed look at the collisional randomness embodied in Boltzmann's equation.

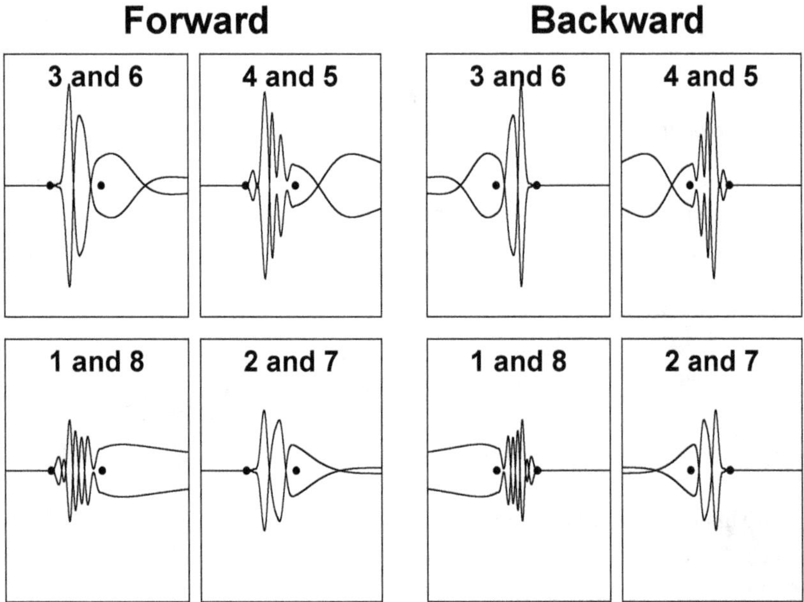

Fig. 8.13 Collision of two quiet diatomics. The exponents all vanish before and, eventually, after the collision. It takes a *long* time for the exponents to vanish, too long to show in the Figure. Again, as in **Figure 8.12** , the time interval occupied *by* the collision is indicated by the two filled circles in each panel.

How seriously should we take the antisymmetry of the Lyapunov vectors? From the formal standpoint it is a striking result in its generality, like Liouville's Theorem. Both have the same source, Hamilton's equations of motion. But unlike Liouville's Theorem it takes a long time for the offset vectors' antisymmetry to assert itself. Another very simple chaotic system, a six-atom ϕ^4 chain, with quartic tethers to the six fixed sites $\{ xo \}$ and with quadratic nearest-neighbor springs,

$$\mathcal{H} = \sum_{i=1}^{6}(1/2)p_i^2 + \sum_{i=1}^{6}(1/4)(x_i - xo_i)^4 +$$

$$\sum_{j=i+1}(1/2)(x_j - 1 - x_i)^2 = 0.167 \; ; \; \{ xo_i \equiv i - 3.5 \} \; ,$$

takes about a billion (10^9) Runge-Kutta timesteps ($dt = 0.001$) for the components of the six vector pairs to agree with three-digit accuracy.

One might well expect that a more complicated system, *twelve* particles with *twelve* vector pairs would take considerably longer for convergence of the vectors. We were surprised to find that the twelve pairs converged just as rapidly as did the six. Trying a variety of initial conditions, including lining up the first Lyapunov vector with the phase-space velocity $\dot{\delta}_1 \propto (\dot{q}_1, \dot{p}_1, \dot{q}_2, \dot{p}_2, \dots \dot{q}_N, \dot{p}_N)$, showed that the convergence time (to three or four significant figures) was on the order of 10^6 . This convergence time for pairing is much larger than the time required for errors due to computer roundoff error, $\simeq 10^{-16} \simeq e^{-37}$, to grow to unit magnitude :

$$\simeq e^{-37} e^{\lambda_1 t} \simeq 1 \ ; \ \lambda_1 \simeq 0.1 \longrightarrow t_{\rm grow} \simeq 370 << t_{\rm pair} \simeq 10^6 \ .$$

8.9 Lyapunov Instability of Many-Body Systems

Two macroscopic bodies, liquid or solid, when flung together generate Lyapunov exponents which are very different going forward and backward in time. We display some results illustrating the dependence of "important particles" on the past. It seems that there can be a symmetry breaking in these collision problems which is different to the usual forward-backward symmetry breaking.

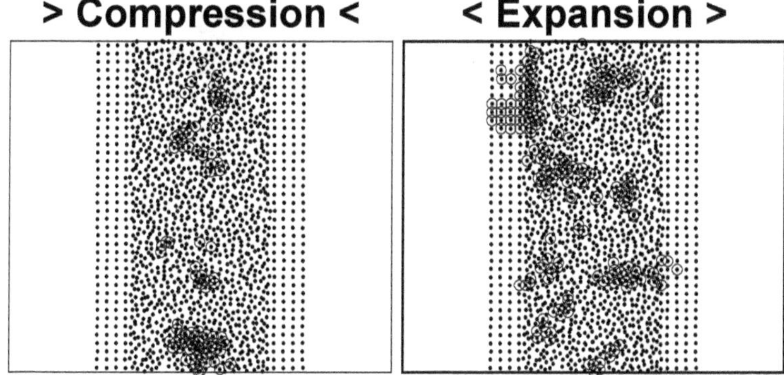

Fig. 8.14 Collision of two 800-particle blocks indicating the important particles. The configurations of the particles are identical in the two snapshots. The "compression" corresponds to a trajectory forward in time while the "expansion" refers to exactly the same trajectory played backward. The "important" particles, from the standpoint of the largest Lyapunov exponent, are very different in the two time directions.

Let us turn away from simple few-body reversible systems to a many-body system with an *irreversible* look, but still with fully-reversible conservative motion equations. Consider a prototypical Hamiltonian process, a bit more elaborate than the usual illustration based on filling *half* of the particles' container with *all* of the particles. Condensed matter trajectories include interactions between pairs of particles as an equilibration mechanism, in addition to their free-streaming motion, which provides the collisional phase-shifts illustrated in **Figure 8.12** for two harmonic molecules.

Figure 8.14 compares two *bit-reversible* snapshots (forward and reversed) from the compressive collision of two fast-moving 800-particle cold blocks. The speeds of the blocks prior to collision were 0.875 with kinetic temperatures of order 0.00001. All pairs of particles interact with a short-ranged soft-disk potential :

$$\{\ (+x,+y,+p_x,+p_y) \longleftrightarrow (-x,-y,-p_x,-p_y)\ \}\ ;\ \phi(r<1) = (1-r^2)^4\ .$$

The container is a periodic square. In the snapshots those particles which are making an above-average contribution to the largest Lyapunov exponent are darkened. Despite the reversibility of all the motion equations the "important" particles in the reversed motion are entirely different. The difference between the important particles in the two directions of time provides possible definitions for the distance of a given state from equilibrium. For more details of this simulation see pages 236-237 of our 2012 book, *Time Reversibility, Computer Simulation, Algorithms, Chaos.*

In a typical equilibrium situation the future and past are not very different from the present. In a strongly nonequilibrium situation such as the two-block collision the lack of time symmetry becomes quite obvious. This asymmetry suggests the possibility of metrics measuring the distance from equilibrium. One might also expect that the local Kolmogorov entropy, the sum of the positive Lyapunov exponents, would describe the opening up of phase space forward in time, or its closing down in the reversed motion. Our cursory explorations of the Kolmogorov entropy showed no systematic behavior in these collision problems.

In the collision of two chunks of cold material, converting kinetic energy to thermal energy changes the stability of the many-body trajectory. In the cold material the Lyapunov exponents are relatively small while in the hot material they are much larger. Thus the forward and backward Lyapunov exponents before and after collision differ qualitatively.

In Chapter 6 we described what happens when hot and cold regions are constrained by using Lagrangian or Hamiltonian mechanics to control the kinetic energy of selected degrees of freedom. Although the constraints function properly, with some particles "hot" and others "cold" in purely classical systems there is no flow of heat from hot to cold. Stationary heat flow necessarily involves dissipation, which shows up mechanically as a *change* in phase volume. No phase-volume change is possible in Hamiltonian flows so that no dissipation and no heat flow take place. "Important Particles" do reflect the past rather than predict the future. It appears to be very timely to explore the differences in stabilities of forward and reversed Hamiltonian motions.

8.10 Antique Dogs and Stochastic Fleas

How does our understanding of Second Law irreversibility compare to that of Boltzmann, Loschmidt, and Zermélo? Our current understanding of irreversibility in terms of zero-volume fractals with divergent entropy is certainly more elaborate and detailed than theirs. But the overall message, both now and then, is simply that nonequilibrium states are rare. We all agree on this but can differ on the details. The main question is how nonequilibrium states are best generated and analyzed. The classic stochastic methods can be modified to do this but deterministic constraints seem simpler, less expensive, and much better suited to analysis and explanation. It is refreshing to look back a century to review the models then current for the understanding of irreversibiility.

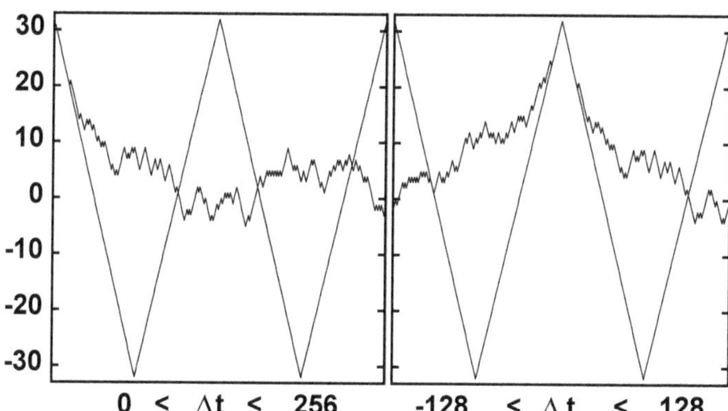

Fig. 8.15 The classic dog-flea model using random number generators with ergodic periods of 2^6 and 2^{22} . The left dog's number of excess fleas is plotted as a function of time.

About a century ago the Ehrenfests provided us with a simple stochastic model for "understanding" the Second Law in terms of the likelihood of states. In their model many individually-numbered fleas were initially sited on two dogs at time zero. Then at each successive time interval (imagine the ticks of a clock at unit values of the time) a randomly chosen flea is moved from one dog to the other. A *deterministic* version of their model can be used to illustrate differences between reversible and irreversible mechanical problems.

Reminded of the checkerboard of **Figure 8.7** we choose to use 64 fleas, putting all of them originally on the "left" dog. We then replace each tick of the clock with a fresh computer-generated deterministic "pseudorandom number" (equally and ergodically distributed in a predetermined order in the range from 1 through 64) to choose the flea to move. All 64 fleas jump from left to right in sequence as the 64 moves are selected. As a consequence the left panel of **Figure 8.15** shows a straight line for the 64 "random numbers" corresponding to times $1 \ldots 64$. At time 64 the righthand dog is fully occupied.

This dynamics is far from random, as the entire "pseudorandom" but deterministic dynamics follows from the initial conditions. *Random* fleas would exhibit fluctuations of the order of $\sqrt{N} = 8$ during a Poincaré recurrence time of order $2^{N/2}$.

In the same **Figure** we include excess-flea data based on a more sophisticated pseudorandom number generator that sequences through the integers $\{ \ \mathtt{I} \ \}$ from 1 to 2^{22} . The number of the selected flea is given by `mod(I,64) + 1` . The sequence repeats after each flea has made 2^{16} jumps. The right panel is centered on time 2^{22} and shows that both the simple and the sophisticated generators agree for a few steps. Except near the start and finish of the 2^{22} jumps the fluctuations in the flea populations are indistinguishable from those generated stochastically.

For the first five steps (left side of the left panel; center of the right panel) the directions of the jumping fleas (but not their identities) are the same. The two approaches agree, with the excess, $N_{\rm left} - 32$ decreasing from 32 to 27. At that point the sophisticated generator happens to choose flea number 1 for the second time, sending him back from right to left, so that from then on the two approaches are mostly different. But as the cycle length of the sophisticated generator is approached (after 2^{22} jumps) the two schemes come to agree again for a few steps, both forward and backward in time. For the long interval in between the start and its recurrence the two schemes are totally different. The understanding of this problem, dating back to Boltzmann's era, and discussed by the Ehrenfests, was for a long time *the* understanding of irreversible behavior. Namely the evolution of a complex system can be described probabilistically as the likely seeking out of more probable situations (roughly equal numbers of fleas on the two dogs). Of course this argument is subject to the familiar objections of Zermélo and Loschmidt. Boltzmann, Gibbs, Maxwell, and Poincaré, and later Lebowitz, viewed this explanation of irreversibility as the Grail, a convenient way of avoiding the embarrassing reversibility of Hamiltonian mechanics.

8.11 Summary

Before computers were developed, models explaining the coexistence of reversible micromechanics and irreversible thermodynamics relied on simple theoretical analyses. Kinetic theory and stochastic differential equations introduced irreversibility at the outset. Gibbs' ensembles of weakly-coupled microstates, more than could ever be explored, even on astrophysical timescales, seem unnecessarily abstract. When computers became available the Central Limit Theorem and the Monte Carlo method, with their promise of eventual success with only small fluctuations in big systems provided the courage to go on. Even so, solving the Boltzmann equation or the Fokker-Planck equation was difficult, even in the absence of realistic boundary conditions. Liouville's Theorem, Boltzmann's H Theorem, and Brownian Motion all occupied separate compartments in our understanding of mechanics. Attempts to go beyond simple linear theories were too complex to attract a real following.

Necessity, the Mother of Invention, led to finite-difference and series expansion techniques suitable for practical problems. Eiffel's Tower, the Wright Brothers' powered flight, and the Depression's Hoover Dam were necessarily based on simple pencil-and-paper designs. World War II brought qualitative change. Navigation, tracking, munitions, and coding efforts required the development of cheap computation. Our heritage today includes pocket-size communication/computation devices *billions of times smaller and faster* than their predecessors, as well as *millions of times less expensive*. Today *formulating* problems to challenge our modern computational tools is the main difficulty.

The present computer age, making it possible to solve complex problems quickly and efficiently, provides a new, different, and more detailed explanation for nonequilibrium systems in contact with thermal (or moving or otherwise perturbing) boundaries. With an *infinite* reservoir available, and represented by no more than a single Nosé-Hoover degree of freedom, a stationary interacting system approaches a fractal phase-space distribution, where dimensionality rather than phase volume is reduced. The Galton Board and the heat-conducting oscillator are prototypical of this situation. The Galton Board shows a phase-space cross section dimension varying from two to zero as the field strength increases. Similarly the N-body heat-conducting ϕ^4 chain shows a dimension varying from $2N + 2$ toward unity as the temperature is lowered while the flow remains nonequilibrium.

The ancient and modern explanations of irreversibility do share a common thread : nonequilibrium states are *rare*. Nonequilibrium systems correspond to many fewer states than do equilibrium ones. In time-reversible models including explicit interactions with external reservoirs the ratio of nonequilibrium to equilibrium states becomes *zero*, not just small. We think that the reader will agree that our time-reversible deterministic models are simpler, both logically and computationally, than their stochastic ancestors. Generating nonequilibrium states with just a few additional control variables (possibly only one) is simpler, elegant, and readily portable from one investigator's laptop to another's. The equations of motion remain time-reversible, recurrent, and often chaotic. This way of managing nonequilibrium systems avoids the pitfalls of Poincaré, Zermélo, and Loschmidt. Those traps are revealed as natural consequences of the conservative Newton-Lagrange-Hamilton model of mechanics. Though much more needs to be revealed, the path toward an enhanced understanding of nonequilibrium systems, through chaos and time-reversible dissipation is well underway.

In the next and final Chapter we consider more complicated state-of-the-art applications of computation to another interesting, and somewhat more complex, problem area, Life on Earth.

8.12 References

There are by now many interesting references dealing with the fractal distributions and Lyapunov Spectrum of the Periodic Lorentz Gas. For an early effort see J. Machta and R. Zwanzig, "Diffusion in a Periodic Lorentz Gas", Physical Review Letters **50** , 1959-1962 (1983). See particularly Bill's work with the same title, B. Moran, W. G. Hoover, and S. Bestiale, Journal of Statistical Physics **48** , 709-726 (1987); Ch. Dellago, L. Glatz, and H. A. Posch, "Lyapunov Spectrum of the Driven Lorentz Gas", Physical Review E **52** , 4817-4826 (1995); Ch. Dellago, H. A. Posch and Wm. G. Hoover, "Lyapunov Instability in a System of Hard Disks in Equilibrium and Nonequilibrium Steady States", Physical Review E **53** , 1485 (1996); and C. Dettmann, "Diffusion in the Lorentz Gas", arχiv: 1402:7010 = Communications in Theoretical Physics **62** , 521-540 (2014).

Some details of the Galton Board's history can be found in A. A. M. Daud's "Mathematical Modelling and Symbolic Dynamics Analysis of Three New Galton Board Models", Communications in Nonlinear Science and Numerical Simulation" **19** , 3476-3491 (2014). Be sure to have a look through the many interesting examples in Clint Sprott's *Chaos and Time-Series Analysis* (Oxford University Press, 2003) .

We have done considerable work on analyzing small-system multifractal distributions. See, for instance Ch. Dellago and Wm. G. Hoover, "Finite-Precision Stationary States At and Away from Equilibrium", Physical Review E **62** , 6275-6281 (2000); Wm. G. Hoover, C. G. Hoover, H. A. Posch, and J. A. Codelli, "The Second Law of Thermodynamics and Multifractal Distribution Functions: Bin Counting, Pair Correlations, and the Kaplan-Yorke Conjecture", Communications in Nonlinear Science and Numerical Simulation **12** , 214-231 (2007); and Wm. G. Hoover, C.G. Hoover, and F. Grond, "Phase-Space Growth Rates, Local Lyapunov Spectra, and Symmetry Breaking for Time-Reversible Dissipative Oscillators", Communications in Nonlinear Science and Numerical Simulation **13** , 1180-1193 (2008).

Much work has helped to elucidate canonical thermostats. See D. Kusnezov, A. Bulgac, and W. Bauer, "Canonical Ensembles from Chaos", Annals of Physics **204** , 155-185 (1990); D. Kusnezov and A. Bulgac, "Canonical Ensembles from Chaos II: Constrained Dynamical Systems", Annals of Physics **214** , 180-218 (1992)]; Wm. G. Hoover and B. L. Holian, Kinetic Moments Method for the Canonical Ensemble, Physics Letters A **211** , 253-257 (1996); and G. J. Martyna, M. L. Klein, and M. Tuckerman, "Nosé-Hoover Chains: The Canonical Ensemble *via* Continuous Dynamics", Journal of Chemical Physics **97** , 2635-2643 (1992).

8.13 Problems

1. Comment on the conservation of angular momentum , ($xp_y - yp_x$) , in the Galton Board problems.

2. Calculate the fluctuation in p^2 according to the three sets of equations of motion that constrain the time-averaged values of $\langle\, p^{\{\ 2,4,6\ \}}\,\rangle$.

3. Compute the average number of random jumps giving recurrence in systems (like a coin and like a die) with two or with six or with Ω equally-likely states.

4. Solve the Galton Board problem with constant friction and determine whether or not the resulting distribution of collisions is fractal.

5. By considering the form of the Leapfrog algorithm formulate a checkerboard algorithm along the lines of **Figure 8.2** which *is* reversible in the algebraic sense.

6. Analyze the averaged behavior of the dog-flea model analytically and compare the analysis to the short-time 64-flea computer simulation using 2^{22} pseudorandom numbers as given in **Figure 8.15** on page 259 .

7. Use Stirling's Approximation, $\ln(N!) \simeq N\ln(N/e) + (1/2)\ln(2\pi N)$ to estimate the minimum and maximum number of fleas expected on one of the dogs during a session of 2^{22} random jumps. Start with all of the fleas on one dog and find the minimum number occurring on that dog during a run of 2^{22} jumps, using what you believe is a good random number generator with a cycle length exceeding 2^{22} . If the answer is far from 12 you might want to find a better generator.

8. Estimate the length of time it would take your computer to find all 64 fleas on one dog again assuming that this *does not* occur in the first 64 jumps.

Chapter 9

Outlook for Progress ; Life on Earth

Topics

Introduction–Water and Real Life / Computational Methods for Water / Bernal and Fowler / Abascal and Vega's Models / Feynman's Path Integrals / Wigner-Kirkwood Quantum Corrections / Car-Parrinello Lagrangian / Density Functional Theory / "Multiscale" Simulations / Karplus, Levitt, and Warshel (2013) / Life / Miller and Urey's Experiment / Shockwave Syntheses / Computational Chemistry and Drug Design /

9.1 Introduction – Modelling Water and Real Life

This final chapter is devoted to a look at the applications of atomistic simulation from the standpoint of understanding the past and predicting the future. Like climate modelling, evolution and biophysics and astrophysics are competing and lobbying for public and government support. We have the optimistic view that Harvard's motto and goal, *Veritas*, or Yale's more elaborate *Lux et Veritas* will eventually win out to the benefit of all.

In this book we have explored our lifelong research interest, classical computer simulations of simple models with relatively complex and chaotic consequences. We have surveyed ideas dating back a century or more, to Boltzmann, to Gibbs, to Lyapunov, and Poincaré. Computer implementations and development of numerical techniques cover the latter half of that period. Applying these methods to selected problems which challenge the intellect has provided us with new points of view and enhanced understanding.

It is a relatively small step from our idealized one- and two-dimensional models to real-world applications. There are many possibilities from which to choose, ranging from manufacturing and materials science, to biology and medicine, and to the origin of life on earth. We illustrate the current state of computer models by reviewing four sorts of "realistic" simulations, water, protein dynamics, shockwave-induced amino-acid chemistry, and drug design. These areas are contemporary samples of man's ongoing effort better to know himself and his Universe.

Water, the basis of human life, is a good place to begin. Microscopic models of water are undergoing rapid development so as to keep pace with discoveries in the laboratory and the capacities of fast computers. More than fifteen different solid phases ! Numerical investigations of water's phase diagram have been pushed to pressures of 50 megabars, motivated or rationalized by the presumed water content of Neptune, Uranus, and the nearly countless planets outside our own solar system. See Hermann, Ashcroft, and Hoffmann's "High Pressure Ices" in the Proceedings of the National Academy of Science (2011) for a simulation-based survey of its title.

Medicine and biology function at more familiar earth's-surface temperatures and pressures. There is a widespread faith that simulations of aqueous amino acids and proteins will improve our self-knowledge and add to the duration and quality of our lives. Let us begin with a review of progress toward microscopic models for water and then turn to their overall usefulness in applications.

9.2 Computational Models for Water

The launch of Bill's undergraduate general chemistry course of 1953-1954 was memorable. The first class began with Professor Luke Eby Steiner turning on a tap at the demonstration table of Chemistry's main lecture hall, northwest of Tappan Square. After watching the water flow for about a minute Professor Steiner asked the hundred or so Oberlin Freshmen a question: "What do you see?". Taking responses from several students soon led to a consensus: "Water is an economical fluid that flows from a pipe". Fluids other than water are quite rare on the earth's surface. We ourselves are about half water.

Though relatively inexpensive, non-toxic, and easy to transport, water is far from simple. In terms of the models of the 1950s the proton and electron ingredients of water and neon are the same, ten of each. Deuterated "heavy water" has also the same number of neutrons, ten, as does neon. Ordinary water has eight neutrons. Evidently the neon and water structures are quite different. Neon's triple point temperature, 25K, is less than a tenth that of (ordinary) water, 273K. The difference stems from the separated charges. Water has three of them, at the two hydrogen atoms and at the oxygen that together make up each water molecule. Instead of neon's spherically symmetric charge distribution water has a "bent" shape. In current electromechanical models of the molecule the ${}^{H}{}_{O}{}^{H}$ angle is nearly tetrahedral, giving relatively long-ranged attractive forces from the resulting electrical dipole and quadrupole moments. See Problem 9.1 .

Hooke's-law Models of quantum harmonic oscillators, when fitted to spectroscopic data, allow the basic geometry of the molecule to be described by three or more centers of charge (the two relatively-positive hydrogens and the relatively-negative oxygen). At the same time, in ordinary ice or liquid water, nearby molecules are able to polarize and orient themselves to "screen out" most of the electrostatic interactions, reducing them by a factor of 80 (water's "dielectric constant"). The strength [$13.6ev \simeq 158,000$kK] of the electron-proton interaction in the hydrogen atom exceeds the vaporization energy of a water molecule [$\simeq$ 4900kK] water by more than an order of magnitude so that the structures of water's condensed phases are relatively subtle. The acid-base properties of water and its solutions are dependent on anion-cation concentrations of the order of one part in ten million, far outside the concentration range of today's practical many-body simulations.

Nevertheless, understanding the properties of water is prerequisite to understanding our bodies, our weather, and the world around us. Most of the earth's surface and most of ourselves are water. Despite its ubiquity, the prospect of microscopic simulations of water looks particularly daunting in view of its complex phase diagram. Unlike "simple" fluids, which coexist with a single solid phase, water exhibits more than fifteen solid phases. The 50-megabar simulations mentioned in the Introduction suggest that this complexity continues on as far as one would care to follow it.

9.2.1 *Bernal-Fowler Mechanical Model for Water (1933)*

John Bernal's handsome Depression-era models of "simple fluids" and of water are illustrated and described with justifiable admiration by his student, John Finney, in the memoir "Bernal and the Structure of Water", Journal of Physics : Conference Series **57**, 40-52 (2007) . Nowadays ball-and-stick physical models, one of Bill's first research efforts with Duward

Shriver ["Inexpensive Stuart-type Molecular Models", Journal of Chemical Education, **38**, 295 (1961)] and inspired by Bernal's work, have been superseded by computer movies of protein structural dynamics based in part on the classical mechanics of masses interacting with empirical force laws.

To reproduce the dipole and quadrupole moments of water requires, at the very least, two positive Coulomb charges neutralized by a negative pair. Today's computational models of water generally incorporate a double negative charge, near the oxygen nucleus, and two compensating positive charges at the locations of the hydrogenic protons. Barker and Watts' "Structure of Water; a Monte Carlo Calculation", in Chemical Physics Letters (1969) and Stillinger and Rahman's "Molecular Dynamics Study of Liquid Water", in the 1971 Journal of Chemical Physics, used tetrahedral arrangements of two negative and two positive charges.

It is surprising that considerable headway can be made treating water with classical mechanics. It is humbling to realize that today's models bear a strong resemblance to Bernal and Fowler's 1933 model, illustrated in **Figure 9.1** , differing only in small few-percent details of the geometry and charge magnitudes. The hidden complexity described by the "dielectric constant" is more humbling still. Whatever electrical charges might describe the water-water interaction in "real" water are almost completely screened out in bulk water (or in ordinary ice) where additional molecules are available to intervene. In the Bernal-Fowler model of a water molecule the nonCoulombic water-water interaction is described by a Lennard-Jones 6-12 potential with a well depth of 157kK .

Earnshaw's Theorem, which is evidently entertaining to check numerically, states that Coulomb forces alone cannot provide an equilibrium arrangement of charges. No motion or confining boundary conditions are permitted by the Theorem, because Coulomb attraction could otherwise be offset by centrifugal forces. Likewise, the instability associated with purely-repulsive Coulomb charges can be offset by constraining the charges to lie within a circle or on the surface of a sphere. Glasser and Every reported on the conformation of from 50 to 100 charges on a sphere's surface in the 1992 Journal of Physics A. Evidently *any* static model for water must provide additional short-ranged forces preventing the Coulomb charges from coalescing. This additional complexity is usually avoided by choosing a *rigid* framework.

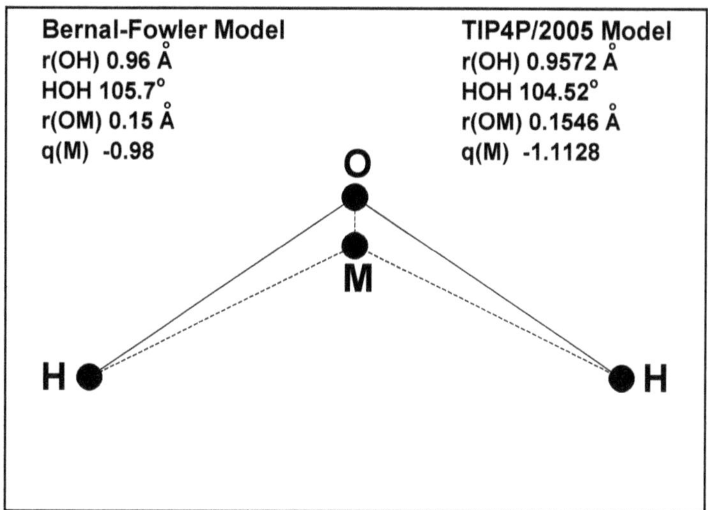

Fig. 9.1 Bernal and Fowler's 1933 model of water and the Abascal-Vega 2005 modification of it. The negative charge neutralizing the two positive charges is displaced downward from the Lennard-Jones central-force site at "O" to "M" in both models.

9.2.2 *Abascal and Vega's Model for Water (2005)*

A model closely related to Bernal and Fowler's 1933 ideas (Volume **1** of the Journal of Chemical Physics) is described by Abascal and Vega in a later volume of that same Journal, **123**, 234505 (2005). In both cases a four-site model for the potential energy involves Coulomb interactions with three charges on each molecule together with a Lennard-Jones 6-12 intermolecular potential with a well depth of ϵ and a collision diameter σ :

$$\phi_{\rm LJ} = 4\epsilon[\ (\sigma/r)^{12} - (\sigma/r)^6\ ] \longrightarrow \phi(r = 2^{1/6}\sigma) = -\epsilon\ .$$

Abascal and Vega adopted a water-model well depth of 93.2kK and a collision diameter 3.1589Å. Their main goal was a semiquantitative fit of the atmospheric-pressure dependence of density on temperature for both water and ice. They used packaged software (both molecular dynamics and Monte Carlo) to compute the phase diagram for liquid water together with ten of its solid phases. The pressure and temperature ranges covered are shown in **Figure 9.2** . The classical rigid-framework calculations give phase boundaries with errors of order 30K in temperature and a kilobar in pressure. The rigid-structure model also provides reasonable values for the surface tension and viscosity (but not the heat capacity). The qualitative features of the model are good, but will need improvement for trustworthy biomolecular simulations, the main interest of atomistic water modellers today.

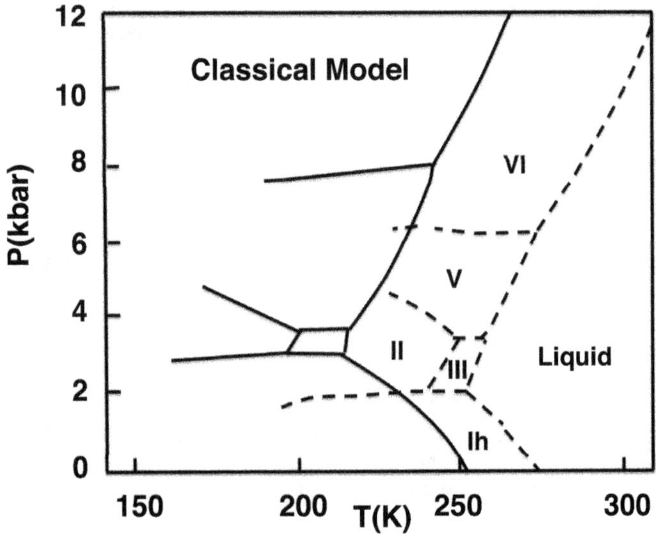

Fig. 9.2 Lines represent the phase diagram for water according to classical simulations with Abascal and Vega's three-charge four-site TIP4P/2005 model. The charges at the "H" locations of **Figure 9.1** are +0.5564 electronic charges. The dashes represent a portion of the experimental phase diagram for water. Ordinary ice is the "Ih" phase.

There are a certain number of flies in the ointment when it comes to using classical mechanics to represent "real" (quantum) molecules. The "Uncertainty Principle" reminds us that quantum coordinates and momenta are coupled and cannot be simultaneously specified or known. This can be visualized as a smearing out of particle coordinates q and momenta p with the product of the uncertainties of order Planck's constant h .

In the special case of a one-dimensional harmonic oscillator with frequency ν the ground-state energy exceeds the classical energy minimum by an additional "zero-point energy" $(h\nu/2)$:

$$\langle\ (p^2/2m)\ \rangle = \langle\ (\kappa q^2/2)\ \rangle = (h\nu/4)\ .$$

Thus the zero-point energy for a harmonic oscillator is $(h\nu/2)$ at zero temperature, half kinetic and half potential. If we imagine that a typical continuum acoustic frequency in water corresponds to the sound speed divided by an intermolecular spacing the corresponding temperature is that of a cold winter's night in Nevada (using cgs units) :

$(h\nu/\mathrm{k}) = 6.6\times 10^{-27}(1.5\times 10^5)/[\ (3\times 10^{-23})^{1/3}(1.38\times 10^{-16})\ ] \simeq 230\mathrm{K}$.

Evidently at "room temperature" (298K) the bulk vibrational modes of water would still have a residual quantum energy of about 0.05kT, exceeding the classical vibrational energy by an average of 15kK each. The zero-point energy associated with the "rigid" OH bonds would be very much greater, corresponding to classical thermal energies at temperatures of hundreds of degrees. This energy omission can be offset, at least roughly, by adjusting the strength of the central-force Lennard-Jones potential.

At low temperature the vibrational degrees of freedom for water are in their ground states. This observation suggests a rigid nonlinear model for the HOH molecule with three degrees of freedom for its center of mass and three additional rotational degrees of freedom for its orientation about the center of mass. This six-degrees-of-freedom model has a classical gas-phase N-molecule heat capacity of only $C_V = [\ (3/2) + (3/2)\ ]N\mathrm{k}$ for three translational degrees of freedom and three more rotational. The actual heat capacity of room-temperature water is about three times larger, roughly $9N\mathrm{k}$, just what a fully-vibrational model would give. The extra energy of the measured heat capacity is usually attributed to "hydrogen bonding". Eventually the rigid-framework models for water will have to give way to more complicated "flexible" models with polarizability and mobile charges.

In fact the agreement of the classical models with experimental data is surprisingly good. The phase diagram would seem to be a particularly searching test of potentials. See the semiquantitative agreement of the classical and experimental phase diagrams shown in **Figure 9.2** taken from Abascal and Vega's 2005 paper. Although the rigid-framework model seems a bit crude, to a large extent errors in free energy due to quantum and vibrational corrections tend to cancel. A calculation of quantum corrections—from the nuclear zero-point motion discussed in the next subsection—to the classical diagram appears in **Figure 9.3** , drawn to resemble McBride, Noya, Aragones, Conde, and Vega's Figure 3 in their 2012 paper in Physical Chemistry Chemical Physics **14**, 10140-10146 = arχiv 1205.5181 .

Models involving long-range electric charges, corresponding to the Madelung sums for perfect Coulomb crystals, caused trouble in the early days of computer simulation. Special "Ewald" periodic boundary conditions were required to evaluate (the Fourier transforms of) the Coulomb interactions. At the same time long-range Coulomb forces are not realistic in condensed phases like water. The dielectric constant is the ratio of the force between two charges in vacuum and the force when the intervening space is filled with the material of interest. In the case of water and ordinary ice where the dielectric constant is about 80, the equilibrated orientation of

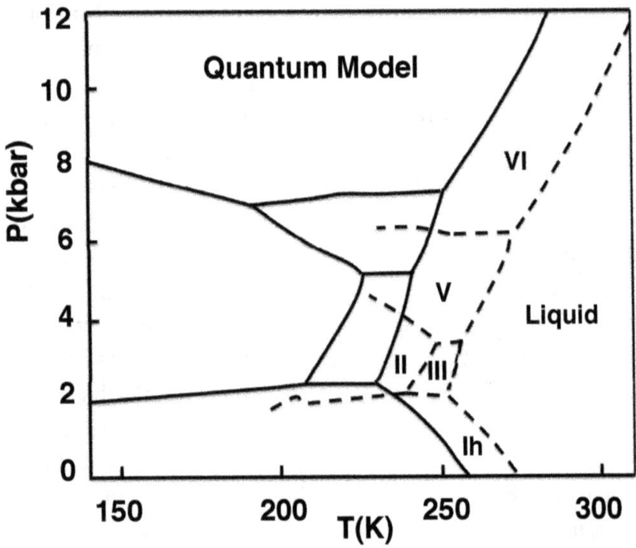

Fig. 9.3 Phase diagram for water according to path-integral quantum simulations of a modified version of Abascal and Vega's four-site TIP4P/2005 model. Here the positive charges were increased by 0.02 , from 0.5564 to 0.5764 electronic charge, and the negative charge was decreased by 0.04 better to fit room-temperature data. The diagram is drawn after Figure 3 in McBride, Noya, Aragones, Conde, and Vega's 2012 paper.

intervening water molecules provides an internal field offsetting nearly all of the applied external field. The dielectric constant is a key property in that some of the rigid-framework water models erroneously show a factor of two difference between the values for water and for ordinary ice "Ih".

Evidently the Coulomb forces in both phases are nearly completely screened so that a shorter-ranged force could explain most of classical water's dynamics. Water models generally contain three or four central Coulomb forces, with additional central Lennard-Jones intermolecular forces applied at or near the location of the oxygen nucleus. The development of "flexible" water models is still in its infancy and is no doubt an activity from which conceptual breakthroughs can soon be expected.

Simulations can never be better than the models incorporated in them. Here we concentrate on the simplest possible models in order to draw conclusions with widespread applicability. The relative simplicity of the theoretical force-law models should make it possible for diligent readers to reproduce results from the literature. Such work typically entails the use of packaged software rather than home-grown computer programs.

9.2.3 *Feynman's Path Integrals (1948)*

Two more-sophisticated computational models, intended to include quantum nuclear-and-electron motion corrections to the classical models, are mainly due to Feynman (for nuclear quantum effects) and to Car and Parrinello (for electronic effects). Feynman developed the "path-integral" approach, which has recently been applied to the water-structure problem. Although the physical picture is highly-cumbersome, involving the summed-up integrals over "all" trajectory paths $\{ q(t) \}$, it is possible to implement it in combination with classical Monte-Carlo path sampling, where each particle is replaced by a few dozen images, representing a series of sampling points along the paths. The contributions of the paths vary in magnitude and phase according to weights based on the classical action integral along closed-loop paths, $\simeq e^{2\pi i \int dt[\ K - \Phi\ ]/h}$.

The effect of Feynman's approach is to smear out the positions of the nuclei over a distance of order the de Broglie wavelength, $\lambda_{\rm dB}$. Jancovici considered this smearing effect for hard spheres in 1969. He found that to first order in $\sqrt{1/T}$ the effect of the quantum smearing is simply to increase the effective diameter of the interacting spheres :

$$\lambda_{\rm dB} \equiv h/\sqrt{2\pi mkT}\ ;\ \sigma_Q = \sigma_C + (\lambda/\sqrt{8})\ .$$

Jancovici's work is published in two papers in the Physical Review. For hydrogen and oxygen the corresponding size increases are about one third and one tenth of an Ångström, respectively. (The oxygen-to-hydrogen spacing in water is about one Ångström , as is indicated in **Figure 9.1**).

In applications of Feynman's work to water it is the molecules which are smeared. Nuclear quantum corrections to the phase diagram for water (using various rigid-framework models) were explored by McBride, Noya, Aragones, Conde, and Vega in their paper, "The Phase Diagram of Water from Quantum Simulations": arχiv 1205.5181. **Figure 9.3** shows a portion of the calculated and experimental phase diagrams for water. It appears that the quantum simulations remove about half of the offset separating the theoretical and experimental phase lines. The phase lines track pairs of coexisting phases in **Figure 9.2** . In the end the temperature errors are about 30K with pressure errors of order 500 bars. Though far from quantitative the overall impression left by these models is one of commendable progress from Bernal's time. Perhaps the inclusion of optimized flexible, as opposed to rigid, models will reduce the remaining discrepancies?

9.2.4 *Wigner-Kirkwood versus Feynman (1932-1948)*

The Correspondence Principle declares that quantum mechanics approaches a "classical limit" at high temperature. For example, the quantum canonical partition function approaches the classical one :

$$Z \equiv e^{-A(N,V,T)/kT} = \sum_{\text{states}} e^{-E_s/kT} \longrightarrow$$

$$\int \ldots \int \prod [\ dqdp\ ] e^{-\mathcal{H}/kT}/(N!h^{DN}) \ .$$

Here D is the dimensionality, almost always 3 . Wigner and Kirkwood, in two papers in the Physical Review (1932 and 1933) made the Correspondence Principle explicit for the partition function by deriving a rapidly-converging series in Planck's constant h . This "Wigner-Kirkwood series" relates the Helmholtz' free energy A_Q to its classical counterpart. By introducing plane-wave momentum states the quantum sum can be converted to the classical partition-function integral, with a series of corrections. For the one-dimensional harmonic oscillator the correspondence is easy to work out explicitly :

$$\sum_n e^{-(n+\frac{1}{2})h\nu/kT} = (kT/h\nu)e^{-(1/24)(h\nu/kT)^2+(1/2880)(h\nu/kT)^4 \ - \ \ldots} \longrightarrow$$

$$\int_{-\infty}^{+\infty} dq \int_{-\infty}^{+\infty} dp e^{-\kappa q^2/2kT} e^{-p^2/2mkT}/h = (kT/h\nu) \ ,$$

where the dots indicate higher even powers of $(h\nu/kT)$. This approach seems *much* simpler than Feynman's path integrals, in that it is only required to add on to the pair potential the quantum correction consistent with the series. In 1969 Hansen and Weis [Physical Review **188**, 314-318] showed that the fourth-order term for triple-point neon is 40 times smaller than the quadratic one, which suggests that isothermal Nosé-Hoover molecular dynamics with a different potential energy function ,

$$\Delta\phi = \phi_Q - \phi_C \simeq (1/12)(\hbar^2/mkT)[\ \phi'' + 2(\phi'/r)\ ] \ ,$$

would provide a good description of neon (and maybe water) at room temperature. Because Nosé-Hoover dynamics is isothermal rather than isoenergetic it would seem that molecular dynamics simulations of biomolecular problems could be carried out much more simply from this viewpoint than from the (equivalent, but more complex) Feynman path-integral approach.

9.2.5 *The Car-Parrinello Lagrangian (1985)*

Roberto Car and Michele Parrinello had a clever idea for sampling the ground-state Lagrangian dynamics of systems containing both nuclei and electrons. This was in 1985, perhaps motivated by the "artificial" Lagrangian used by Nosé to carry out canonical-ensemble dynamics rather than the usual microcanonical dynamics. Car and Parrinello invented a fictitious dynamics for individual ground-state electronic wavefunctions. The resulting nuclear motion is not fictitious but real. Nuclei move subject to interactions with the other nuclei as well as with the much more-rapidly-changing electronic wave functions. The electronic energy is itself determined by density functional theory (described in the next section). The Hellman-Feynman Theorem guarantees that the nuclear motions correctly model nature provided that they move in response to the correct electronic distribution.

Simulations based on the Car-Parrinello model for more than a few atoms are relatively time-consuming for three reasons. First, it is usual that *several* electrons need to be followed for each nucleus. Second, the electronic motion is rapid. The timescales of the nuclear and electronic motions differ by about two orders of magnitude. Third, the electronic wave functions are typically complex superpositions of hundreds of orthogonal basis functions. The resulting *ab initio* simulations furnish great detail for the time-dependent electronic density (and hence the forces) along the ground state (or zero-point) energy surface for the nuclei.

9.2.6 *Thomas-Fermi (1927) → Density Functional Theory*

The Hellman-Feynman Theorem allows the ground state nuclear motion to be computed from the classical electrostatics of the electronic density which provides the energy surface for the nuclei associated with that density. The energy of a uniform free-electron gas can be estimated [See Berni Alder and David Ceperley's 1980 Physical Review Letter, "Ground State of the Electron Gas by a Stochastic Method"], as can also be "gradient-dependent" corrections, taking into account that natural electronic density functions are not uniform on an atomic scale. Today's density-functional theories include estimates for electronic density based on energies for a free-electron gas in a uniform neutralizing background.

Thousands of investigators have worked on the dynamical properties of water. Even so, discrepancies in the diffusion coefficient and in the differences between heavy and ordinary water, are at the level of a factor of two. Kohn and Pople shared the 1998 Nobel Prize in chemistry for their contributions to density functional theory and computational chemistry.

9.2.7 *Modern "Multiscale" Simulations*

A somewhat different approach has been followed by those interested in biomolecular dynamics. Their work necessarily includes water, because biomolecular compounds (protein, enzymes, amino acids, viruses, ...) operate *in* water within the human body. Among these workers too it is usual to repeat Dirac's assertion that all of chemistry, and quite a bit of physics, depends upon solving the Schrödinger Equation. It is interesting to see that the 2013 winners of the Nobel Prize for Chemistry, Martin Karplus, Michael Levitt, and Arieh Warshel, were lauded for combining classical and quantum models so as to attack practical problems involving thousands of atoms.

Not even an atomistic-scale, let alone an electronic-scale, description of biomolecular problems is possible. Biolmolecular problems necessarily involve a "multiscale" approach. A typical multiscale picture links together small, medium, and large descriptions—from the millimeter-scale Kolmogorov length to the size of an airplane or ship in turbulent flows ; from the size of a star to a galaxy to the Universe in astrophysics, and from the size of an interstitial impurity, to a crystal grain, to a rivet, to an aircraft window in fracture mechanics. The simulation of biomolecular problems is intrinsically multiscale, involving the dynamics of Ångström-sized entities, composing small molecules, up to the micron scale, and for relatively long times, perhaps up to microseconds.

The most elaborate biomolecular models require an outer boundary, surrounding a continuous dielectric medium, within which rigid water molecules, and one or more idealized biomolecules interact. These biomolecules themselves may be an enzyme or a virus or a protein or a drug. The challenging dynamics of such problems necessarily combines a variety of algorithms, classical and quantum. It piqued the interests of many scientists some thirty years ago. The successful efforts of three of these men were rewarded by a Nobel Prize just this last year, 2013.

9.3 2013 Chemistry Prize: Karplus, Levitt, Warshel

One might guess that classical mechanics is worthless for the lightest atom, hydrogen. But in Karplus' 2013 Nobel lecture he emphasizes that accurate calculations of the potential surface make it possible to treat fairly well, classically [!] , even the hydrogen exchange reaction :

$$H + HH \longleftrightarrow HH + H \ .$$

Karplus pioneered the use of classical molecular dynamics for simulations of protein dynamics in aqueous solution. The result was an elaborate widely-distributed software package called CHARMM [an acronym for **C**hemistry, **HAR**vard, **M**acromolecular **M**echanics], a classical molecular dynamics program. The dynamics uses the leapfrog integrator, augmented by thermostating options such as Nosé-Hoover temperature control. The software includes interaction potentials describing the various central, angular, and torsional interactions needed to describe water and a variety of small molecules including amino acids, and larger structures, proteins for example. Amino acids contain a distinctive middle portion joining the amino (NH_2) and acid (COOH) groups together. Glutamic acid is a typical amino acid :

$$(HO)(CO)(CH_2)(CH_2)(CHNH_2)(CO)(OH) = C_5H_9NO_4 \ .$$

Proteins are a carbon chain with from dozens to hundreds of such "side-chain" amino acids attached to it. Proteins are key ingredients of living cells with lifetimes on the order of days. It is a difficult modelling challenge to find the "native" (optimally-folded) structure of a protein from a knowledge of its sequence of side chains. The native structure is the topology of the working protein in solution, crucial to its ability to interact with the other cell components required for life. The proper topology is apparently a daunting computational problem.

Levinthal's Paradox is just another version of the ergodic problem: for more than just a few atoms there are too many topological states to sample for the protein to find its "native state" in a reasonable time. Compare the situation to a gas with all of the atoms on one side of their container being "right" and all on the other side "wrong". Evidently the time to get "right" is of the order of 2^N . But unlike states in a constant-energy microcanonical ensemble the protein states are *isothermal* and cover a wide distribution of energies. At constant temperature, interacting with their surroundings as they do, these states are far from equally likely. Zwanzig, with Szabo and

Bagchi, considered a more realistic model taking the environment of the thermalized protein into account. With a *thermostated* heat bath the *rate* of making an improvement can be quite different to the reversed rate of making things worse, say by a factor of two or three. This idea, appropriate to a canonical distribution of energies, can be used to resolve the paradox.

In their prize-winning work Karplus, Levitt, and Warshel developed and implemented detailed models to simulate "protein folding", the surprisingly quick process by which extended carbon chains of amino acids adopt their "native" folded-up structure. To reduce the size and time of simulating such problems the amino-acid side chains can be replaced by single torsional degrees of freedom describing the orientation of a side chain relative to the neighbors hindering its rotation. Simulations were developed which began with a relaxation phase: a damped side-chain dynamics represented effects of the immersion of the protein in water. The damped dynamics then was followed up with a period of thermal equilibration. To model the thermal fluctuations accompanying that equilibration random forces were added. See Levitt and Warshel's "Computer Simulation of Protein Folding" in Nature (1975) and Karplus' autobiographical and enjoyable "Spinach on the Ceiling" from the 2006 Annual Review of Biophysics and Biomolecular Structure, seven years prior to their shared Nobel award.

9.4 How Did the Building Blocks of Life Emerge and Grow?

Physicists and chemists interested in the origin and history of life, not just its biomedical improvement and extension, are as motivated to use the fruits of quantum chemistry as are the biologists, medical doctors, and pharmicists. They would like to understand how "life" evolved to reach its current state. The chronology of the Universe rests on evidence dating back to the Big Bang, some fourteen billion years ago. Though the data and the tentative explanatory theories are both most interesting and stimulating, they resemble classical simulation codes, explaining a bit, but always in the company of the unexplained. Current gaps include explanations for dark matter and energy as well as mechanisms for the formation of the first one-celled "animals". We are invariably far from a clear and complete understanding of how life on earth began and how it ended up with us.

Because half a century is a short time in the earth's 4.5 billion years, one can predict with confidence that these problems will "soon" be understood better, and that the practice of medicine will likewise become more and more intricately detailed and expensive, as will also be our prognoses. As we find out more about the relative importance of environment and genetics we will better understand the functions and structure of our bodies and our brains.

So let us back up from our survey of water structure and protein folding to the questions of how and where these life-giving ingredients originated. Georges Lemaître's Big Bang (1927) gave us a start long ago, with plenty of hydrogen and helium. Gravitation then gave us stars, burning hydrogen to make more helium, continuing on to lithium, and succeeding heavier members of the Periodic Table. Following paths converting mass to energy, along came carbon and oxygen as well as the other elements up to the most stable nucleus, iron. Oxygen and the other elements less massive than iron were born in burning stars. After iron comes gravitational collapse, along with more nuclear fusion, creating the supernovae explosions that have distributed our bodies' building blocks : carbon, nitrogen, oxygen, ... throughout the Universe. This astrophysical background eventually led to the earth, to water, to ammonia, and to other simple small-molecule building blocks needed for life.

Life, though relatively hard to define, depends in its present form upon about twenty amino (NH_2) acids (COOH). Glycine (NH_2CH_2COOH) is the simplest. Amino acids are the basic building blocks for proteins and for the genetic information that distinguishes one species from another. How were these building blocks formed from the elements and the "prebiotic" molecules – the molecules needed for life – (H_2, H_2O, CO_2, CH_4, N_2 ...) available in the early days of a recently solidified earth?

There are several possible sources of the energy necessary to molecule formation. Ultraviolet radiation in space, electrical discharges on the earth's surface, kinetic energy of meteorites and comets striking the earth, and geothermal reactions under the early sea. At present there is a good correspondence between quantum chemical simulations of many-atom mechanisms for their formation and corresponding experiments both slow and fast. Let us consider some of each.

9.5 Stanley Miller and Harold Urey's Experiment (1953)

After hearing Harold Urey lecture on the possibility of making life-forming chemicals by adding energy to simple mixtures of methane, ammonia, hydrogen, and water, a new student at the University of Chicago, Stanley

Miller, volunteered to try the idea. Using an electric discharge to simulate lightning, in the vapor above boiling liquid, he produced many of the amino acids necessary to life on a timescale of several days. Heat and high pressure can be found at the bottom of the ocean and in meteoric or cometary collisions with the earth's surface. Any or all of these possibilities might be responsible for the precursors of presentday life on earth.

In 1969 a relatively large meteorite (at least 100 kilograms, with individual pieces weighing up to seven kilograms) landed near Murchison, Australia. This "Murchison meteorite" bore a mixture of nearly 20 amino acids, similar to those of Miller and Urey's work, reinforcing the idea that early solar system materials, plus energy, could lead to organic compounds forming a basis for life. Because organic compounds are stable over a limited temperature range there is no need for prolonged high temperatures in these syntheses.

Higher-pressure shockwaves and undersea vulcanism are two additional common energy sources that could contribute to a life-forming mechanism. In sorting out this wealth of ideas and ingredients chemical simulations, along with corroborating experiments, have together confirmed the potential of shockwaves for the formation of amino acids.

9.6 Computer Simulations and Corroborating Experiments

Computer simulations along the lines of the Miller-Urey experiments are possible now. Although real life on the atomic level is quantum-mechanical we have seen that considerable progress has been made in modelling chemistry in systems with a few hundred atoms, principally hydrogen, carbon, nitrogen, and oxygen. Nir Goldman, at the Livermore Laboratory, collaborated with Evan Reed, William Kuo, Laurence Fried, Christopher Mundy, Alesandro Curioni, and Amitesh Maiti, on high-pressure small-molecule simulations of ground-state chemistry. They studied states which can be reached in shockwave experiments under conditions similar to those reached in the many comets impacting the earth.

Rather than simulating all the details of a shockwave with a steady-state region adjoining an entrance and an exit, the simulation models of Goldman *et alii* are instead based on following the "Rayleigh line" the linear relationship between P_{xx} and the volume V :

$$\rho u = A \text{ and } P_{xx} + \rho u^2 = B \rightarrow P_{xx} = B - (A^2/\rho) \ .$$

In Goldman's work the compression is carried out homogeneously, with periodic boundaries, over a time somewhat longer than the shockwave rise time.

This computational technique was first validated with simulations of strongly shocked water. It is described in the Journal of Chemical Physics **130**, 124517 (2009). Classical molecular dynamics simulations were carried out on the quantum ground-state potential surface obtained with density-functional theory. **Figure 9.4** is taken from this work. It shows the initial state, 64 water molecules at unit density and room temperature, as well as the end state following 2.36-fold shockwave compression. The final shockwave pressure and temperature were 680 kilobars and 3650K. The shockwave velocity was 11 km/sec .

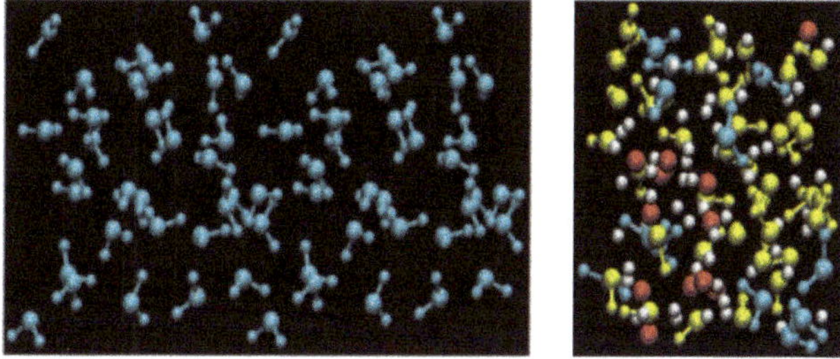

Fig. 9.4 Before and after renderings of a 64-molecule density-functional theory simulation of the quantum chemistry of water. Here the shockwave has dissociated 90% of the molecules, generating mainly white unbonded hydrogens, red unbonded oxygens, and yellow bonded hydrogen-oxygens. The simulations are in good agreement with corresponding experimental data for the ionic conductivity of water.

This same shock-simulation technique was then used to study the high-temperature high-pressure chemistry of an initial mixture of 210 atoms with a typical cometary composition :

$$20\,H_2O \;+\; 10\,CH_3OH \;+\; 10\,NH_3 \;+\; 10\,CO \;+\; 10\,CO_2\;.$$

The resulting compounds included glycine and alanine, amino acids of the kind required for life on earth.

The 210-atom mixture was taken to temperatures as high as 4000K and pressures of a few hundred kilobars corresponding to collisions of cometary material with the earth at a relative velocity of 30 km/sec . It is evident that these shockwave simulations model well the potential surface on which the nuclei move. The electronic part of the energy is based on a uniform electron gas with exchange and correlation corrections included (density functional theory) to model the interactions of the electrons with one another and to provide an energy surface for the nuclei. Soon after the shockwave simulations were published experiments consistent with them were reported in Nature Geoscience **6**, 1045 (2013) by Martins, Price, Goldman, Sephton, and Burchell.

9.7 Computational Biochemistry and Drug Design

In his own prize lecture 15 years earlier Walter Kohn (1998) preceded Karplus (2013) in crediting Dirac (1929) with the condescending observation that all of chemistry amounts to the "details" of solving Schrödinger's (impossibly difficult) equation for the appropriate nuclei and electrons. Some 85 years later prizes are still being awarded for some of those details, including density functional theory. Karplus emphasized the ability of classical molecular dynamics to solve intrinsically quantum problems by approximating Schrödinger's equation for the electrons. The truth in his prediction is borne out in the field of computational biochemistry.

The successful advances in computational chemistry in the last fifty years have led to computation's significant rôle today in the structural design of drugs. The results of *ab initio* quantum calculations are combined with experimental data to parameterize the forces and potentials that are vital components of molecular mechanics programs. Molecular mechanics computer programs are used for the structural exploration of interactions between biological macromolecules (such as proteins, hormones, and peptides) and smaller molecules ("ligands" : molecules that bind to other molecules) that can affect the biological functions of the macromolecule. Macromolecules are composed of anywhere from a few hundred to several thousand atoms. They are designed with highly sophisticated graphics packages that allow for three-dimensional visualization of the molecular structure and physical properties. Finding the equilibrium configuration of large proteins with molecular mechanics is a slow and computationally expensive calculation. Understanding the isothermal mechanisms for protein folding of proteins has been and still is a supercomputer "Grand Challenge" problem.

There are hundreds of thousands of molecules that can be used as ligands. Databases of families of biological molecules that will interact with our body chemistry to modify abnormal physiology or improve immunity have been developed in government laboratories, private companies, and research institutions around the world. Computational chemists use the databases to find ligands that fit into the *pockets* of proteins. **Figure 9.5** shows the docking result from "Homology Model of $SARS - CoV\ M^{pro}$ Protease", by E. Ashley Wiley and Ghislain Deslongchamps, at the University of New Brunswick, in Canada. The databases are also used by chemical companies synthesizing ligands for drug research and development.

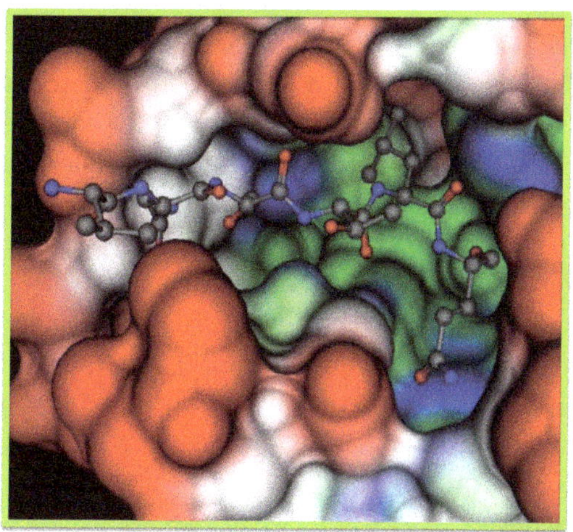

Fig. 9.5 A homology model (protein with a bound ligand) of the $SARS - CoVM^{pro}$ virus. The model is an analytic Connolly surface representation with molecular dynamics relaxation of the $SARS - CoV\ M^{pro}$ active site with bound peptidomimetic chloromethyl ketone inhibitor. It was derived using two crystal structures deposited into the Protein Data Bank. The homology model was computed using the Molecular Docking program, Molecular Operating Environment (*MOE*, www.chemcomp.comp). The color correspondences using the "Pocket" option in *MOE* are red for exposed surfaces, green for the hydrophobic regions, and blue for the polar regions.

Computer programs ("docking programs") are available to search through the databases to find sets of molecules that best match the input characterization of the macromolecule-ligand interaction. Any qualifying ligand must meet two criteria : [i] The macromolecule-ligand interaction must result in inhibiting, activating, or enhancing the biological function of the macromolecule ; [ii] The size of the ligand and its binding affinity must be selected to fit into the interaction surface and chemical properties of the macromolecule. Docking programs are designed to calculate a numerical rating or *score* for the ligands and to select the best set based on this rating. The ratings include factors such as molecular size and shape, solvent accessible surface, total energy, cluster analysis, molecular rotations and conformations.

9.7.1 *Drug Treatments for Diabetes*

Computational biochemistry has been a strong contributor in the development of treatments for diabetes. Diabetes is a disease resulting from high blood sugar levels. It is caused when insulin secreted from the pancreas is at levels that are too low to produce normal sugar (glucose) levels in the blood stream. Serious long term effects of diabetes include heart disease, stroke, kidney failure, foot ulcers, and damage to the eyes. There are two forms of diabetes. Type 1 diabetes, childhood diabetes, must be treated with insulin shots because the pancreas is unable properly to regulate the production of insulin on its own. Drug treatment is not currently a possibility for childhood diabetes. Type 2 diabetes represents about 90 percent of the cases of diabetes in the United States. With this disease type pancreatic functions are impaired but can be improved with drugs.

In September of 2014 the Food and Drug Administration announced the approval of a weekly drug injection for Type 2 diabetes. This weekly injection greatly benefits diabetes patients because the prior treatments were either a daily injection or oral treatments. These earlier treatments also produced undesirable side effects – weight gain and hypoglycemia – low blood sugar levels – with long term use. The new weekly injection drug has not only reduced the frequency of injections. It has also has been shown in clinical trials to reduce patient weight and eliminate the prospect for hypoglycemia in the longer term.

Type 2 diabetes alters the normal *incretin effect* that keeps insulin production at a steady level. Incretins are a group of gastrointestinal hormones that regulate insulin secretion in the pancreas and thus decrease glucose levels in the blood in the case of diabetics. The incretins are peptides with a signaling capability that transmits biological activation beyond the cells in the gastrointestinal system in which they are located. Peptides are composed of amino acid chains surrounded by a carboxyl end (COOH) and an amino end (NH_2). They are usually distinguished from proteins by being composed of roughly forty or fewer amino acids.

Medical research has shown that gastrointestinal peptides are released when a healthy patient eats. One example is the **G**lucagon-**L**ike **P**eptide GLP-1 which is secreted into the small intestine. GLP-1 performs four signaling functions :

[i] it delays emptying of the stomach ;

[ii] it initiates the neural signal that the stomach is full ;

[iii] it stimulates the release of insulin from the pancreas ; and

[iiii] it inhibits glucagon release in the pancreas.

This last function prevents the breakdown of glycogen into glucose in the liver and then into the blood. Under normal blood-glucose conditions the release of glucose from the liver responds to vigorous physical activity.

GLP-1 is also referred to as GLP-1 *receptor* because it has a target pocket for ligand binding. An agonist is a drug that increases the stimulation of a biological action over its natural activity. In particular, the GLP-1 receptor agonist drug enhances the signaling capability of the GLP-1 receptor. The first two functions of the GLP-1 receptor agonist help with weight control reaction as a benefit of the weekly injection.

GLP-1 is metabolized by the enzyme DPP-4 in the liver. As a result the half life of physiological GLP-1 is about 2 minutes! DPP-4 inhibitors have been developed to counterbalance this effect. Recently it has been shown that GLP-1 receptor agonists result in a five-fold increase in the peak stimulation level above the normal physiological level compared to two-fold increases with DPP-4 inhibitors. Furthermore the GLP-1 receptor agonists have shown weight loss compared to the weight-neutral effect with DPP-4 inhibitors in clinical trials.

The biomedical journals detail several research efforts directed toward producing a GLP-1 receptor agonist. In the journal Evidence-Based Complementary and Alternative Medecine **2014**, 385120 Hsin-Chieh Tang and Calvin Yu-Chian Chen report their successful search for ligands in the Traditional Chinese Medicine database. Traditional Chinese Medicine has evolved over several thousand years and complements herbal remedies with meditation, acupuncture, and *tai chi*, as a means to treat or prevent health problems. The Tang-Chen article is an interesting example of the holistic treatment approach in Chinese medicine. Although the English is a bit difficult the article illustrates all the computational steps involved in drug design. The authors conclude that they have found three ligands that are good prospects for producing a commercial GLP-1 receptor agonist drug.

Computational biochemistry is now an integral part of research efforts to develop drug treatments for many diseases including HIV virus, some cancers, and influenza. The effort in this field will grow as the computational chemistry programs are redesigned for cluster and massively-parallel computers. An exciting and affordable combination of Graphical Processing Units (GPUs) and Central Processing Units (CPUs) has demonstrated performance that is better than that achieved on cluster systems. The cost of the GPU/CPU systems, a few thousand dollars, is affordable by individual university researchers. The massively parallel computers open up new possibilities for larger scale more-nearly-accurate and more-detailed quantum calculations, including the explicit molecular modelling of water and other solvent molecules. Computational chemistry software is available to all researchers. Comprehensive lists of free and commercial computer programs with major capabilities flagged are listed in the references below.

9.8 Summary

This short tour shows that many-body simulation is very much in the mainstream of theoretical and applied computational physics. The subject is far from maturity, as is evidenced by the gaps between water's properties and the results of simulation. The design of drugs and the crumpling of cars are two applications with long histories and bright futures, especially with self-driving cars on the horizon.

From the academic perspective the difficulties in visualizing flows in so few as four space dimensions shows too that analyses of simple models will be surprising us for decades. Although these complex visualization problems and the biomedical applications of simulation are both rather far from our own research interests there is a powerful cross-fertilization linking the fruits of small-scale investigations of chaos and nonequilibrium systems to new algorithms for mesoscopic and macroscopic simulations devoted to improving the quality of our lives.

9.9 References

For references to technical publications describing quantum chemistry see the millions of Google listings. The arχiv is also a useful source. The several Nobel Laureates awarded prizes for their applications of computational chemistry have recorded their thoughts and popular writings on the internet and in hundreds of popular as well as technical publications. Karplus' "Spinach on the Ceiling" is specially recommended. It is next to impossible to enter this field as a researcher without joining a team and investing years of study becoming familiar with the details of packaged software and current research.

Wikipedia references useful in locating biomolecular software can be found using the phrases "List of quantum chemistry and solid-state physics software", "List of software for molecular mechanics modelling", "List of molecular graphics systems", and "Molecular Docking".

9.10 Problems

1. A dipole can be thought of as two charges located at $x = \pm(\delta/2)$. Evidently the large-$x \equiv r$ potential follows an inverse-square law :

$$\frac{1}{[\, r + (\delta/2)\,]} - \frac{1}{[\, r - (\delta/2)\,]} \simeq -(\delta/r^2) + O(\delta^3/r^4) \ .$$

Explore and characterize the structures and the power laws corresponding to quadrupoles (four charges) and octupoles.

2. Confirm the high-temperature expansion of the oscillator partition function that appears on page 274 .

3. What is the quantum correction to the Coulomb pair potential according to the Wigner-Kirkwood series of Section 9.2.4 ?

4. As temperature drops at what temperature does the de Broglie wavelength of hydrogen, $\lambda = h/\sqrt{2\pi mkT}$, first exceed Bohr's electron-proton hydrogen radius, $0.53\mathring{A}$?

5. Discuss methods for determining the distance dependence of the dielectric constant.

Index

Biographical Information

Bill and Carol Hoover met in 1975 at Edward Teller's Department of Applied Science, a graduate department of the University of California's Davis Campus, adjacent to the Lawrence Livermore National Laboratory. Bill worked as a physicist and Carol as a computer scientist at the Laboratory. Carol was Bill's student in graduate courses in Statistical Mechanics and Kinetic Theory. They married in the Fall of 1988 in preparation for their research leaves to work with Toshio Kawai and Shuichi Nosé at Keio University in Yokohama, 1989-1990 . Bill stayed on at Livermore for ten years after his early retirement in 1994 until Carol retired in 2005. The Hoovers built a new home in Ruby Valley Nevada in 2004-2005 to escape the crowding and high taxes of California. The loving friends and neighbors of Ruby Valley have provided the Hoovers an ideal environment for retirement, relaxation, and an active research life.

This photograph of the Hoovers was taken in Australia's Victorian Alps in December 2014 just prior to the Ian Snook Memorial Conference at the Royal Melbourne Institute of Technology.